Springer Series in Computational Physics

Editors

H. Cabannes   M. Holt   H. B. Keller
J. Killeen   S. A. Orszag   V. V. Rusanov

Springer Series in Computational Physics

Editors: R. Glowinski M. Holt P. Hut H. B. Keller J. Killeen
S. A. Orszag V. V. Rusanov

Roger Peyret
Thomas D. Taylor

Computational Methods for Fluid Flow

With 125 Figures

Springer-Verlag   New York   Berlin   Heidelberg
London   Paris   Tokyo   Hong Kong   Barcelona

Roger Peyret
CNRS
Département de Mathématiques
Université de Nice
F-06000 Nice
France

Thomas D. Taylor
Applied Physics Laboratory
Johns Hopkins University
Laurel, Maryland
USA

Editors

Library of Congress Cataloging in Publication Data
Peyret, Roger.
    Computational methods for fluid flow.

    (Springer series in computational physics)
    Bibliography: p.
    Includes index.
    1. Fluid dynamics—Mathematics. I. Taylor,
Thomas D. (Thomas Darwin), 1935–   . II. Title.
III. Series.
TA357.P525   532'.05'01515   82-757
                                       AACR2

9 8 7 6 5 4 3 (Corrected Third Printing, 1990)

ISBN 0-387-13851-X Springer-Verlag New York Heidelberg Berlin
ISBN 3-540-13851-X Springer-Verlag Berlin Heidelberg New York

Preface

In developing this book, we decided to emphasize applications and to provide methods for solving problems. As a result, we limited the mathematical developments and we tried as far as possible to get insight into the behavior of numerical methods by considering simple mathematical models.

The text contains three sections. The first is intended to give the fundamentals of most types of numerical approaches employed to solve fluid-mechanics problems. The topics of finite differences, finite elements, and spectral methods are included, as well as a number of special techniques. The second section is devoted to the solution of incompressible flows by the various numerical approaches. We have included solutions of laminar and turbulent-flow problems using finite difference, finite element, and spectral methods. The third section of the book is concerned with compressible flows. We divided this last section into inviscid and viscous flows and attempted to outline the methods for each area and give examples.

The completion of this book was accomplished because of a number of organizations and special people in both the United States and France. A large number of scientists furnished material for the text and we have attempted to acknowledge each and everyone in the book. Should we have unintentionally missed anyone, we express our regrets. In addition, we wish to express special thanks to Henri Viviand and the management at O.N.E.R.A. who made it possible for both authors to work with the aerodynamic computational group at O.N.E.R.A. in Chatillon, near Paris. In the United States, special appreciation is extended to Mort and Ralph Cooper, formerly of the Office of Naval Research, Al Loeb of the U.S. Army, Robert Moore and Richard Hoglund, formerly of DARPA, and Phil Selwyn of the U.S. Navy for the supporting research that made the book possible.

We also express our thanks to Wolf Beiglböck and Henri Cabannes for allowing us to publish in this Springer-Verlag series and to Jill Owens for typing the manuscript.

Lastly, we wish to thank our wives, Anne Peyret and Françoise Taylor, for their encouragement in this venture.

Contents

PART I
NUMERICAL APPROACHES

CHAPTER 1

Introduction and General Equations

The introduction of the computer into engineering has resulted in the growth of a completely new field termed computational fluid dynamics. This field has led to the development of new mathematical methods for solving the equations of fluid mechanics. The improved methods have permitted advanced simulations of flow phenomena on the computer for wide and varied applications. The areas range from aircraft and missile design to large-scale simulations of the atmosphere and ocean. A measure of the current state-of-the-art in missile simulation capabilities is demonstrated by the complex missile launch geometry shown in Fig. 1.1. This figure shows a two-dimensional simulation of a Titan missile launch indicating the nature of the flow about the missile and in the launch duct. The flow was calculated by solving the unsteady inviscid equations by the method of Godunov (1959). In the material that follows we will describe how one can conduct such a calculation.

In addition to the advances in system simulations there have been large strides in research predictions of transition and turbulence. One of the most recent is simulation of a turbulent shear layer by Riley and Metcalfe (1980)

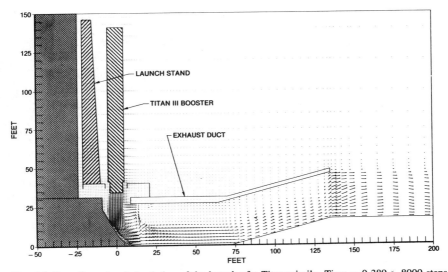

Fig. 1.1 Two-dimensional simulation of the launch of a Titan missile. Time = 0.389 s, 8000 steps.

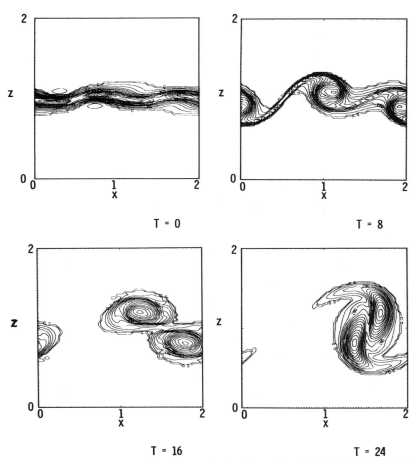

Fig. 1.2 Simulation of the turbulent behavior of a shear layer. (Furnished by J. Riley and R. Metcalfe.)

shown in Fig. 1.2. The results were generated by solving the complete two-dimensional unsteady Navier–Stokes equations by a *Spectral method*. Shown are mean flow results obtained after time averaging the turbulent fluctuations. The calculation was initiated using primarily a theoretically derived instability mode (Michalke, 1964). Note the prediction of the vortex roll up in the subsequent flow times. These examples demonstrate the complexity of the flows that are being modeled today. It is clear, however, that even more complicated cases will become commonplace simulations for the future.

The progress in computer simulation has, in fact, generated a competition between the experimentalists and the computer modelers. The first real controversy regarding this matter was generated by a paper entitled "Computers versus Wind Tunnels" authored in 1975 by Chapman, Mark, and Pirtle. This paper brought into focus the trend of things to come in the area of flow simulation on the computer. The drawbacks at the time of the paper were both the speed of the computer and the methods.

Subsequently, however, there has been a significant improvement in both computers and methods. In a ten-year period beginning in 1965, two orders of magnitude improvement has occurred in both areas (Chapman, 1979). This trend in efficiency is continuing and is leading to increased use of computer simulations in preliminary design to avoid increasing experimental costs.

In the computer area past improvements have been primarily in the speed of computation, but recently the trend has taken another direction. The speed is no longer dropping at an exponential rate, instead the speed is being held about constant and the cost of computing components is now dropping rapidly due to the new integrated circuits. The result is that in the near future computers with 10-ns cycle times and a million words of memory will cost about $300,000. This will revolutionize scientific computing, and the large computing centers will be complemented by smaller low cost units that can be employed by small groups to perform sophisticated fluid dynamic calculations. Currently, units that will fulfill this prediction are under construction commercially and under government support. Table 1 shows a comparison of speeds, capabilities, and estimated costs of some units currently under development in the United States. Other units are under development in England and France. Figure 1.3 shows the declining cost of computer memory and the commercial availability of a million words of memory at a moderate cost. The result is that one can now visualize the construction of a very high speed computer with large memory, as shown in Fig. 1.4. Each processor could have the cycle time of 10 ns and a core memory that could contain millions of words. This progress in computers will no doubt lead to more growth in computational fluid mechanics.

Table 1 Estimates of computer performance.

	MIPS	MFLOPS	MTTF	Approximate relative hardware cost
FPS	36[b]	12 (6)	3000 h	0.03
CDC 7600	40	40 (10)	days	1.00
CRAY[a]	80	80 (25)[c]	7 h	0.60[d]

MIPS—million instructions per second.

MFLOPS—million floating point operations per second.

MTTF—mean time to failure.

[a] Taken from LASAL CRAY-1 evaluation.

[b] FPS instruction contains six independent operations: floating point add, floating point multiply, indexing, branching, register transfer, and memory transfer. The FPS-AP modular units may be configured into a larger system that can increase the thruput.

[c] ( ) Typical and/or usually obtained speeds.

[d] Without system control computer.

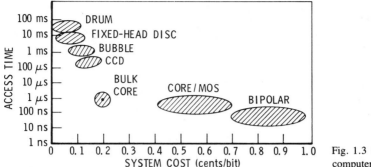

Fig. 1.3 Cost of various computer memories.

The impact will cause a surge in large-scale flow simulations and well founded attacks on the fundamental understanding of turbulence and transition. Also, many groups will no longer require the aid of a major government-supported computer center to perform large-scale flow simulations.

A recent example of the power of the use of the new low cost processors has been published by Enselme, Brochet, and Boisseau (1981). These authors

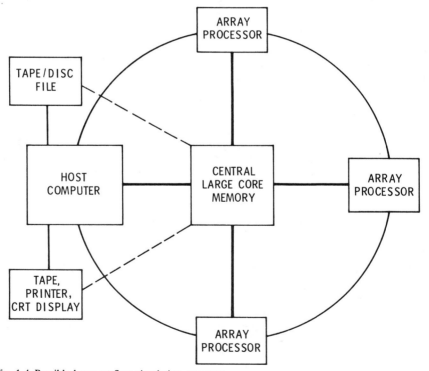

Fig. 1.4 Possible low-cost flow-simulation computer.

computed flow in a supersonic compressor and, using a dedicated computer with a high speed processor, obtained computation speeds of two times the CDC 7600. The cost of the system was about $200,000.

The advance in computers will also promote improvement of computational methods. The trend in this area will be to move away from methods that rely on linear and quadratic function fits to reduce the differential equations to difference equations. This advance is already beginning to take place since problems are now being attacked by employing Hermitian techniques (Hirsh, 1975; Peyret, 1978) and Spline Interpolation (Rubin and Khosla, 1977). In addition, Fourier and Chebyshev expansions are being used in conjunction with fast Fourier transforms to solve the nonlinear fluid-mechanics equations to high accuracy (Orszag and Israeli, 1974). This advance offers the possibility of reducing computation times by more than an order of magnitude over finite differences for two- and three-dimensional problems, the reason being that these types of expansions need fewer grid points to obtain the same resolution as a second- or third-order finite-difference method. In recent publications by Haidvogel, Robinson, and Schulman (1980) as well as Myers, Taylor, and Murdock (1981) these types of improvements are suggested.

Even with this expected progress in computational capability, a number of technical areas offer obstacles to be overcome in simulation of flows. Difficult areas include two-phase flow simulation and simulation of complicated gas dynamic flows with strong embedded shocks and expansions. An example of such a flow is shown in Fig. 1.5. This figure displays the transient flow that

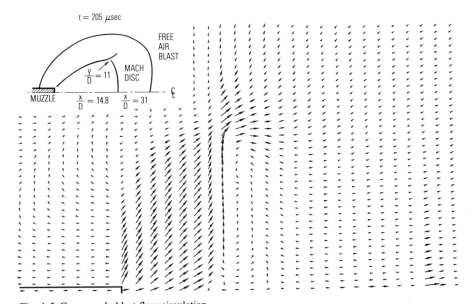

Fig. 1.5 Gun muzzle blast flow simulation.

occurs at the exit of a gun after firing. This type of flow has very strong expansions at the muzzle lip and also strong shock reflections if muzzle brakes are employed. Other examples of difficult simulations include high Reynolds number laminar flows with separation, transition to turbulent flow, fully turbulent flow, transonic flows, stratified flow and radiation coupled flows, as occur in gas lasers.

In the material that follows, an attempt has been made to provide the reader with tools for attacking complicated flow problems. We begin by giving the basis for the computational methods and then proceed to applications that will guide the reader in the use of the methods. In the applications, the pitfalls of the methods are pointed out. The book is meant to be practical and application oriented. As a result, we have intentionally avoided extensive mathematical arguments and attempted to complement other books on the subject such as those by Gottlieb and Orszag (1977), Holt (1977), Richtmyer and Morton (1967), and Roache (1972).

We begin the discussion by first introducing the flow equations that will be considered in various forms in the book.

1.1 The General Navier–Stokes Equations

The motion of a continuous medium is governed by the principles of classical mechanics and thermodynamics for the conservation of mass, momentum, and energy. Application of these principles in a Galilean (absolute) frame of reference leads to the following conservation equations in integral form for mass, momentum, and energy, respectively,

$$\frac{d}{dt} \int_v \rho \, dv + \int_\Sigma \rho \mathbf{V} \cdot \mathbf{N} \, d\Sigma = 0 \tag{1.1}$$

$$\frac{d}{dt} \int_v \rho \mathbf{V} \, dv + \int_\Sigma [(\mathbf{N} \cdot \mathbf{V})\rho \mathbf{V} - \mathbf{N}\boldsymbol{\sigma}] \, d\Sigma = \int_v \mathbf{f}_e \, dv \tag{1.2}$$

$$\frac{d}{dt} \int_v \rho E \, dv + \int_\Sigma \mathbf{N} \cdot [\rho E \mathbf{V} - \boldsymbol{\sigma}\mathbf{V} + \mathbf{q}] \, d\Sigma = \int_v \mathbf{f}_e \cdot \mathbf{V} \, dv \tag{1.3}$$

In these equations, t is the time, ρ the density, $\mathbf{V}$ the velocity of a material particle in the frame of reference, E the total specific energy:

$$E = e + \tfrac{1}{2}\mathbf{V}^2 \tag{1.4}$$

where e is the specific internal energy; $\boldsymbol{\sigma}$ is the stress tensor, $\mathbf{q}$ the heat-flux vector, $\mathbf{f}_e$ the external force per unit volume, and $\mathbf{N}$ is the unit outward normal to the boundary Σ of the fixed control volume v. The energy equation is valid under the assumption that there is no source or sink of energy in v.

The properties of the medium need not be continuous functions of space and time; if they are continuous and sufficiently differentiable in some domain of space and time, then the conservation equations in integral form (1.1)–(1.3) can be transformed into an equivalent set of partial differential equations through the divergence theorem:

$$\frac{\partial \rho}{\partial t} + \nabla \cdot (\rho \mathbf{V}) = 0 \tag{1.5}$$

$$\frac{\partial}{\partial t}(\rho \mathbf{V}) + \nabla \cdot (\rho \mathbf{V}\mathbf{V} - \boldsymbol{\sigma}) = \mathbf{f}_e \tag{1.6}$$

$$\frac{\partial}{\partial t}(\rho E) + \nabla \cdot (\rho E \mathbf{V} - \boldsymbol{\sigma}\mathbf{V} + \mathbf{q}) = \mathbf{f}_e \cdot \mathbf{V} \tag{1.7}$$

One thus obtains the equations in *divergence*, or *conservative*, form. Equivalent *nonconservative* forms are

$$\frac{D\rho}{Dt} + \rho \nabla \cdot \mathbf{V} = 0 \tag{1.8}$$

$$\rho \frac{D\mathbf{V}}{Dt} - \nabla \cdot \boldsymbol{\sigma} = \mathbf{f}_e \tag{1.9}$$

$$\rho \frac{De}{Dt} - \boldsymbol{\sigma} \cdot \nabla \mathbf{V} + \nabla \cdot \mathbf{q} = 0 \tag{1.10}$$

where $D/Dt = \partial/\partial t + \mathbf{V} \cdot \nabla$ is the material derivative.

The above equations are based on the Eulerian approach for the description of the continuum motion: the characteristic properties of the medium (ρ, $\mathbf{V}$, etc.) are considered as functions of time and space in the frame of reference. An alternative description is provided by the Lagrangian formulation in which the dependent variables are the characteristic properties of material particles that are followed in their motion: these properties are thus functions of time and of parameters used to identify the particles, such as the particle coordinates at some fixed initial time. The Lagrangian description, or more precisely mixed Lagrangian–Eulerian formulations, are particularly interesting for problems involving different media with interfaces; however, they are not as widely used in fluid mechanics as the Eulerian formulation and will not be considered here.

The basic dependent variables in Eqs. (1.1)–(1.3), or (1.5)–(1.10), are ρ, $\mathbf{V}$, and E (or e). Constitutive relationships for the stress tensor $\boldsymbol{\sigma}$ and for the heat-flux vector $\mathbf{q}$ must be added to these equations in order to obtain a closed system. We are concerned here with the case of Newtonian fluids, i.e., by definition, fluids such that the stress tensor is a linear function of the velocity gradient. From this definition, excluding the existence of distributed force couples, results Newton's law, also called the Navier–Stokes law, for $\boldsymbol{\sigma}$:

$$\boldsymbol{\sigma} = -p\mathbf{I} + \boldsymbol{\tau}$$
$$\boldsymbol{\tau} = \lambda(\nabla \cdot \mathbf{V})\mathbf{I} + 2\mu \ \text{def } \mathbf{V} \tag{1.11}$$

with def $\mathbf{V} = \frac{1}{2}[\nabla\mathbf{V} + (\nabla\mathbf{V})^t]$, the superscript t denoting the transpose of a tensor. In these relations, p is the pressure, $\boldsymbol{\tau}$ the viscous stress tensor, λ and μ are the two coefficients of viscosity, and def $\mathbf{V}$ is the tensor of rates of deformation. Furthermore, the fluid is assumed to obey Fourier's law of heat conduction for $\mathbf{q}$:

$$\mathbf{q} = -k\,\nabla T \tag{1.12}$$

where T is the absolute temperature, and k is the thermal conductivity coefficient. Many fluids, in particular air and water, follow Newton's law and Fourier's law.

Introducing Newton's law into the momentum and energy equations, for example, in the forms (1.9) and (1.10), one obtains, respectively:

$$\rho\frac{D\mathbf{V}}{Dt} + \nabla p = \mathbf{f}_e + \mu\,\nabla^2\mathbf{V} + (\lambda + \mu)\nabla(\nabla\cdot\mathbf{V})$$
$$+ (\nabla\cdot\mathbf{V})\,\nabla\lambda + 2(\text{def }\mathbf{V})\,\nabla\mu \tag{1.13}$$

and

$$\rho\frac{De}{Dt} + p\,\nabla\cdot\mathbf{V} = \Phi - \nabla\cdot\mathbf{q} \tag{1.14}$$

where Φ is the dissipation function:

$$\Phi = \boldsymbol{\tau}\cdot\nabla\mathbf{V} = \lambda(\nabla\cdot\mathbf{V})^2 + 2\mu(\text{def }\mathbf{V})\cdot(\text{def }\mathbf{V}) \tag{1.15}$$

An alternate form of the energy equation is obtained by introducing the specific entropy s:

$$\rho T\frac{Ds}{Dt} = \Phi - \nabla\cdot\mathbf{q} \tag{1.16}$$

The state variables ρ, e, T, and p are connected by thermodynamic relationships (assuming local thermodynamic equilibrium). We consider the case of a simple fluid such that all its thermodynamic properties can be deduced from a single fundamental relationship which, for a compressible fluid, can be chosen of the type

$$s = s\,(\rho, e)$$

From this relationship, the pressure p and temperature T are obtained in terms of the basic variable ρ and e from

$$p = -\rho^2 T\left(\frac{\partial s}{\partial\rho}\right)_e, \qquad T = \frac{1}{(\partial s/\partial e)_\rho} \tag{1.17}$$

An important special case is a perfect gas with constant specific heats c_p and c_v. For such a gas the laws of state are

$$p = (\gamma - 1)\rho e, \qquad \gamma = \frac{c_p}{c_v} \tag{1.18a}$$

$$e = c_v T \tag{1.18b}$$

The viscosity and thermal conductivity coefficients depend on the local thermodynamic state; in most conditions they depend only on the temperature:

$$\lambda = \lambda(T), \qquad \mu = \mu(T), \qquad k = k(T) \tag{1.19}$$

From the second law of thermodynamics the dissipation function Φ cannot be negative. It can be shown that this leads to the conditions

$$3\lambda + 2\mu \geq 0, \qquad \mu \geq 0 \tag{1.20}$$

and, in the absence of internal relaxation phenomena which would involve departure from local thermodynamic equilibrium, the Stokes relationship

$$3\lambda + 2\mu = 0 \tag{1.21}$$

is generally accepted as a valid approximation.

1.2 Various Forms of the Navier–Stokes Equations

1.2.1 Dimensionless form

First we consider the dimensionless form of these flow equations. To define dimensionless variables, characteristic values of all the variables entering the Navier–Stokes equations can be constructed from the following reference quantities: a reference length L, a reference velocity V^*, a reference density ρ^*, and reference values μ^* and k^* of the coefficients of viscosity and thermal conductivity.

All other characteristic quantities can be derived from these basic ones; we choose L/V^* for t, $\rho^* V^{*2}$ for $\boldsymbol{\sigma}$, $\rho^* V^{*2}/L$ for f_e, V^{*2} for e and E, and $\rho^* V^{*3}$ for $\mathbf{q}$, so that Eqs. (1.1)–(1.10) remain unchanged in dimensionless form (denoting the dimensionless quantity by means of the same symbols as the corresponding dimensional variables). Furthermore, we use as reference values $\rho^* V^{*2}$ for p and $\mu^* V^*/L$ for $\boldsymbol{\tau}$ so that the constitutive relationships (1.11) become in dimensionless form:

$$\boldsymbol{\sigma} = -p\mathbf{I} + \frac{1}{\mathrm{Re}}\,\boldsymbol{\tau}, \qquad \boldsymbol{\tau} = \lambda(\nabla \cdot \mathbf{V})\mathbf{I} + 2\mu\,\mathrm{def}\,\mathbf{V} \tag{1.22}$$

where $\mathrm{Re} = V^* L \rho^*/\mu^*$ is a characteristic Reynolds number, and, considering the case of a perfect gas, relationship (1.12) becomes

$$\mathbf{q} = -\frac{\gamma}{\mathrm{Re}\,\mathrm{Pr}}\,k\,\nabla e \tag{1.23}$$

where $\text{Pr} = \mu^* c_p / k^*$ is the Prandtl number. The energy equation in the form (1.14) remains unchanged in dimensionless form if Φ is now defined as $(\boldsymbol{\tau} \cdot \nabla \mathbf{V}) / \text{Re}$. Let us introduce dimensionless temperature and entropy by using characteristic values T^* and s^* such that $s^* T^* = V^{*2}$. Then the energy equation (1.16) and the law of state (1.17) remain unchanged in dimensionless form. For a perfect gas with constant specific heats, the law of state (1.18a) is unchanged, while (1.18b) becomes $e = T / [\gamma(\gamma - 1)M^{*2}]$, where $M^* = V^* / \sqrt{\gamma R T^*}$ is a characteristic Mach number. If we choose $s^* = c_v$, that is, $c_v T^* = V^{*2}$, then we get simply $e = T$.

1.2.2 Orthogonal curvilinear coordinates

We now consider the Navier–Stokes equations written in an orthogonal curvilinear coordinate system fixed with respect to the absolute frame of reference. Denote x_i ($i = 1, 2, 3$) as these coordinates and let $\boldsymbol{\alpha}$, $\boldsymbol{\beta}$, and $\boldsymbol{\gamma}$ be the local unit vectors, forming an orthogonal triad, tangent to the coordinate lines x_1, x_2, and x_3, respectively. An elementary displacement can be written as

$$d\mathbf{s} = h_1 \, dx_1 \boldsymbol{\alpha} + h_2 \, dx_2 \boldsymbol{\beta} + h_3 \, dx_3 \boldsymbol{\gamma}$$

where h_1, h_2, and h_3 are the metric coefficients. The velocity components in the local axis are u_1, u_2, u_3:

$$\mathbf{V} = u_1 \boldsymbol{\alpha} + u_2 \boldsymbol{\beta} + u_3 \boldsymbol{\gamma}$$

The continuity equation (1.5) is then written

$$\frac{\partial \rho}{\partial t} + \frac{1}{h_1 h_2 h_3} \frac{\partial}{\partial x_j} \left(\frac{1}{h_j} h_1 h_2 h_3 \, \rho u_j \right) = 0 \tag{1.24}$$

where the summation convention is used.

We write the equation for the component of momentum in the $\boldsymbol{\alpha}$ direction, i.e., ρu_1; the two other equations for ρu_2 and ρu_3 follow by cyclic permutation. From the momentum equation in the form (1.6) we get

$$\frac{\partial}{\partial t} (\rho u_1) + \frac{1}{h_1 h_2 h_3} \frac{\partial}{\partial x_j} \left(\frac{1}{h_j} h_1 h_2 h_3 \, \mathcal{T}_{j_1} \right) + \frac{1}{h_1 h_2} \left(\mathcal{T}_{12} \frac{\partial h_1}{\partial x_2} - \mathcal{T}_{22} \frac{\partial h_2}{\partial x_1} \right)$$
$$+ \frac{1}{h_1 h_3} \left(\mathcal{T}_{13} \frac{\partial h_1}{\partial x_3} - \mathcal{T}_{33} \frac{\partial h_3}{\partial x_1} \right) = \mathbf{f}_e \cdot \boldsymbol{\alpha} \tag{1.25}$$

where $\mathcal{T}_{ij}$ are the components of the tensor $\mathcal{T}$:

$$\mathcal{T} = \rho \mathbf{V} \mathbf{V} - \boldsymbol{\sigma} = \rho \mathbf{V} \mathbf{V} + p \mathbf{I} - \frac{1}{\text{Re}} \boldsymbol{\tau}$$

$$\mathcal{T}_{ij} = \rho u_i u_j + p \delta_{ij} - \frac{1}{\text{Re}} \tau_{ij}, \qquad \tau_{ij} = \lambda (\nabla \cdot \mathbf{V}) \delta_{ij} + 2\mu \, (\text{def } \mathbf{V})_{ij},$$

where δ_{ij} is Kronecker's symbol.

$$\nabla \cdot \mathbf{V} = \frac{1}{h_1 h_2 h_3} \frac{\partial}{\partial x_l} \left(\frac{1}{h_l} h_1 h_2 h_3 u_l \right)$$

$$(\text{def } \mathbf{V})_{11} = \frac{1}{h_1} \left(\frac{\partial u_1}{\partial x_1} + \frac{u_2}{h_2} \frac{\partial h_1}{\partial x_2} + \frac{u_3}{h_3} \frac{\partial h_1}{\partial x_3} \right),$$

with $(\text{def } \mathbf{V})_{22}$ and $(\text{def } \mathbf{V})_{33}$ following by cyclic permutation.

$$(\text{def } \mathbf{V})_{ij \ (i \neq j)} = \frac{1}{2} \left[\frac{1}{h_i} \frac{\partial u_j}{\partial x_i} + \frac{1}{h_j} \frac{\partial u_i}{\partial x_j} - \frac{1}{h_i h_j} \left(u_i \frac{\partial h_i}{\partial x_j} + u_j \frac{\partial h_j}{\partial x_i} \right) \right] \tag{1.26}$$

The energy equation (1.7) becomes

$$\frac{\partial}{\partial t} (\rho E) + \frac{1}{h_1 h_2 h_3} \frac{\partial}{\partial x_j} \left\{ \frac{1}{h_j} h_1 h_2 h_3 \right.$$

$$\left. \times \left[(\rho E + p) u_j - \frac{1}{\text{Re}} \tau_{ji} u_i - \frac{\gamma k}{\text{Re Pr}} \frac{1}{h_j} \frac{\partial e}{\partial x_j} \right] \right\} = \mathbf{f}_e \cdot \mathbf{V} \tag{1.27}$$

The same transforms can be applied to the other equations presented in the previous section.

1.2.3 Plane flows

Here we consider in more detail the equations for two-dimensional plane flows. These equations are deduced from the above equations by choosing the planes that contain the flow trajectories as coordinate surface $x_3 = \text{const}$, so that $u_3 = 0$, $h_3 = 1$, $\partial / \partial x_3 = 0$; x_1 and x_2 being the orthogonal coordinates in these planes.

First, we write the Navier–Stokes in Cartesian coordinates x, y ($x_1 = x$, $x_2 = y$, and $u_1 = u$, $u_2 = v$). The system of the four conservation equations without external forces can be written in the following form:

$$\frac{\partial f}{\partial t} + \frac{\partial F}{\partial x} + \frac{\partial G}{\partial y} = 0 \tag{1.28}$$

where f, F, G are four-component vectors. The vectors F and G can be decomposed as

$$F = F_\text{I} - \frac{1}{\text{Re}} F_\text{V}, \qquad G = G_\text{I} - \frac{1}{\text{Re}} G_\text{V} \tag{1.29}$$

where F_I and G_I correspond to the inviscid part of the terms F, G and F_V, G_V to the viscous part. More precisely,

$$f = \begin{pmatrix} \rho \\ \rho u \\ \rho v \\ \rho E \end{pmatrix}, \qquad F_\text{I} = \begin{pmatrix} \rho u \\ \rho u^2 + p \\ \rho u v \\ (\rho E + p) u \end{pmatrix}, \qquad G_\text{I} = \begin{pmatrix} \rho v \\ \rho u v \\ \rho v^2 + p \\ (\rho E + p) v \end{pmatrix}, \tag{1.30}$$

$$F_V = \begin{pmatrix} 0 \\ \tau_{xx} \\ \tau_{xy} \\ u\tau_{xx} + v\tau_{xy} + \dfrac{\gamma}{\mathrm{Pr}} k \dfrac{\partial e}{\partial x} \end{pmatrix}, \qquad G_V = \begin{pmatrix} 0 \\ \tau_{xy} \\ \tau_{yy} \\ u\tau_{xy} + v\tau_{yy} + \dfrac{\gamma}{\mathrm{Pr}} k \dfrac{\partial e}{\partial y} \end{pmatrix}$$

With the Stokes relation $3\lambda + 2\mu = 0$, we have

$$\tau_{xx} = \tfrac{2}{3}\mu\left(2\frac{\partial u}{\partial x} - \frac{\partial v}{\partial y}\right), \qquad \tau_{xy} = \mu\left(\frac{\partial u}{\partial y} + \frac{\partial v}{\partial x}\right),$$

$$\tau_{yy} = \tfrac{2}{3}\mu\left(2\frac{\partial v}{\partial y} - \frac{\partial u}{\partial x}\right)$$

In addition to these equations, other forms of the Navier–Stokes equation can be considered in arbitrary curvilinear coordinates (which may depend on time). Let, τ, ξ, η be independent variables:

$$\tau = t, \qquad \xi = \xi(x, y, t), \qquad \eta = \eta(x, y, t)$$

then a *fully conservative form* of the transformed equations can be derived as shown by Viviand (1974). These equations are written

$$\frac{\partial \overline{f}}{\partial \tau} + \frac{\partial \overline{F}}{\partial \xi} + \frac{\partial \overline{G}}{\partial \eta} = 0 \tag{1.31}$$

where $\overline{f} = f/D$, $D = \partial(\xi, \eta)/\partial(x, y)$ is the Jacobian of the transformation, and

$$\overline{F} = \frac{1}{D}\left(f\frac{\partial \xi}{\partial t} + F\frac{\partial \xi}{\partial x} + G\frac{\partial \xi}{\partial y}\right)$$

$$\overline{G} = \frac{1}{D}\left(f\frac{\partial \eta}{\partial t} + F\frac{\partial \eta}{\partial x} + G\frac{\partial \eta}{\partial y}\right)$$

The first-order derivatives included in F_V and G_V can be transformed in a similar manner.

A *quasiconservative form* is obtained from Eq. (1.28) by the straightforward use of the chain rule:

$$\left(\frac{\partial}{\partial \tau} + \frac{\partial \xi}{\partial t}\frac{\partial}{\partial \xi} + \frac{\partial \eta}{\partial t}\frac{\partial}{\partial \eta}\right)f$$

$$+ \left(\frac{\partial \xi}{\partial x}\frac{\partial}{\partial \xi} + \frac{\partial \eta}{\partial x}\frac{\partial}{\partial \eta}\right)F + \left(\frac{\partial \xi}{\partial y}\frac{\partial}{\partial \xi} + \frac{\partial \eta}{\partial y}\frac{\partial}{\partial \eta}\right)G = 0 \tag{1.31a}$$

It is obvious that any derivative involved in the viscous terms F_V and G_V must also be differentiated with the same chain rule.

1.3 The Navier–Stokes Equations for Incompressible Flow

1.3.1 The primitive-variable formulation
An incompressible flow is characterized by the condition that

$$\mathbf{\nabla} \cdot \mathbf{V} = 0 \tag{1.32}$$

This condition when introduced into the continuity equation (1.5) implies that

$$\frac{\partial \rho}{\partial t} + \mathbf{V} \cdot \mathbf{\nabla} \rho = 0 \quad \text{or} \quad \frac{D\rho}{Dt} = 0 \tag{1.33}$$

This condition states that the density is constant along a fluid particle trajectory. In most cases one usually assumes ρ to be uniform so that this condition is satisfied identically everywhere. For stratified flows Eq. (1.33) is frequently employed to compute a small perturbation in density, ρ', about a mean constant density ρ_0 so that $\rho = \rho_0 + \rho'$. However, one must take care in employing such an approach to be certain that it is consistent with all transport equations.

In the event that $\mu = \text{const}$, then the momentum equation (1.6) reduces to the form

$$\rho \left[\frac{\partial \mathbf{V}}{\partial t} + \mathbf{\nabla} \cdot (\mathbf{VV}) \right] + \mathbf{\nabla} p - \mu \, \nabla^2 \mathbf{V} = \mathbf{f}_e \tag{1.34}$$

For this case the unknowns are the velocity field and pressure. These can be determined from Eqs. (1.32) and (1.34), but this presents some difficulty in incompressible flow problems since boundary conditions typically only exist for the velocity field. The most direct way to solve for the pressure is to combine Eqs. (1.32) and (1.34) to obtain a Poisson equation for the pressure of the form

$$\nabla^2 p = g(u, v, w)$$

where $g(u, v, w)$ is a function of the components of the velocity vector. This equation must be solved subject to Eq. (1.34) applied at each boundary for the boundary condition on $\partial p / \partial N$, the normal pressure gradient. If, however, the viscosity vanishes and the flow is steady and irrotational, then one has

$$p + \tfrac{1}{2}\rho(u^2 + v^2 + w^2) = \text{const}$$

for determination of the pressure.

Equation (1.34) can be transformed by using condition (1.32) to obtain

$$\rho \left[\frac{\partial \mathbf{V}}{\partial t} + (\mathbf{V} \cdot \mathbf{\nabla})\mathbf{V} \right] + \mathbf{\nabla} p - \mu \, \nabla^2 \mathbf{V} = \mathbf{f}_e \tag{1.35}$$

This form is called the *nonconservative* form (or *convective* form) of the Navier–Stokes equations while Eq. (1.34) is in *conservative* form (or *divergence* form).

1.3.2 The stream-function vorticity formulation

Another formulation of the Navier–Stokes equations makes use of the vorticity vector

$$\boldsymbol{\omega} = \boldsymbol{\nabla} \times \mathbf{V} \qquad (1.36)$$

An equation for $\boldsymbol{\omega}$ is obtained by applying the curl operator to Eq. (1.35) so that the pressure term disappears. The result is

$$\frac{\partial \boldsymbol{\omega}}{\partial t} + (\mathbf{V} \cdot \boldsymbol{\nabla})\boldsymbol{\omega} - (\boldsymbol{\omega} \cdot \boldsymbol{\nabla})\mathbf{V} - \nu\, \boldsymbol{\nabla}^2 \boldsymbol{\omega} = \frac{1}{\rho} \boldsymbol{\nabla} \times \mathbf{f}_e \qquad (1.37)$$

where $\nu = \mu/\rho$. This equation is usually associated with an equation for a solenoidal *stream-function vector* $\boldsymbol{\Psi}$ such that

$$\mathbf{V} = \boldsymbol{\nabla} \times \boldsymbol{\Psi} \qquad (1.38)$$

which automatically satisfies incompressibility condition (1.32). The equation satisfied by $\boldsymbol{\Psi}$ is derived by applying the curl operator to Eq. (1.38) and using definition (1.36) to obtain

$$\nabla^2 \boldsymbol{\Psi} + \boldsymbol{\omega} = 0 \qquad (1.39)$$

Such a formulation becomes most interesting when the vector $\boldsymbol{\Psi}$ has only one component. This is true, in particular, for plane flows where

$$\mathbf{V} = \boldsymbol{\nabla} \times (\Psi \mathbf{k}) \qquad (1.40)$$

where $\mathbf{k}$ is the unit vector normal to the plane of flow and Ψ is a scalar function. In this case, the vorticity $\boldsymbol{\omega} = \omega \mathbf{k}$ and Eqs. (1.37) and (1.39) become scalar equations:

$$\frac{\partial \omega}{\partial t} + (\mathbf{V} \cdot \boldsymbol{\nabla})\omega - \nu\, \boldsymbol{\nabla}^2 \omega = \frac{1}{\rho} \boldsymbol{\nabla} \times \mathbf{f}_e \qquad (1.41)$$

$$\nabla^2 \Psi + \omega = 0 \qquad (1.42)$$

As for the equations in primitive variables, Eq. (1.41) is said to be in *nonconservative* form (or *convective* form) and the *conservative* form (or *divergence* form) of the vorticity equation is

$$\frac{\partial \omega}{\partial t} + \boldsymbol{\nabla} \cdot (\mathbf{V}\omega) - \nu\, \boldsymbol{\nabla}^2 \omega = \frac{1}{\rho} \boldsymbol{\nabla} \times \mathbf{f}_e \qquad (1.43)$$

If the stream-function vorticity equations are made dimensionless by scaling vorticity ω by V^*/L, the stream function by V^*L, the space dimensions by L, and t by L/V^*, then in Eqs. (1.37), (1.41), and (1.43), the quantity $\nu = \mu/\rho$ must be replaced by the reciprocal Reynolds number, i.e., $1/\mathrm{Re} = \nu/V^*L$. Also in the right-hand sides of the equations f_e/ρ must be replaced by $f_e L/\rho V^{*2}$.

Note also that an important feature of the stream-function vorticity formulation is that the pressure is no longer explicit in the equations. If, however, one needs the pressure along a line or a boundary, one can integrate the tangential derivative of pressure obtained from the momentum equation written in coordinates along the boundary or line. For the pressure in the complete field, however, one must solve the Poisson equation presented earlier.

References

Chapman, D. R., Mark, H., and Pirtle, M. W. *Astronaut. Aeronaut.* 22–35 (1975).

Chapman, D. R. AIAA Paper 79-0129 (1979).

Enselme, M., Brochet, J., and Boisseau, J. P. Low-Cost Three-Dimensional Flow Computations using a Mini-system; 5th AIAA Computational Fluid Dynamics Conference, Palo Alto, June 1981.

Godunov, S. K. *Mat. Sb.* **47**, 271–306 (1959).

Gottlieb, D., and Orszag, S. A. *Numerical Analysis of Spectral Methods.* SIAM, Philadelphia, 1977.

Haidvogel, D. B., Robinson, A. R., and Schulman, E. E. *J. Comput. Phys.* **34**, 1–53 (1980).

Hirsh, R. *J. Comput. Phys.* **19**, 90–109 (1975).

Holt, M. *Numerical Methods in Fluid Dynamics.* Springer, New York, 1977.

Michalke, A. *J. Fluid Mech.* **19**, 543–556 (1964).

Myers, R., Taylor, T. D., and Murdock, J. *J. Comput. Phys.* **34**, 180–188 (1981).

Orszag, S. A., and Israeli, M. *Ann. Rev. Fluid Mech.* **6**, 281–317 (1974).

Peyret, R. ONERA T. P. No. 1978-29, Chatillon, France (1978) and *Proceedings 1st International Conf. on Numerical Methods in Laminar and Turbulent Flow,* pp. 43–54. Pentech Press, Plymouth, U.K., 1978.

Richtmyer, R. D., and Morton, K. W. *Difference Methods for Initial-Value Problems.* Interscience, New York, 1967.

Riley, J., and Metcalfe, R. AIAA Paper 80-0274 (1980).

Roache, P. J. *Computational Fluid Dynamics.* Hermosa, Albuquerque, 1972.

Rubin, S. G., and Khosla, P. *J. Comput. Phys.* **24**, 217–244 (1977).

Viviand, H. *Rech. Aerosp.* No. 1974-1, 65–78 (1974).

CHAPTER 2

Finite-Difference Methods

Until recently, numerical methods for solving fluid-flow problems have been dominated by finite-difference approximations. These methods are powerful and play a major role in problem solutions. In this chapter we attempt to present the fundamental advances and insight into these methods. We begin by discussing the concept of discrete pointwise approximation of functions.

2.1 Discrete Approximations

Let $f(x)$ be a function defined in the range $a \le x \le b$. The interval $[a,b]$ is discretized by considering the set $x_0 = a, x_1, \ldots, x_i, \ldots, x_{N+1} = b$, and the discrete representation of $f(x)$ is the set $\{f(a), f(x_1), \ldots, f(x_i), \ldots, f(b)\}$. Generally, the value $f(x_i)$ is denoted by f_i. When $f(x)$ is known as a solution of some mathematical problem, say the solution of a differential equation, the values $f(x_i)$ are not known exactly but are the result of some approximation and, in this case, $\{f_i\}$ is a discrete approximation of $f(x)$ and we note $f_i \cong f(x_i)$.

The quantity $(x_{i+1} - x_i)$ is the mesh size and we shall assume, for the sake of simplicity, that this mesh size is a constant: $\Delta x = (b - a)/(N + 1)$ and $x_i = a + i\Delta x, i = 0, \ldots, N + 1$.

The mth derivative of $f(x)$ at point x_i is approximated in the form

$$\frac{d^m f(x_i)}{dx^m} \cong \sum_{j=-J_1}^{j=J_2} \alpha_j f_{i+j} \tag{2.1.1}$$

where the α_j's are determined by means of Taylor expansions of f_{i+j} around x_i and J_1, J_2 are integers depending on the order m of the considered derivative and also on the degree of accuracy of the approximation. If for $m = 1$ we consider an approximation using three values of f_i, i.e., $J_1 + J_2 = 2$, and we take $J_1 = J_2 = 1$, we can write a general expression

$$\frac{df(x_i)}{dx} \cong \frac{(1 - \alpha) f_{i+1} + 2\alpha f_i - (1 + \alpha) f_{i-1}}{2 \Delta x} \tag{2.1.2}$$

where α is an arbitrary constant. The error of this approximation is

$$- \frac{\alpha \, \Delta x}{2} \frac{d^2 f}{dx^2} - \frac{\Delta x^2}{6} \frac{d^3 f}{dx^3} + O\,(\Delta x^3)$$

By specifying the value of α we obtain the standard differences:

Centered: $\alpha = 0$

$$\frac{df(x_i)}{dx} \cong \frac{f_{i+1} - f_{i-1}}{2 \, \Delta x} \equiv \Delta_x^0 f_i, \quad \text{error} = O(\Delta x^2) \tag{2.1.3a}$$

Backward: $\alpha = 1$

$$\frac{df(x_i)}{dx} \cong \frac{f_i - f_{i-1}}{\Delta x} \equiv \Delta_x^- f_i, \quad \text{error} = O(\Delta x) \tag{2.1.3b}$$

Forward: $\alpha = -1$

$$\frac{df(x_i)}{dx} \cong \frac{f_{i+1} - f_i}{\Delta x} \equiv \Delta_x^+ f_i, \quad \text{error} = O(\Delta x) \tag{2.1.3c}$$

Now, if we choose $J_1 = 2$ and $J_2 = 0$ and employ a Taylor series to find the α_j's we obtain the second-order accurate approximation.

$$\frac{df(x_i)}{dx} \cong \frac{1}{2 \, \Delta x} \, (3 \, f_i - 4 \, f_{i-1} + f_{i-2}) \tag{2.1.4a}$$

Similarly if $J_1 = 0$, $J_2 = 2$, we obtain the second-order accurate approximation

$$\frac{df(x_i)}{dx} \cong \frac{-3 \, f_i + 4 \, f_{i+1} - f_{i+2}}{2 \, \Delta x} \tag{2.1.4b}$$

A fourth-order accurate result is obtained if $J_1 = J_2 = 2$, i.e.,

$$\frac{df(x_i)}{dx} \cong \frac{-f_{i+2} + 8 \, f_{i+1} - 8 \, f_{i-1} + f_{i-2}}{12 \, \Delta x} \tag{2.1.5}$$

In the same way, we can define approximations for all the derivatives. For example, the classical difference approximation of the second derivative is

$$\frac{d^2 f(x_i)}{dx^2} \cong \frac{f_{i+1} - 2 \, f_i + f_{i-1}}{\Delta x^2} \equiv \Delta_{xx} f_i \tag{2.1.6}$$

which is accurate to $O(\Delta x^2)$.

The use of the difference operators Δ_x^0, Δ_x^+, Δ_{xx}, . . . is very useful, and it is easy to verify the following relationships:

$$\tfrac{1}{2}(\Delta_x^+ + \Delta_x^-) = \Delta_x^0$$

$$\Delta_x^+ - \Delta_x^- = \Delta x \, \Delta_{xx}$$

$$\Delta_x^+ \, \Delta_x^- = \Delta_{xx}$$

However, note that $\Delta_x \Delta_x \neq \Delta_{xx}$ because

$$\Delta_x \Delta_x f_i = \frac{f_{i+2} - 2f_i + f_{i-2}}{4 \, \Delta x^2}$$

2.2 Solution of an Ordinary Differential Equation

Let us consider as a model the simple differential equation

$$A \frac{df}{dx} - \nu \frac{d^2 f}{dx^2} = 0, \qquad 0 < x < 1 \tag{2.2.1a}$$

$$f(0) = U_0, \qquad f(1) = U_1 \tag{2.2.1b}$$

where $\nu = \text{const} > 0$ and $A = A(x)$. The interval $(0,1)$ is discretized with $x_i = i \, \Delta x, \, i = 0, \ldots, N + 1$, with $\Delta x = 1/(N + 1)$. If we denote by f_i the numerical approximation of $f(x_i)$ at point x_i, a finite-difference approximation of (2.2.1) is

$$A_i \Delta_x^0 f_i - \nu \Delta_{xx} f_i = 0, \qquad i = 1, \ldots, N \tag{2.2.2a}$$

$$f_0 = U_0, \qquad f_{N+1} = U_1 \tag{2.2.2b}$$

The use of centered differences leads to an error of second order with respect to the mesh size Δx. Equation (2.2.2) produces a linear algebraic system whose solution yields the N values of f_i. In the present case the associated matrix is tridiagonal, so it is very efficient to use a direct method of solution. The general method of factorization leads to a simple solution algorithm described in the following section.

2.2.1 The method of factorization

This method to solve a linear algebraic system,

$$\mathcal{A}F = G \tag{2.2.3}$$

where $\mathcal{A}$ is a tridiagonal matrix, F the vector of unknowns, and G a given vector, is a particular form of the general method of factorization which expresses $\mathcal{A}$ as a product of two triangular matrices $\mathcal{L}$ and $\mathcal{M}$, i.e., $\mathcal{A} = \mathcal{L}\mathcal{M}$. The solution of Eq. (2.2.3) is then obtained in two steps: First a vector V is computed by $\mathcal{L}V = G$, then the final vector F by $\mathcal{M}F = V$. This decomposition has the advantage that triangular matrices are easily inverted. When applied to a tridiagonal matrix, the method is applied according to the following algorithm.

Let the difference equations corresponding to Eq. (2.2.3) be

$$\begin{aligned} a_i f_{i-1} + b_i f_i + c_i f_{i+1} &= d_i, \qquad i = 1, \ldots, N \\ f_0 = U_0, \qquad f_{N+1} &= U_1. \end{aligned} \tag{2.2.4}$$

For this equation there exists a recurrence relation

$$f_i = X_i f_{i+1} + Y_i, \qquad i = 1, \ldots, N \tag{2.2.5}$$

where the coefficients X_i and Y_i are easily found (by identification) to be

$$X_i = \frac{-c_i}{b_i + a_i X_{i-1}}, \qquad Y_i = \frac{d_i - a_i Y_{i-1}}{b_i + a_i X_{i-1}}, \qquad i = 1, \ldots, N \tag{2.2.6}$$

The coefficients X_0 and Y_0 are determined by considering Eqs. (2.2.4) and (2.2.5) with $i = 1$. Identification of these two equations gives

$$\frac{b_1}{1} = \frac{c_1}{-X_1} = \frac{d_1 - a_1 U_0}{Y_1}$$

which, compared to Eq. (2.2.6) written for $i = 1$, yields

$$X_0 = 0, \qquad Y_0 = U_0$$

The method is applied by first calculating X_i, Y_i ($i = 1, \ldots, N$) with Eq. (2.2.6). Then formula (2.2.5) is applied with i varying from N to 1 for the calculation of f_i.

It can be proven (Isaacson and Keller, 1966) that the algorithm gives accurately bounded results if the following conditions are satisfied:

$$|b_1| > |c_1|, \qquad |b_i| \geq |a_i| + |c_i|, \qquad i = 2, \ldots, N - 1$$

$$|b_N| > |a_N| \tag{2.2.7a}$$

This set of conditions is a special case of the diagonally dominant property of a matrix $\mathcal{A} = [\alpha_{l,m}]$ which states

$$|\alpha_{l,l}| \geq \sum_{m=1}^{N} |\alpha_{l,m}|, \qquad l = 1, \ldots, N \text{ with } l \neq m \tag{2.2.7b}$$

In the case of Eqs. (2.2.2) the conditions (2.2.7a) are satisfied if and only if

$$\frac{|A| \, \Delta x}{v} < 2$$

where $|A| \, \Delta x / v$ is called the *mesh Reynolds number*.

2.2.2 Iterative methods

As we have seen in the previous section, in the case of a tridiagonal matrix, the direct method of solution of the system (2.2.3) is very efficient. However, because iterative methods are often used when the matrix is no longer tridiagonal, it seems interesting to describe briefly the standard iterative methods. *Jacobi method*: The iterative procedure gives the value of f_i at iteration $m + 1$ by the formula ($b_i \neq 0$):

$$f_i^{m+1} = \frac{1}{b_i} (-a_i f_{i-1}^m - c_i f_{i+1}^m + d_i), \qquad i = 1, \ldots, N \tag{2.2.8}$$

Gauss–Seidel method: Here, f_i^{m+1} is calculated by using the values of the present iteration $m + 1$ which have just been computed. So, if the calculation is made with the index i increasing, the Gauss–Seidel method is

$$f_i^{m+1} = (-a_i f_{i-1}^{m+1} - c_i f_{i+1}^m + d_i)/b_i, \quad i = 1, \ldots, N \tag{2.2.9}$$

Successive-relaxation method: This method defines a provisional value $\tilde{f}_i^{m+1}$ with the Gauss–Seidel technique, then the final value f_i^{m+1} is computed with a relaxation characterized by the positive parameter ω.

$$\begin{aligned} \tilde{f}_i^{m+1} &= (-a_i f_{i-1}^{m+1} - c_i f_{i+1}^m + d_i)/b_i \\ f_i^{m+1} &= \omega \tilde{f}_i^{m+1} + (1 - \omega)f_i^m \end{aligned} \tag{2.2.10}$$

The convergence of these methods is obtained if $\text{Max}_i \, |f_i^{m+1} - f_i^m| \to 0$, when $m \to \infty$, whatever the starting value f_i^0. For the methods of *Jacobi* and *Gauss–Seidel*, a sufficient condition of convergence is that the matrix $\mathcal{A}$ is strictly diagonally dominant; i.e., (2.2.7b) is satisfied with a strict inequality. This condition is not always satisfied; for instance, it is not satisfied in Eq. (2.2.2a). On the other hand, it would be satisfied if a nonderivative term Bf ($B > 0$) was included in the left-hand side of Eq. (2.2.1a).

Therefore, another sufficient condition of convergence, more easily satisfied, is that the matrix is irreducible* and strongly diagonally dominant**, i.e., (2.2.7b) is satisfied with a strict inequality for one value of l at least.

For the *successive-relaxation method*, the known condition of convergence is more stringent: If the matrix $\mathcal{A}$ is symmetrical and definite positive, the method converges if and only if $0 < \omega < 2$. But in fact the method converges for nonsymmetrical matrices. We refer the reader to specialized books (Varga, 1962; Young, 1971) for a detailed study of the convergence of the iterative procedures and in particular for the determination of the optimal value of the relaxation parameter ω.

Remark: The iterative procedures just described are expressed in matrix notation in the following manner. Assuming $b_i \neq 0$, Eq. (2.2.4), after division by b_i, yields a system (2.2.3) in which the main diagonal entries are 1. The matrix $\mathcal{A}$ is then decomposed according to

$$\mathcal{A} = I - \mathcal{M} - \mathcal{N}$$

where $\mathcal{M}$ and $\mathcal{N}$ are, respectively, lower and upper strictly triangular matrices. (Here, they have only one diagonal not null.) The *Jacobi method* is defined by

$$F^{m+1} = (\mathcal{M} + \mathcal{N})F^m + G$$

the *Gauss–Seidel method* by

---

*A matrix is irreducible if the solution of the algebraic system associated cannot be reduced to the successive solution of two systems of lower order. Simple conditions of irreducibility are found from the theory of graphs [see Varga (1962)].

**Note this terminology differs from Young's (1971).

$$(I - \mathcal{M})F^{m+1} = \mathcal{N} F^m + G$$

or

$$F^{m+1} = (I - \mathcal{M})^{-1}\mathcal{N}F^m + (I - \mathcal{M})^{-1}G$$

and the *successive-relaxation method* by

$$\tilde{F}^{m+1} = \mathcal{M}F^{m+1} + \mathcal{N}F^m + G$$
$$F^{m+1} = \omega\tilde{F}^{m+1} + (1 - \omega)F^m$$

2.2.3 Analogy between iterative procedures and equations of evolution

As was shown by Garabedian (1956), there exists a close analogy between some iterative procedures and equations of evolution. If we identify formally the index m defining a time $m\Delta t$ we can easily find the analog of each procedure described above by using a Taylor expansion in space as well as in time. This is the general technique used to determine the truncation error of a finite-difference scheme for a time-dependent equation and is described in detail in Section 2.6.

Therefore, if Taylor expansions are made in each of the equations (2.2.8), (2.2.9), or (2.2.10), we obtain an equation of the form

$$K \frac{\partial f}{\partial t} = \nu \frac{\partial^2 f}{\partial x^2} - A \frac{\partial f}{\partial x} \tag{2.2.11}$$

where only the significant terms have been retained. The coefficient K depends on the iterative procedure considered. The convergence of the iterative procedure is equivalent to the existence of a steady solution of Eq. (2.2.11). A necessary condition for this existence is that Eq. (2.2.11) be parabolic in the direction, $t > 0$. For the Jacobi method,

$$K = K_J = 2\nu \frac{\Delta t}{\Delta x^2} > 0$$

and for the Gauss–Seidel method

$$K = K_{GS} = \frac{\Delta t \nu}{\Delta x^2} \left(1 - \frac{A \Delta x}{2\nu} \right) > 0$$

if $A \Delta x < 2\nu$. Finally, for the successive-relaxation method

$$K = K_{SR} = \frac{\Delta t \nu}{\Delta x^2} \left(\frac{2 - \omega}{\omega} - \frac{A \Delta x}{2\nu} \right)$$

The coefficient K_{SR} is positive if $A \Delta x/\nu < (2 - \omega)/\omega$. However, the condition $K > 0$ is not sufficient to ensure that numerical solution of (2.2.11), given by schemes (2.2.8)–(2.2.10), has a steady limit. A supplementary condition must be considered—that is, the stability of the scheme used to approximate Eq. (2.2.11). The stability of such numerical schemes will be

considered in Section 2.6. If an analysis of stability is made for schemes
(2.2.8)–(2.2.10), conditions equivalent to some of the conditions of convergence mentioned in the previous section are obtained.

Comparisons of the various K coefficients make clear the relative properties
of convergence of the iterative procedures. For example, if $A = 0$, we obtain
$K_{GS} = \frac{1}{2} K_J$ and $\omega K_{SR} = (2 - \omega) K_{GS}$. From Eq. (2.2.11), it is clear that the
smaller K is the more rapid the convergence of the iterative procedure. In
particular, we see that the successive-relaxation method converges more rapidly than the Gauss–Seidel method if $1 < \omega < 2$.

Note that an approximation of the optimal value ω can be obtained by a
slightly more sophisticated analysis (Garabedian, 1956).

2.3 Analytical Solution of the Finite-Difference Problem

We consider now the solution of the differential equation

$$A \frac{df}{dx} - \nu \frac{d^2f}{dx^2} = 0, \qquad 0 < x < 1 \tag{2.3.1a}$$

with

$$f(0) = U_0 \qquad \text{and} \qquad f(1) = U_1 \tag{2.3.1b}$$

where A and ν are constants ($\nu > 0$). The solution is

$$f = U_0 + (U_1 - U_0) \frac{e^{\delta x} - 1}{e^{\delta} - 1} \tag{2.3.2}$$

where $\delta = A/\nu$. Figure 2.3.1 shows the graph of the solution. The essential
feature is the existence of a boundary layer of thickness $O(|\delta|)$ when $\delta \to 0$.
This boundary layer is located near $x = 0$ when $\delta < 0$ and near $x = 1$ where
$\delta > 0$.

Assume Eq. (2.3.1) to be approximated with centered differences: Δ_x^0 for the
first derivative and Δ_{xx} for the second derivative. The resulting finite-difference
problem is

$$(2 - R)f_{i+1} - 4 f_i + (2 + R)f_{i-1} = 0, \qquad i = 1, \ldots, N \tag{2.3.3a}$$

$$f_0 = U_0, \qquad f_{N+1} = U_1 \tag{2.3.3b}$$

where R is the algebraic mesh Reynolds number

$$R = \delta \, \Delta x = \frac{A \, \Delta x}{\nu}$$

It is possible, in simple cases like this one, to calculate analytically the solution
of the difference problem. The theory of finite differences is similar to the

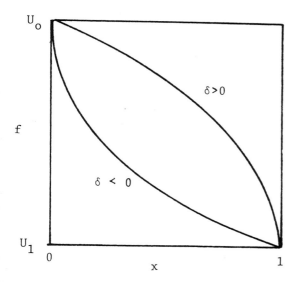

Fig. 2.3.1 Graph of the function f.

theory of ordinary differential equations [see, for example, Godunov and Ryabenski (1964)].

The general solution of Eq. (2.3.3a) is

$$f_i = C_1 \, q_1^i + C_2 \, q_2^i \qquad\qquad (2.3.4)$$

where q_1 and q_2 are the roots of the characteristic equation associated with Eq. (2.3.3a),

$$(2 - R)q^2 - 4\,q + (2 + R) = 0 \qquad\qquad (2.3.5)$$

and C_1, C_2 are constants to be determined by the boundary conditions (2.3.3b). The roots of (2.3.5) are

$$q_1 = 1, \qquad q_2 = \frac{2 + R}{2 - R}$$

and the constants

$$C_1 = U_0 - \frac{U_1 - U_0}{q_2^{N+1} - 1}, \qquad C_2 = \frac{U_1 - U_0}{q_2^{N+1} - 1}$$

so that the solution (2.3.4) is

$$f_i = U_0 + (U_1 - U_0)\,\frac{q_2^i - 1}{q_2^{N+1} - 1} \qquad\qquad (2.3.6)$$

Now consider the limit $\Delta x \to 0$, with $x_i = i\,\Delta x$ fixed. Assuming $\delta = O(1)$, we can show that

$$q_2^i = \left(\frac{2 + \delta \Delta x}{2 - \delta \Delta x}\right)^i = \left(\frac{2 + \delta \Delta x}{2 - \delta \Delta x}\right)^{x_i/\Delta x}$$

$$= e^{\delta x_i}\left[1 + \frac{\delta^3 x_i \Delta x^2}{12} + O(\Delta x^4)\right] \tag{2.3.7}$$

Therefore, we remark that q_2^i is an approximation to second order of the basic solution $e^{\delta x}$ of the differential equation (2.3.1a). At the same time, we note that the other solution $q_1^i = 1$ is exactly the second basic solution of the differential equation.

Finally, by combining Eqs. (2.3.6) and (2.3.7) we obtain

$$f_i = f(x_i) + \Delta x^2 (U_1 - U_0) \Phi(x_i) + O(\Delta x^4) \tag{2.3.8a}$$

$$\Phi(x_i) = -\frac{\delta^3}{12(e^\delta - 1)^2}\left[e^\delta(e^{\delta x_i} - 1) - (e^\delta - 1)x_i e^{\delta x_i}\right] \tag{2.3.8b}$$

where $f(x)$ is the exact solution and the error term is exhibited.

At this point it is interesting to note that expansion (2.3.8) can be derived another way by considering the differential equation obtained from the finite-difference equation (2.3.3) with Taylor's series (construction of the truncation error). This equation is

$$A\frac{df}{dx} - \nu\frac{d^2f}{dx^2} + \frac{\Delta x^2}{6}\left(A\frac{d^3f}{dx^3} - \frac{\nu}{2}\frac{d^4f}{dx^4}\right) + O(\Delta x^4) = 0 \tag{2.3.9}$$

If we look for the solution of this equation in the form of an asymptotic expansion

$$F(x, \Delta x) = F_0(x) + \Delta x F_1(x) + \Delta x^2 F_2(x) + O(\Delta x^3) \tag{2.3.10}$$

with

$$F_0(0) = U_0, \qquad F_0(1) = U_1, \qquad F_j(0) = F_j(1) = 0 \qquad \text{for } j = 1,2$$

we obtain a set of equations determining the various F_j, $j = 0, 1, 2$, where

$$F_0(x) = f(x) = \text{the exact solution}$$

$$F_1(x) = 0$$

$$F_2(x) = (U_1 - U_0)\Phi(x) = \text{the } \Delta x^2 \text{ error term}$$

The formal nature of this analysis does not make it of general application, however, it does make clear the relationship between the solution of the finite-difference scheme (2.3.3) and the *expanded differential equation* (2.3.9).

Now, we assume $\Delta x \ll 1$ but δ is not necessarily of the order of unity, so that the mesh Reynolds number $|R| = |\delta| \Delta x$ can be $O(1)$. We remark that $q_2 < 0$ if $|R| > 2$, which means q_2^i is positive for even values of the index i and negative for the odd values. This fact (Roache, 1972) explains some of the wiggles which appear in the numerical solution when $|R| > 2$. It is recalled that $|R| < 2$ ensures the strong diagonal dominance of the matrix associated with the linear algebraic system defined by the finite-difference equations (2.3.3).

We assume now that the first-order derivative is approximated with the weighted scheme

$$\frac{df}{dx} \cong \tfrac{1}{2}[(1 - \alpha)\Delta_x^+ + (1 + \alpha)\Delta_x^-]f_i \tag{2.3.11}$$

where α is a constant. If $\alpha = 0$ we obtain the centered approximation. If $\alpha = \pm 1$ we get the noncentered approximations. The classical *upwind scheme* is obtained if

$$\alpha = \epsilon \qquad \text{where } \epsilon = \text{sign}(R).$$

When the above approximation is used in Eq. (2.3.1), the numerical solution f_i is again given by Eq. (2.3.4), but here

$$q_2 = \frac{2 + (1 + \alpha)R}{2 - (1 - \alpha)R} \tag{2.3.12}$$

We note that $q_2 > 0$ if

$$\alpha > \left(1 - \frac{2}{R}\right) = \alpha_0 \qquad \text{or} \qquad \alpha < \left(-1 - \frac{2}{R}\right) = \alpha_1 \tag{2.3.13}$$

and no oscillations occur in the solution. These conditions also ensure the diagonal dominance of the associated matrix.

When $\Delta x \ll 1$ and $\delta = O(1)$, the mesh Reynolds number $R = \delta \Delta x$ is $\ll 1$ and conditions (2.3.13) are easily satisfied. In this case, the expansion analogous to (2.3.10) is obtained in the form

$$f_i = f(x_i) - 6 \Delta x (U_1 - U_0) \frac{\alpha}{\delta} \Phi(x_i) + O(\Delta x^2) \tag{2.3.14}$$

and the expanded differential equation analogous to (2.3.9) is

$$A \frac{df}{dx} - \nu \frac{d^2f}{dx^2} - \frac{\alpha \Delta x}{2} A \frac{d^2f}{dx^2} + \frac{\Delta x^2}{6} \left[A \frac{d^3f}{dx^3} - \frac{\nu}{2} \frac{d^4f}{dx^4} \right]$$
$$+ O(\Delta x^3) = 0 \tag{2.3.15}$$

We note that the use of the weighted upwind scheme introduces an error of order $\alpha \Delta x$ and this error is of a dissipative nature if $\alpha A > 0$, i.e., if the sign of α is the same as A (or R). When $|R| > 2$ (this is the case where the utilization of upwind schemes is justified) $\alpha_0 > 0$ and $\alpha_1 < 0$ whatever the sign of R. Therefore, it seems necessary to use the condition $\alpha > \alpha_0$ if $R > 0$ and $\alpha < \alpha_1$ if $R < 0$. This last result can be obtained more rigorously than by the previous heuristic argument based on the nature of the main part of the truncation error.

We shall consider successively two cases: $R = \delta \Delta x \rightarrow \pm \infty$ and $R = O(1)$, $\Delta x \rightarrow 0$.

In the first case, we look for a special value of α which gives the *exact* solution of the problem (2.3.1) (Raviart, 1979). Let α^* be this value. It follows that $q_2^i = e^{\delta x_i}$, i.e., $q_2 = e^{\delta \Delta x}$ and

$$\frac{2 + (1 + \alpha^*)\delta \Delta x}{2 - (1 + \alpha^*)\delta \Delta x} = e^{\delta \Delta x} \tag{2.3.16}$$

Then

$$\alpha^* = -\frac{2}{\delta \Delta x} - \frac{1 + e^{\delta \Delta x}}{1 - e^{\delta \Delta x}} \tag{2.3.17}$$

In the case where $A > 0$, $\delta > 0$ and $\delta \Delta x \rightarrow + \infty$, so that

$$\alpha^* \sim -\frac{2}{\delta \Delta x} + 1 + \text{exponentially small terms}$$

But then $\alpha_0 = 1 - 2/\delta \Delta x$, and we can conclude that $(\alpha^* - \alpha_0)$ tends exponentially towards zero when $\delta \Delta x \rightarrow + \infty$. And, since α^* is the value of α which gives the exact solution, we conclude that, in this case where $A > 0$, we must choose $\alpha > \alpha_0$.

It is easy to verify that an analogous result is obtained when $A < 0$ by replacing α_0 by α_1. In conclusion, if $A < 0$, we must choose $\alpha < \alpha_1$.

Now, we consider the most common case in applications: $\Delta x \ll 1$ but $R = (A/\nu) \Delta x = \delta \Delta x = O(1)$. From (2.3.12) we deduce that $q_2 > 1$, whatever the sign of R, if $\alpha > \alpha_0 = 1 - 2/R$ and $0 < q_2 < 1$, whatever the sign of R, if $\alpha < \alpha_1 = -1 - 2/R$. Moreover, $q_2^i = e^{i \ln q_2} = e^{(x_i/\Delta x) \ln q_2}$. Therefore if $\alpha > \alpha_0$, q_2^i tends exponentially toward infinity when $\Delta x \rightarrow 0$, for any value of $i \in [1,N]$ and $x_i = i \Delta x$ fixed. And it is easy to see that $f_i \sim U_0 + \text{exponentially small terms}$. Now if $\alpha < \alpha_1$, we have, under the same limit $\Delta x \rightarrow 0$, $f_i \sim U_1 + \text{exponentially small terms}$. If we recall (Fig. 2.3.1) that a boundary layer exists near $x = 0$ when $A < 0$ and near $x = 1$ when $A > 0$, we may conclude again that $\alpha > \alpha_0$ must be chosen if $A > 0$ and $\alpha < \alpha_1$ if $A < 0$.

Moreover, an interesting conclusion concerns the following case. When $|\delta|\Delta x = O(1)$, Δx is then $O(|\delta|^{-1})$; that is to say the first point of computation is located at the edge of the boundary layer. In this case, if the choice of α has been made as described above, the numerical solution, although not able to describe the boundary layer, will be able, on the other hand, to give a correct approximation of the solution outside the boundary layer.

Finally we note without discussion that some interesting results for analytical solution of upwind finite-difference approximations of nonlinear differential equations have been reported by Cheng and Shubin (1978).

2.4 Upwind Corrected Schemes

Let us assume that problem (2.3.1) is approximated by using the upwind difference (2.1.2) with $\alpha = \text{sign}(R)$ for the first derivative. Such an approximation leads to a strongly diagonally dominant matrix which allows the use of an iterative solution procedure. The resulting finite-difference equation is only first-order accurate. However, second-order accuracy can be recovered if a correction term evaluated at a previous iteration is included into the general procedure in order to give, at convergence, a centered approximation. Such a technique of correction has been considered, in various forms and for various equations, by Dennis and Chang (1969), Khosla and Rubin (1974), Ta Phuoc Loc (1975), Veldman (1973) and Peyret (1971). When applied to the Navier–Stokes equations, the technique has been found to be successful.

But the point is, if the solution oscillates due to a central difference scheme, the correction strategy described above at convergence still does not eliminate the oscillations. In the principle of correction the term must be chosen so that the resulting finite-difference equation at convergence is (i) second-order accurate and (ii) sufficiently dissipative to damp possible oscillations. This second point means that an artificial viscosity term must be introduced into the scheme.

To demonstrate the approach we consider a nonlinear differential equation for which an iterative procedure of solution is natural. Let us consider the solution of

$$\nu \frac{d^2f}{dx^2} - g(f)\frac{df}{dx} = 0, \qquad \nu = \text{const} > 0 \tag{2.4.1}$$

with associated boundary conditions of type (2.3.1b). Equation (2.4.1) is approximated in the form

$$\nu \Delta_{xx} f_i - g_i \Delta_x^* f_i = 0 \tag{2.4.2}$$

where $g_i = g(f_i)$ and Δ_x^* is any finite-difference operator approximating the first-order derivative. This nonlinear problem is solved with the iterative procedure

$$\nu \Delta_{xx} f_i^{m+1} - g_i^m \Delta_x^* f_i^{m+1} = 0 \tag{2.4.3}$$

As it was shown in Section 2.2.1 for the case where $\Delta_x^* = \Delta_x^0$, the matrix associated with Eq. (2.4.3) is strongly diagonally dominant if and only if

$$|g_i^m|\frac{\Delta x}{\nu} < 2 \tag{2.4.4}$$

The upwind approximation that ensures strong diagonal dominance is defined by

$$\Delta_x^* f_i = \tfrac{1}{2}[(1 - \epsilon)\Delta_x^+ + (1 + \epsilon)\Delta_x^-]f_i \equiv (\Delta_x^0 - \tfrac{1}{2}\Delta x\,\epsilon\Delta_{xx})f_i \tag{2.4.5}$$

where $\epsilon = \text{sign } (g_i^m)$. The technique of correction consists of considering the iterative equation

$$\nu \Delta_{xx} f_i^{m+1} - g_i^m \Delta_x^* f_i^{m+1} = \Lambda_x(g_i^m) f_i^m \tag{2.4.6}$$

where the difference operator $\Lambda_x(g_i^m)$ is defined so that

$$[g_i \Delta_x^* + \Lambda_x(g_i)] f_i = \frac{df}{dx} + O(\Delta x^2)$$

A straightforward procedure for defining Λ_x, if Δ_x^* is given by (2.4.5), is simply

$$\Lambda_x(g_i) f_i = \tfrac{1}{2} \Delta x |g_i| \Delta_{xx} f_i \tag{2.4.7}$$

This, however, can produce an oscillating numerical solution. A more systematic way to construct the correction term is to look for an operator of the general form:

$$\Lambda_x(g_i) f_i = \sum_{j=-1}^{1} L_j f_{i+j} \tag{2.4.8}$$

with $L_j = a_j g_{j+1} + b_j g_j + c_j g_{j-1}$. The arbitrary coefficients are determined by requiring damping and second-order accuracy. If we require these conditions we obtain

$$\Lambda_x(g_i) f_i = \tfrac{1}{2} \Delta x \epsilon g_i \Delta_{xx} f_i - \Delta x^2 \, \alpha \bar{\epsilon} (\Delta_x^0 g_i) \cdot (\Delta_{xx} f_i)$$
$$- \Delta x^3 \, \beta \bar{\bar{\epsilon}} (\Delta_{xx} g_i) \cdot (\Delta_{xx} f_i) \tag{2.4.9}$$

with $\bar{\epsilon} = \text{sign } (\Delta_x^0 g_i)$, $\bar{\bar{\epsilon}} = \text{sign } (\Delta_{xx} g_i)$, and α, β are positive constants that measure the magnitude of the artificial viscosity. With this operator the resulting finite-difference equation (2.4.6) at convergence $(f_i^{m+1} = f_i^m)$ approximates the following differential equation:

$$\nu \frac{d^2 f}{dx^2} - g \frac{df}{dx} + \Delta x^2 \left(\frac{\nu}{12} \frac{d^4 f}{dx^4} - \frac{g}{3} \frac{d^3 f}{dx^3} + \alpha \left| \frac{dg}{dx} \right| \frac{d^2 f}{dx^2} \right)$$
$$+ \Delta x^3 \beta \left| \frac{d^2 g}{dx^2} \right| \frac{d^2 f}{dx^2} + O(\Delta x^4) = 0.$$

The artificial viscosity is proportional to the first and second derivatives of g, but a term of type $|g_i| \, d^2 f/dx^2$ could be included also. Note that the terms involving g are considered here as coefficients of derivatives of f. This is rather artificial and should be reconsidered depending on the form of the function $g(f)$. In the case of the vorticity equation to be studied later, f would be the vorticity and g the velocity.

Figure 2.4.1 shows some results obtained for the case where $g(f) = f$ and where the solution f satisfies the boundary conditions $f(0) = 1$, $f(1) = 0$. In this case, the exact solution is

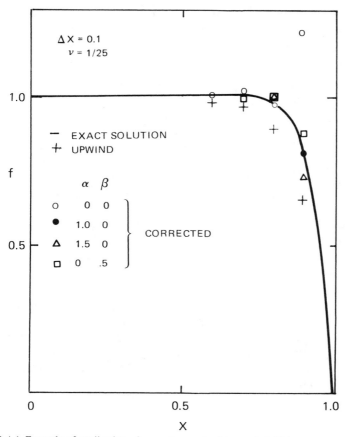

Fig. 2.4.1 Example of application of correction method to upwind difference approximations.

$$f(x) = C_1 \tanh \left(C_2 - \frac{C_1}{2\nu} x \right)$$

with $C_1 = 1$ and $C_2 = 12.5$ if $\nu = 1/25$.

2.5 Higher-Order Methods

In many problems the accuracy of second order is not sufficient and it is best to use a higher-order method. Classical finite-difference approximations can be constructed by using difference formulas of type (2.1.1). However, the higher the approximation the higher the number of discretization points involved. In this way, a fourth-order accurate method, Eq. (2.1.5), leads to algebraic systems with a pentadiagonal matrix. Another way to construct difference equations is the use of "compact" methods. Their purpose is to obtain higher-order (generally of fourth order) methods leading to the solution of algebraic systems with tridiagonal matrices. Among these types of methods, we select here the *Hermitian*, the *Mehrstellen* and the *OCI methods* (see Collatz, 1966, 1972).

2.5.1 Hermitian method

This method considers as unknowns at each discretization point not only the value of the function itself f_i, but also those of its derivatives $f'_i, f''_i, \ldots$. The system is closed by considering relationships between the function and its derivatives in three successive discretization points. These relationships are called *Hermitian formulas*. The idea of such a method is rather old, as explained by Collatz (1966), but only recently has it found application to flow problems.

The general three-point Hermitian formula for first- and second-order derivatives is written as

$$H_i = \sum_{j=-1}^{1} (a_j f_{i+j} + b_j f'_{i+j} + c_j f''_{i+j}) = 0 \tag{2.5.1}$$

where the terms a_j, b_j, and c_j are constants determined by requiring relationship (2.5.1) to represent (to some order of accuracy related to Δx) the correct relationship that exists between any regular function $f(x)$ and its two first derivatives $f'(x)$, $f''(x)$ at three points $x_{i-1} = (i-1)\,\Delta x$, $x_i = i\,\Delta x$, and $x_{i+1} = (i+1)\,\Delta x$. Practically, Taylor expansions around x_i are made in Eq. (2.5.1) and we obtain

$$\sum_{k=0}^{\infty} A_k \frac{d^k f}{dx^k} = 0 \tag{2.5.2}$$

If formula (2.5.1) was exact, all the A_k terms would be identically zero. In fact, only a finite number, K, of A_k can be made equal to zero;

$$A_k = 0, \qquad k = 0, \ldots, K \tag{2.5.3}$$

The maximal possible value of K depends essentially on the number of arbitrary parameters involved in Eq. (2.5.1).

If $K = 3$, the satisfaction of the four homogenous equations (2.5.3) leaves five undetermined coefficients and after a change of notation of the parameters we have

$$a_{-1} = \frac{15\beta - 8\gamma}{\Delta x^2} + \frac{3\rho - 2\theta}{\Delta x} \qquad\qquad b_{-1} = \frac{3\alpha + 7\beta - 5\gamma}{\Delta x} + \rho - \theta$$

$$a_0 = \frac{16\gamma}{\Delta x^2} + \frac{4\theta}{\Delta x} \qquad\qquad b_0 = \frac{16\beta}{\Delta x} + 4\rho$$

$$a_1 = -\frac{15\beta + 8\gamma}{\Delta x^2} - \frac{3\rho + 2\theta}{\Delta x} \qquad\qquad b_1 = -\frac{3\alpha - 7\beta - 5\gamma}{\Delta x} + \rho + \theta$$

$$c_{-1} = \alpha + \beta - \gamma, \qquad c_0 = 4\alpha, \qquad c_1 = \alpha - \beta - \gamma$$

where the arbitrary parameters α, β, γ, ρ, and θ are constants independent of Δx. The above expressions are substituted into formula (2.5.1) which will be written henceforth in the symbolic form

$$H_i [\alpha, \beta, \gamma, \rho, \theta] = 0 \tag{2.5.4}$$

This formula depends homogeneously on the parameters, so it is possible to express four of these parameters in terms of the remaining parameter which could be taken equal to 1. The error $\mathcal{E}$ associated with formula (2.5.4) is obtained by performing Taylor expansions as above and is found to be

$$\mathcal{E} = \frac{\Delta x^3}{6} \theta f^{IV} + \frac{\Delta x^4}{90} [3\rho f^{V} + (3\alpha - 2\gamma) f^{VI}]$$

$$+ \frac{\Delta x^5}{630} (7\theta f^{VI} - 2\beta f^{VII}) + O(\Delta x^6) \tag{2.5.5}$$

where superscripts refer to derivatives.

Various higher-order methods make use of formulas that can be derived directly from the general formula (2.5.4). Some of these will be described below. Other methods can be related to this general formula, for example, the method based on spline approximations (Rubin and Khosla, 1977) or the OCI method (Ciment et al., 1978).

Also note that it is possible to derive Hermitian formulas for unequally spaced meshes (Adam, 1975).

We next present applications of the Hermitian method to the solution of problem (2.3.1). Equation (2.3.1a) is approximated by

$$A f_i' - \nu f_i'' = 0, \qquad i = 1, \ldots, N \tag{2.5.6}$$

where f_i' and f_i'' are the respective approximations of df/dx and d^2f/dx^2 at the discretization point $x_i = i \, \Delta x$ with $\Delta x = 1/(N + 1)$. In Eq. (2.5.6), the unknowns are the derivatives f_i' and f_i''. Two supplementary equations have to be considered in order to connect these derivatives with the function itself. In the usual finite-difference method f_i' and f_i'' are explicitly expressed by formulas of the type (2.1.1). Here we consider instead two independent Hermitian formulas:

$$\begin{aligned} H_i [\alpha_1, \beta_1, \gamma_1, \rho_1, \theta_1] = 0 \\ H_i [\alpha_2, \beta_2, \gamma_2, \rho_2, \theta_2] = 0 \end{aligned} \qquad i, = 1, \ldots, N \tag{2.5.7}$$

Moreover, in order to close the system, it is necessary to add to the boundary conditions

$$f_0 = U_0, \qquad f_{N+1} = U_1 \tag{2.5.8}$$

Hermitian relations written at the boundaries or the differential equation evaluated at the boundary points associated with some Hermitian relations.

The resulting algebraic system is generally a 3×3 block tridiagonal system, the unknowns being the vectors $\psi_i = (f_i, f_i', f_i'')$. This system is solved by a matrix factorization method identical to the method described in Section 2.1 for a simple tridiagonal system.

Let us consider now, some specific Hermitian methods.

Implicit formula for second-order derivatives: The method has been studied and applied by Hirsh (1975) and Adam (1975). It corresponds to a special choice of the Hermitian formula (2.5.7), i.e.,

$$\alpha_1 = \beta_1 = \gamma_1 = 0, \qquad \rho_1 = 1, \qquad \theta_1 = 0$$

and

$$\alpha_2 = \frac{5}{3}, \qquad \beta_2 = 0, \qquad \gamma_2 = 1, \qquad \rho_2 = \theta_2 = 0$$

which respectively give

$$f_{i+1}' + 4 f_i' + f_{i-1}' - \frac{3}{\Delta x} (f_{i+1} - f_{i-1}) = 0 \qquad (2.5.9)$$

and

$$f_{i+1}'' + 10 f_i'' + f_{i-1}'' - \frac{12}{\Delta x^2} (f_{i+1} - 2f_i + f_{i-1}) = 0 \qquad (2.5.10)$$

In this case, the system is closed in the following manner: (i) boundary conditions (2.5.8), (ii) Eq. (2.5.6) written at $i = 0$ and $i = N + 1$, and (iii) noncentered Hermitian formulas:

$$H_1[0, 0, 0, 1, -1] = 0$$
$$H_N[0, 0, 0, 1, 1] = 0 \qquad (2.5.11)$$

The relationships (2.5.9) and (2.5.10) are fourth-order accurate, while the relationships (2.5.11) are only third-order accurate. However, an increase in the truncation error near a limit where Dirichlet conditions are imposed does not necessarily destroy the accuracy of the numerical solution. It has been shown (Peyret, 1978b) that the error of the numerical solution with respect to the exact solution is in the case where $U_0 = 1$, $U_1 = 0$:

$$E_i = f_i - f(x_i) = \Delta x^4 \frac{\delta^2}{60} \Phi (x_i) + O(\Delta x^5),$$

$$i = 1, \ldots, N \qquad (2.5.12)$$

where $\Phi(x_i)$ has been defined by Eq. (2.3.8b) and $\delta = A/\nu$.

The solution of the problem requires inversion of a 3×3 block tridiagonal matrix. However, it is possible to reduce the dimensions of the blocks by using (Adam, 1975) a linear combination of (2.5.6) at $i - 1, i$, and $i + 1$ along with

(2.5.10) in order to eliminate the second-order derivatives f''_{i-1}, f''_i, and f''_{i+1}. In this case, the blocks are 2×2, with the unknown being the vector $\psi_i = (f_i, f'_i)$.

The method has been applied to the solution of the Navier–Stokes equations in stream-function vorticity variables by Hirsh (1975), Mehta (1977), Daube and Ta Phuoc Loc (1978), Ta Phuoc Loc (1980) (see Section 6.6.2), Bontoux et al. (1978), Lecointe and Piquet (1981), and also by Rubin and Khosla (1977) in the *spline approximation* formulation. For the Navier–Stokes equations in primitive variables, the method has been used by Elsaesser and Peyret (1979), Elsaesser (1980), and Ghia et al. (1979).

When applying the method to the Navier–Stokes equations in primitive variables, it has been found (Elsaesser and Peyret, 1979) more practical not to close the system by considering the momentum equation at the boundaries as mentioned above for the model equation (2.5.6). The reason is to avoid evaluation of the pressure gradient at the boundary. In the present differential equation problem (2.3.1), the technique is to replace Eq. (2.5.6) written at points $i = 0$ and $i = N + 1$ by the fourth-order accurate Hermitian formulas:

$$H_1 [1, \tfrac{1}{4}, \tfrac{3}{4}, 0, 0] = 0, \qquad H_N [1, -\tfrac{1}{4}, \tfrac{3}{4}, 0, 0] = 0 \qquad (2.5.13)$$

The main part of the error E_i corresponding to this choice is again given by (2.5.12).

Explicit formula for second-order derivatives: The second-order derivative f''_i can be expressed in terms of the values of the function and of the first-order derivative at neighboring points (Adam, 1977). Such an expression is obtained from the general Hermitian formula (2.5.4) with $\alpha = \gamma = 1$ and $\beta = \rho = \theta = 0$, so that

$$f''_i = -\frac{f'_{i+1} - f'_{i-1}}{2 \Delta x} + 2\frac{f_{i+1} - 2f_i + f_{i-1}}{\Delta x^2} \qquad (2.5.14)$$

This expression is fourth-order accurate. The unknowns $\psi_i = (f_i, f'_i)$ are then solutions of the 2×2 block tridiagonal system obtained from Eqs. (2.5.6), relations (2.5.9) and (2.5.14), the boundary conditions (2.5.8), and the relations (2.5.11).

For problem (2.3.1) such a method leads to an error $E_i = f_i - f(x_i)$ of the order Δx^4 but slightly larger in magnitude than (2.5.12). However, when applied to a complex problem as the solution of the Navier–Stokes equations in velocity–pressure variables (Elsaesser and Peyret, 1979), the explicit formula (2.5.16) has produced results with an accuracy comparable to those given by the implicit formula (2.5.10).

2.5.2 Mehrstellen and OCI methods

The Mehrstellen method, used by Krause et al. (1976), for the solution of the boundary layer equations is slightly different from the two above methods in the sense that four independent Hermitian formulas are considered instead of the two formulas (2.5.7). More precisely, the four relations

$$H_i [0, -1, 1, 0, 0] = 0, \qquad H_i [1, 0, 1, 0, 0] = 0$$

$$H_i [0, 1, 1, 0, 0] = 0 \tag{2.5.15a}$$

$$H_i [0, 0, 0, 1, 0] = 0 \tag{2.5.15b}$$

associated with Eq. (2.5.6) written simultaneously at $i - 1$, i, and $i + 1$, leads to a system of seven algebraic equations from which the derivatives f'_{i-1}, f'_i, f'_{i+1}, and $f''_{i-1}, f''_i, f''_{i+1}$ are eliminated. Therefore, there results a finite-difference equation involving only the values of the unknown f_{i-1}, f_i, f_{i+1}, which along with the boundary conditions (2.5.8) forms a tridiagonal system that is easily solved. The accuracy of such a method is fourth order and the error $E_i = f_i - f(x_i)$ is comparable in magnitude with the error given by the explicit elimination technique of the previous section.

The standard *operator compact implicit* (OCI) method developed, in particular, by Ciment et al. (1978) leads to a finite-difference equation identical to the equation given by the above Mehrstellen method. However, the way to derive this equation is different and can be used to construct generalized OCI schemes (Berger et al., 1980; Lecointe and Piquet, 1981). Let us describe the method for the general linear differential second-order problem:

$$L(f) = G, \qquad 0 < x < 1 \tag{2.5.16a}$$

$$f(0) = U_0, \qquad f(1) = U_1 \tag{2.5.16b}$$

The construction is based upon the linear form between $L(f)$ and the function f:

$$\beta_1 L(f_{i+1}) + \beta_0 L(f_i) + \beta_{-1} L(f_{i-1}) = \alpha_1 f_{i+1} + \alpha_0 f_i + \alpha_{-1} f_{i-1}. \tag{2.5.17}$$

We assume that parameters α_j and β_j with $j = -1, 0, 1$ can be found so that Eq. (2.5.17) represents, to some order, the correct relationship between the function and its derivatives through the differential equation (2.5.16a). Then, by replacing the quantity $L(f)$ by G in (2.5.17), one obtains the finite-difference equation

$$\alpha_1 f_{i+1} + \alpha_0 f_i + \alpha_{-1} f_{i-1} = \beta_1 G_{i+1} + \beta_0 G_i + \beta_{-1} G_{i-1} \tag{2.5.18}$$

which, with the boundary conditions (2.5.16b), leads to a linear algebraic system with a tridiagonal matrix.

The determination of the coefficients α_j, β_j is accomplished the same way as for the Hermitian formula (2.5.1)—by employing Taylor expansions in (2.5.17). An equation analogous to (2.5.2) results.

The standard OCI scheme is obtained by considering Eqs. (2.5.3) with $K = 4$. This system of five homogeneous equations determines five of the coefficients α_j, β_j, in terms of one unknown as previously noted. The resulting finite-difference equation is fourth-order accurate; it is identical to the equation given by the Mehrstellen method.

The generalized OCI schemes (Berger et al., 1980) are constructed by using three equations $A_k = 0$, $k = 0$, 1, 2 and two equations $A_k = O(\Delta x^4)$, $k = 3$, 4. By proceeding in this manner fourth-order accuracy is conserved and some arbitrariness remains which can be used to require special supplementary properties: diagonal dominance of the associated matrix and existence of a discrete maximum principle for example.

2.6 Solution of a One-Dimensional Linear Parabolic Equation

Let us consider the linear parabolic equation

$$\frac{\partial f}{\partial t} + A \frac{\partial f}{\partial x} - \nu \frac{\partial^2 f}{\partial x^2} = 0, \qquad A = \text{const}, \ \nu = \text{const} > 0 \qquad (2.6.1)$$

The simplest scheme that can be used to solve this equation is the *explicit scheme*:

$$\frac{f_i^{n+1} - f_i^n}{\Delta t} + A \frac{f_{i+1}^n - f_{i-1}^n}{2 \Delta x} - \nu \frac{f_{i+1}^n - 2f_i^n + f_{i-1}^n}{\Delta x^2} = 0 \qquad (2.6.2)$$

where Δx and Δt are, respectively, the mesh size and the time step so that the plane (x, t) is discretized as $x = i \Delta x$, $t = n \Delta t$ where i and n are integers. In Eq. (2.6.2), $f_i^n \cong f(i \Delta x, n \Delta t)$. The numerical solution f_i^n, given by Eq. (2.6.2), must converge to the exact solution $f(i \Delta x, n \Delta t)$ of Eq. (2.6.1), when Δx and $\Delta t \to 0$ at a given point $(i \Delta x, n \Delta t)$. The Lax *equivalence theorem* states that, for a well-posed initial-value problem associated with a linear equation of evolution and approximated with a *consistent* scheme, *stability* is a necessary and sufficient condition for *convergence*.

We now define briefly consistency and stability in a simplified manner. For a rigorous analysis, see the book by Richtmyer and Morton (1967).

Equation (2.6.1) is written in the general form:

$$\left(\frac{\partial}{\partial t} + L \right) f = 0 \qquad (2.6.3)$$

and scheme (2.6.2) as

$$\Lambda_1 f_i^{n+1} + \Lambda_0 f_i^n = 0 \qquad (2.6.4)$$

where Λ_1 and Λ_0 are two difference operators defined here by

$$\Lambda_1 f_i = I f_i = f_i$$
$$\Lambda_0 f_i = [-I + \Delta t (A \Delta_x^0 - \nu \Delta_{xx})] f_i \qquad (2.6.5)$$

where I is the identity operator and

$$\Delta_x^0 f_i = \frac{1}{2\,\Delta x}(f_{i+1} - f_{i-1})$$

$$\Delta_{xx} f_i = \frac{1}{\Delta x^2}(f_{i+1} - 2f_i + f_{i-1}) \qquad (2.6.6)$$

Consistency is the property of scheme (2.6.4) to represent correctly Eq. (2.6.3). That is to say, any regular solution of (2.6.3) must satisfy Eq. (2.6.4) with an error tending towards zero when Δx, $\Delta t \to 0$. This error is called the *truncation error* and defined by

$$\mathcal{E} = \frac{1}{\Delta t}[\Lambda_1 f(i\,\Delta x,\,(n+1)\,\Delta t) + \Lambda_0 f(i\,\Delta x,\,n\,\Delta t)] \qquad (2.6.7)$$

where f is a regular solution of Eq. (2.6.3). The consistency is

$$|\mathcal{E}| \to 0 \qquad \text{when } \Delta x \to 0 \text{ and } \Delta t \to 0 \qquad (2.6.8)$$

The practical study of consistency is carried out by means of Taylor expansion around point $i\,\Delta x$, $n\,\Delta t$ so that

$$\Lambda_1 f(i\,\Delta x,\,(n+1)\,\Delta t) = f + \Delta t\,\frac{\partial f}{\partial t} + \frac{\Delta t^2}{2}\,\frac{\partial^2 f}{\partial t^2} + O(\Delta t^3)$$

$$\Lambda_0 f(i\,\Delta x,\,n\,\Delta t) = -f + A\,\Delta t\left[\frac{\partial f}{\partial x} + \frac{\Delta x^2}{6}\,\frac{\partial^3 f}{\partial x^3} + O(\Delta x^4)\right]$$

$$-\nu\,\Delta t\left[\frac{\partial^2 f}{\partial x^2} + \frac{\Delta x^2}{12}\,\frac{\partial^4 f}{\partial x^4} + O(\Delta x^4)\right]$$

and

$$\mathcal{E} = \left(\frac{\partial f}{\partial t} + A\,\frac{\partial f}{\partial x} - \nu\,\frac{\partial^2 f}{\partial x^2}\right) + \frac{\Delta t}{2}\,\frac{\partial^2 f}{\partial t^2} + \frac{\Delta x^2}{6}\left(A\,\frac{\partial^3 f}{\partial x^3} - \frac{\nu}{2}\,\frac{\partial^4 f}{\partial x^4}\right)$$
$$+ O(\Delta t^2,\,\Delta x^3) \qquad (2.6.9)$$

The first term on the right-hand side of this equation is zero because f is a solution of (2.6.1). The truncation error is $O(\Delta t,\,\Delta x^2)$; therefore the scheme is consistent and said to be of first-order accuracy in time and second-order in space.

The definition of stability used in this book is that if the initial data associated with Eq. (2.6.1) is decomposed in Fourier space as

$$f(x,\,0) = \sum_{k=-\infty}^{\infty} F^0(k)\,\exp(ikx), \qquad \text{where } i = \sqrt{-1} \qquad (2.6.10)$$

any component $F^{n+1}(k)$ computed from scheme (2.6.4) must not be amplified when time is increasing. Let us consider one component of (2.6.10). The finite-difference equation (2.6.4) admits a solution of the form

$$f_i^n = F^n(k) \exp(iki\,\Delta x) \qquad (2.6.11)$$

so that the numerical solution, approximating the exact solution, can be written

$$f_i^n = \sum_{k=-\infty}^{\infty} F^n(k) \exp(iki\,\Delta x) \qquad (2.6.12)$$

with F^0 given from (2.6.10). The stability indication is that $|F^n(k)|$ remains bounded for any k and $n \to \infty$. Introducing (2.6.11) into (2.6.4), we obtain an expression of the form

$$g_1 F^{n+1} + g_0 F^n = 0$$

or

$$F^{n+1} = g F^n \qquad \text{with } g = \frac{-g_0}{g_1} \qquad (g_1 \neq 0) \qquad (2.6.13)$$

The coefficient g is called the *amplification factor*. From (2.6.13), we obtain $F^{n+1} = g^{n+1} F^0$ (where the superscript $n+1$ in g means a power). Therefore, the condition of stability (strict Von Neumann condition*) is

$$|g| \leq 1 \qquad (2.6.14)$$

for any value of $k\,\Delta x$ so that $0 \leq k\,\Delta x \leq 2\pi$.

Instead of carrying out a special analysis of the amplification factor g corresponding to scheme (2.6.2), we shall now study the stability of the more general two-level, three-point scheme approximating Eq. (2.6.1). This scheme is written

$$\alpha_1 f_{i+1}^{n+1} + \alpha_0 f_i^{n+1} + \alpha_{-1} f_{i-1}^{n+1} = a_1 f_{i+1}^n + a_0 f_i^n + a_{-1} f_{i-1}^n \qquad (2.6.15)$$

where the coefficients α_j and a_j must satisfy the minimal condition

$$(\alpha_1 + \alpha_0 + \alpha_{-1}) - (a_1 + a_0 + a_{-1}) = 0$$

expressing the fact that Eq. (2.6.1) has only derivatives (and no term f). Without the loss of generality we can prescribe

$$\alpha_1 + \alpha_0 + \alpha_{-1} = a_1 + a_0 + a_{-1} = 1 \qquad (2.6.16)$$

The supplementary conditions necessary to ensure consistency will not be specified in the general case. Introducing (2.6.11) into (2.6.15), we obtain the amplification factor:

$$g = \frac{1 - (a_1 + a_{-1}) + (a_1 + a_{-1})\cos k\,\Delta x + i\,(a_1 - a_{-1})\sin k\,\Delta x}{1 - (\alpha_1 + \alpha_{-1}) + (\alpha_1 + \alpha_{-1})\cos k\,\Delta x + i\,(\alpha_1 - \alpha_{-1})\sin k\,\Delta x}$$

$$(2.6.17)$$

---

*The proper Von Neumann condition is $|g| \leq 1 + O(\Delta t)$. It must be used if the exact solution grows exponentially with time (this is not the case here).

After straightforward calculations, we obtain

$$|g|^2 = g\bar{g} = 1 - X\left[\frac{(B' - B)X + C' - C}{B'X^2 + C'X + 1}\right]$$

with

$$X = \tfrac{1}{2}(1 - \cos k\,\Delta x), \qquad B = 16a_1 a_{-1}$$

$$C = 4[(a_1 - a_{-1})^2 - (a_1 + a_{-1})]$$

$$B' = 16\,\alpha_1\,\alpha_{-1}, \qquad C' = 4[(\alpha_1 - \alpha_{-1})^2 - (\alpha_1 + \alpha_{-1})]$$

The condition $|g| \le 1$ is satisfied if and only if

$$(B' - B)\,X + C' - C \ge 0 \qquad \text{for } 0 \le X \le 1$$

which is obtained if

$$(C' - C) \ge 0, \qquad B' - B + C' - C \ge 0$$

Or in terms of the coefficients of (2.6.15),

$$(\alpha_1 - \alpha_{-1})^2 - (a_1 - a_{-1})^2 - (\alpha_1 + \alpha_{-1} - a_1 - a_{-1}) \ge 0$$

$$4(\alpha_1\alpha_{-1} - a_1 a_{-1}) + (\alpha_1 - \alpha_{-1})^2 - (a_1 - a_{-1})^2$$
$$- (\alpha_1 + \alpha_{-1} - a_1 - a_{-1}) \ge 0 \tag{2.6.18}$$

Now, for the particular scheme (2.6.2), we have

$$\alpha_1 = 0, \qquad \alpha_0 = 1, \qquad \alpha_{-1} = 0$$

$$a_1 = -\frac{A\,\Delta t}{2\,\Delta x} + \frac{\nu\,\Delta t}{\Delta x^2}, \qquad a_0 = 1 - \frac{2\nu\,\Delta t}{\Delta x^2}, \qquad a_{-1} = \frac{A\,\Delta t}{2\,\Delta x} + \frac{\nu\,\Delta t}{\Delta x^2}$$

and the conditions (2.6.18) reduce to

$$\frac{A^2\,\Delta t}{2\nu} \le 1 \qquad \text{and} \qquad \frac{2\nu\,\Delta t}{\Delta x^2} \le 1 \tag{2.6.19}$$

If we introduce the algebraic Reynolds number of the mesh, R, defined by

$$R = \frac{A\,\Delta x}{\nu}$$

conditions (2.6.19) can be written

$$S \le \frac{2}{R^2} \qquad \text{and} \qquad S \le \frac{1}{2} \tag{2.6.20}$$

where $S = \nu\,\Delta t/\Delta x^2$ and the domain of stability is shown in Fig. 2.6.1 (the domain lies below the curve ①).

Note that, if $S = \tfrac{1}{2}$, scheme (2.6.2) identifies with the Jacobi procedure for solution of the algebraic system resulting from the difference equation

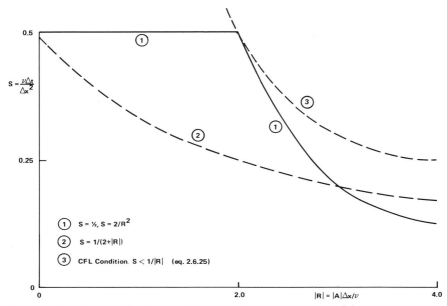

Fig. 2.6.1 Domain of stability from conditions (2.6.20) and (2.6.24) in the $|R|$,S plane.

$$A \, \Delta_x^0 f_i - \nu \, \Delta_{xx} f_i = 0 \tag{2.6.21}$$

Figure 2.6.1 shows that if $S = \frac{1}{2}$ then $|R| \leq 2$, and this is nothing more than the condition of diagonal dominance for the matrix associated with (2.6.21).

2.6.1 Effect of instability

Let us assume that computations are made with scheme (2.6.2) and a given value of Δx. If Δt is chosen satisfying condition (2.6.19), the results are correct and the error of the numerical solution is of the order of the truncation error. On the other hand, if Δt is chosen too large to satisfy condition (2.6.19), the numerical results exhibit oscillations that grow very rapidly, and after a few time steps their amplitude is infinite so that an "overflow" is registered by the computer. The phenomenon is characteristic of instability. In other cases, for example, in the case of a nonlinear equation, the instability could be more difficult to discover because no overflow appears and the amplitude of oscillations can remain bounded. (This phenomenon is experienced with the leapfrog scheme for the nonviscous equation.)

2.6.2 Noncentered schemes

The first condition, (2.6.19), shows that for a fixed value of A, the allowed time step Δt is very small when ν is small. As a result the *centered* scheme Eq. (2.6.2) is difficult to apply when ν/A^2 is too small and impossible to use for the pure advection equation ($\nu = 0$).

A remedy for this problem of stability is to use a noncentered approximation for the first-order derivative—the direction of this noncentered difference being determined by the sign of A. Let us assume that $\nu = 0$. Equation (2.6.1) is then a pure advection equation and the general solution is $f = \phi\,(x - At)$, where ϕ is an arbitrary function (determined by the initial condition). The form of the exact solution $f = \phi\,(x - At)$ shows that f remains constant along the *characteristic line* $x - At = ct$; so that, at a fixed point (x, t) the information comes from a well-determined direction characterized by the value A. Therefore, if $A > 0$, the information propagates in the direction $x > 0$ and it is easily understandable that a backward difference will be used. On the other hand, if $A < 0$, the information propagates in the direction $x < 0$ and a forward difference should be used. Such an approximation, which is called an *upwind* difference, is then closely related to the notion of *domain of dependence*.

The simpler noncentered difference scheme for approximating (2.6.1) can be written

$$\frac{1}{\Delta t}\,(f_i^{n+1} - f_i^n) + \frac{A}{2\,\Delta x}\,[(1 - \epsilon)\,(f_{i+1}^n - f_i^n) + (1 + \epsilon)\,(f_i^n - f_{i-1}^n)]$$

$$- \frac{\nu}{\Delta x^2}\,(f_{i+1}^n - 2f_i^n + f_{i-1}^n) = 0 \tag{2.6.22}$$

where $\epsilon = \text{sign}\,(A)$. This scheme is of the general form (2.6.15), and the criterion of stability is derived from conditions (2.6.18). In this case we find the only criterion

$$\Delta t \leq \frac{\Delta x^2}{2\,\nu + \Delta x\,|A|} \tag{2.6.23}$$

or, by introducing the mesh Reynolds number $|R| = |A|\,\Delta x/\nu$

$$S = \frac{\nu\,\Delta t}{\Delta x^2} \leq \frac{1}{2 + |R|} \tag{2.6.24}$$

the domain of stability (below the curve ②) in the plane $(|R|, S)$, is shown in Fig. 2.6.1. When $\nu = 0$, criterion (2.6.23) becomes (curve ③ of Fig. 2.6.1)

$$\Delta t \leq \frac{\Delta x}{|A|} \tag{2.6.25}$$

which is called *Courant–Friedrichs–Lewy condition* (CFL) and is characteristic of the explicit discretization of a hyperbolic equation (Courant et al., 1928).

From (2.6.23) we note that scheme (2.6.22) cannot be stable if the CFL condition is not satisfied. Figure 2.6.2 shows the various domains of stability in the plane $S = \nu\,\Delta t/\Delta x^2$, $|T| = |A|\,\Delta t/\Delta x$ which is more appropriate than the plane $(|R|, S)$ for the case where $\nu \to 0$. Regions I and II in Fig. 2.6.2 are respectively the domains of stability set by conditions (2.6.19) and (2.6.23).

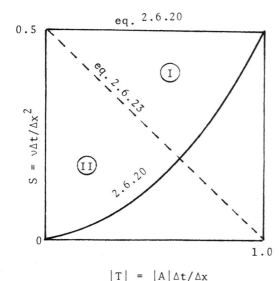

$$|T| = |A|\Delta t/\Delta x$$

Fig. 2.6.2 Domain of stability set by conditions (2.6.20) and (2.6.23) in the $|T|$, S plane.

As we did in Section 2.3 we can introduce an *upwind-weighted scheme* by using the differences described in that section. For Eq. (2.6.1), the scheme is written

$$\frac{1}{\Delta t}(f_i^{n+1} - f_i^n) + \frac{A}{2\,\Delta x}[(1 - \alpha)(f_{i+1}^n - f_i^n) + (1 + \alpha)(f_i^n - f_{i-1}^n)]$$

$$-\frac{\nu}{\Delta x^2}(f_{i+1}^n - 2f_i^n + f_{i-1}^n) = 0 \tag{2.6.26}$$

where α is an arbitrary constant. If $\alpha = \epsilon = \text{sign}\,(A)$, we recover scheme (2.6.22), and if $\alpha = 0$ we recover the centered scheme (2.6.2). The conditions of stability, obtained from (2.6.18), are

$$0 \leqslant \frac{\Delta t}{\Delta x^2}(2\nu + \alpha A\Delta x) \leqslant 1 \tag{2.6.27a}$$

$$A^2\Delta t \leqslant 2\nu + \alpha A\Delta x \tag{2.6.27b}$$

When the conditions (2.6.27a) are satisfied, the condition (2.6.27b) is also satisfied provided $2\nu + (\alpha A - |A|)\,\Delta x \geqslant 0$.

The truncation error associated with scheme (2.6.26) is

$$\frac{\Delta t}{2}\frac{\partial^2 f}{\partial t^2} - \frac{\Delta x}{2}\alpha A\frac{\partial^2 f}{\partial x^2} + O\,(\Delta t^2, \Delta x^2)$$

so that the term in Δx has a dissipative effect if $\alpha A > 0$.

The interest of the weighted scheme (2.6.26) is its stability when ν is small and its truncation error (proportional to α) which can be minimized with a value of α such that $\alpha = \gamma \operatorname{sign}(A)$, with $0 < \gamma < 1$.

The explicit schemes that have been considered are easy to employ, but the condition of stability is restrictive when the diffusive coefficient ν is small. This restriction can be slightly reduced by using upwind schemes but in this case the accuracy in space is reduced except if a three-point noncentered difference [of type (2.1.4a) and (2.1.4b)] is used. The use of such a difference is possible but it necessitates a special treatment of the boundary points.

We shall see later when considering schemes for a nonlinear parabolic equation, that it is possible to derive an explicit scheme of second-order accuracy in space as well as in time with a stability criterion of the CFL type when $\nu \to 0$. However, next we shall describe a scheme that is often used to calculate steady-state solutions: the leapfrog DuFort–Frankel scheme. It is an explicit scheme with good stability properties.

2.6.3 Leapfrog DuFort–Frankel scheme

This is a three-level time-difference scheme which for Eq. (2.6.1) is written

$$\frac{1}{2\,\Delta t}\,(f_i^{n+1} - f_i^{n-1}) + \frac{A}{2\,\Delta x}\,(f_{i+1}^n - f_{i-1}^n)$$

$$- \frac{\nu}{\Delta x^2}\,[f_{i+1}^n - (f_i^{n+1} + f_i^{n-1}) + f_{i-1}^n] = 0 \tag{2.6.28}$$

It is easy to see that the scheme is effectively explicit because f_i^{n+1} can be expressed in terms of the solution at previous times. However, the scheme is not unconditionally consistent with the original Eq. (2.6.1). Scheme (2.6.28) can be written

$$\frac{1}{2\,\Delta t}\,(f_i^{n+1} - f_i^{n-1}) + \frac{A}{2\,\Delta x}\,(f_{i+1}^n - f_{i-1}^n) - \frac{\nu}{\Delta x^2}\,(f_{i+1}^n - 2f_i^n + f_{i-1}^n)$$

$$+ \frac{\nu}{\Delta x^2}\,(f_i^{n+1} - 2f_i^n + f_i^{n-1}) = 0 \tag{2.6.29}$$

and it is easy to see that the last term corresponds to an approximation of the time-derivative term:

$$\frac{\nu\,\Delta t^2}{\Delta x^2}\,\frac{\partial^2 f}{\partial t^2}$$

Therefore, scheme (2.6.28) is consistent with (2.6.1) if and only if $\Delta t/\Delta x \to 0$ where $\Delta t \to 0$ and $\Delta x \to 0$. The study of the stability of (2.6.28) is more complicated because three levels in time are involved. The scheme is not of the general form (2.6.15).

To study the stability, we consider solutions of the form $f_i^n = F^n \exp(iik\,\Delta x)$, so that Eq. (2.6.28) yields

$$(1 + 2S)\,F^{n+1} - 2\,(2S\cos k\,\Delta x - iT\sin k\,\Delta x)F^n - (1 - 2S)\,F^{n-1} = 0$$

Assuming that $F^{n+1} = gF^n = g^2 F^{n-1}$, we obtain the quadratic equation satisfied by the amplification factor g:

$$(1 + 2S)g^2 - 2\,(2S\cos k\,\Delta x - i\,T\sin k\,\Delta x)g - (1 - 2S) = 0$$
$$(2.6.30)$$

This equation can be obtained by a more general approach shown in Appendix A on stability.

The stability requires that the roots g of (2.6.30) satisfy the condition $|g| \leq 1$ for $0 \leq k\,\Delta x \leq 2\pi$. It can be shown (using theorem of Appendix A) that stability of the leapfrog DuFort–Frankel scheme is obtained if the CFL condition $|T| \leq 1$ is satisfied. That is to say, the stability is independent of the viscosity.

2.7 Solution of One-Dimensional Nonlinear Parabolic and Hyperbolic Equations

We now consider finite-difference methods for the solution of

$$\frac{\partial f}{\partial t} + \frac{\partial G(f)}{\partial x} - \nu\frac{\partial^2 f}{\partial x^2} = 0 \qquad (2.7.1)$$

where $\nu = \text{const} > 0$. When $\nu = 0$, Eq. (2.7.1) becomes a hyperbolic equation of conservation law type:

$$\frac{\partial f}{\partial t} + \frac{\partial G(f)}{\partial x} = 0 \qquad (2.7.2)$$

In this case, solution of (2.7.2) admits discontinuities (a *weak solution*). If the discontinuity satisfies an inequality called the *entropy condition* (Oleinik, 1957; Lax, 1957), the weak solution of an initial-value problem in $-\infty < x < \infty$ is unique. Moreover, it has been proven (Oleinik, 1957) that when $\nu \to 0$ the viscous solution of an initial-value problem associated with (2.7.1) tends toward the weak solution of Eq. (2.7.2) with the same initial condition.

Therefore, when ν is small, the solution of Eq. (2.7.1) will have the same behavior as the solution of Eq. (2.7.2) for regions not close to a boundary. For example, in the case of compressible viscous flow, when the Reynolds number is large, there can exist thin regions in the main flow where the gradients are very large. If the fluid is not viscous, these regions will be shock waves, i.e., true mathematical discontinuities. When viscosity is small but not zero, the thickness of a region with large gradients is so small that it generally cannot be described by a finite-difference mesh, and the numerical viscous solution behaves like a nonviscous solution. As a result, a numerical scheme devised to solve Eq. (2.7.1) with small values of ν must be able to calculate the discontinuous solution. For this reason, the methods generally used are direct extensions of methods devised for (2.7.2). When ν is not small, so that the solution is viscous everywhere, the methods in the previous section can be applied. The schemes which follow will focus on one equation, but they can be extended to systems of equations without significant difficulty.

2.7.1 Inviscid methods

Some explicit schemes for the solution of the nonviscous equation (2.7.2) are briefly presented here. We first comment on the first-order methods in space as well as in time. These methods are generally based upon a noncentered discretization of the space derivative. Such schemes include *upwind schemes* (Courant et al., 1952) which were described in the previous section and *majorant schemes* which are a generalization of upwind schemes for a system of equations. More precisely, assume that (2.7.1) is a hyperbolic system, so that the eigenvalues of the matrix $A(f) = dG(f)/df$ are all real. The matrix $A(f)$ can then be decomposed into a sum of two matrices $A^+(f)$ and $A^-(f)$ whose eigenvalues are all nonnegative and nonpositive, respectively. A backward difference is then used for the derivative associated with A^+ and a forward difference for the derivative associated with A^-. Majorant schemes used for equations written in nonconservative form have been proposed and studied by Anucina (1964) and Yanenko and Shokin (1969). These schemes have been considered by Warming and Beam (1978) and Steger and Warming (1981) for equations written in conservative form.

Among other first-order methods, we note the Lax scheme (1954) which was one of the first methods able to handle discontinuities. It has been shown by Lax and Wendroff (1960) that discontinuous solutions can be computed without special treatment of the discontinuity if the differential equation is considered in a conservative form (2.7.2) and if the scheme is *conservative*. A two-level explicit scheme in time is said to be *conservative* if it is of the general form

$$f_i^{n+1} = f_i^n - \frac{\Delta t}{\Delta x} [H(f_{i+j}^n, \ldots, f_{i-j+1}^n) - H(f_{i+j-1}^n, \ldots, f_{i-j}^n)]$$

$$(2.7.3a)$$

where H is a function of $2j$ arguments which must for consistency satisfy the requirement

$$H(f, \ldots, f) = G(f) \qquad \text{for any } f \qquad (2.7.3b)$$

Only methods of this type (*shock-capturing*) are described in this book. We refer to Richtmyer and Morton (1967) and Moretti (1974) for methods (*shock-*

fitting) in which the shock waves are considered in a special manner using explicitly the Rankine–Hugoniot conditions.

A second-order method for inviscid equations was proposed by Lax and Wendroff in 1960. This scheme is based upon a Taylor series expansion with respect to time and limited to the third order, i.e.,

$$f(x, t + \Delta t) = f(x, t) + \frac{\partial f}{\partial t} \Delta t + \frac{1}{2} \frac{\partial^2 f}{\partial t^2} \Delta t^2 + O(\Delta t^3) \tag{2.7.4}$$

The derivatives $\partial f / \partial t$ and $\partial^2 f / \partial t^2$ are calculated from the original Eq. (2.7.2):

$$\frac{\partial f}{\partial t} = -\frac{\partial G}{\partial x}$$

$$\frac{\partial^2 f}{\partial t^2} = -\frac{\partial^2 G}{\partial x\, \partial t} = -\frac{\partial}{\partial x}\left(\frac{\partial G}{\partial t}\right) = -\frac{\partial}{\partial x}\left(\frac{dG}{df} \frac{\partial f}{\partial t}\right) = \frac{\partial}{\partial x}\left(A(f) \frac{\partial G}{\partial x}\right)$$

By introducing these expressions into (2.7.4) and approximating derivatives with centered differences, we obtain the Lax–Wendroff scheme:

$$f_i^{n+1} = f_i^n - \frac{\Delta t}{2\,\Delta x}(G_{i+1}^n - G_{i-1}^n)$$

$$+ \frac{\Delta t^2}{2\,\Delta x^2}[A_{i+1/2}^n(G_{i+1}^n - G_i^n) - A_{i-1/2}^n(G_i^n - G_{i-1}^n)] \tag{2.7.5}$$

where $A_{i+1/2}^n$ is generally defined by

$$A_{i+1/2}^n = A\left(\frac{f_{i+1}^n + f_i^n}{2}\right)$$

Note that scheme (2.7.5) is of the general conservative form (2.7.3) with $j = 1$ and

$$H(a, b) = \tfrac{1}{2}[G(a) + G(b)] - \frac{\Delta t}{2\,\Delta x} A\left(\frac{a + b}{2}\right)[G(a) - G(b)]$$

The stability of such a scheme is studied by assuming $A(f) = $ const and the criterion is the CFL condition:

$$|A(f)| \frac{\Delta t}{\Delta x} \le 1 \tag{2.7.6a}$$

In the case where Eq. (2.7.2) is a vector equation, condition (2.7.6a) is replaced by

$$|\lambda_k(f)| \frac{\Delta t}{\Delta x} \le 1, \qquad k = 1, \dots, m \tag{2.7.6b}$$

where the λ_k are the eigenvalues of the matrix $A(f)$. This last condition can be expressed in geometrical terms: The *numerical domain of dependence* of scheme (2.7.5) must contain the domain of dependence of the differential equation (2.7.2).

In the case where (2.7.2) is a vector equation, scheme (2.7.5) needs the evaluation of matrices $A_{i+1/2}$ and $A_{i-1/2}$. In order to avoid this evaluation, Richtmyer (1962) proposed a two-step version of the Lax–Wendroff scheme. This version, which is often called the *two-step Lax–Wendroff scheme*, is identical to the original scheme (2.7.5) in the one-dimensional case with $A = \text{const}$. After Richtmyer, several two-step schemes were proposed. Among them, the scheme proposed in 1969 by MacCormack quickly became popular because of its simplicity and attractive properties when computing shock waves.

In 1973 Lerat and Peyret (1973b, 1974b, 1975) proposed a generalization of these two-step schemes. More precisely, they looked for the most general schemes with the following characteristics:

1. explicit conservative form;
2. two-step *predictor–corrector* type;
3. three-point, two-level (the solution of f_i^{n+1} depends only on three values of f at level n);
4. second-order accurate in time and in space.

The schemes having all these properties form a general class depending on the two parameters α and β which characterize the location of the point where the predictor $\tilde{f}_i$ is calculated (Fig. 2.7.1). These schemes, called S_β^α, are written

$$\tilde{f}_i = (1 - \beta)f_i^n + \beta f_{i+1}^n - \alpha \frac{\Delta t}{\Delta x}(G_{i+1}^n - G_i^n) \qquad (2.7.7a)$$

$$f_i^{n+1} = f_i^n - \frac{\Delta t}{2\,\alpha\,\Delta x}\big[(\alpha - \beta)\,G_{i+1}^n + (2\beta - 1)\,G_i^n$$

$$+ (1 - \alpha - \beta)\,G_{i-1}^n + \tilde{G}_i - \tilde{G}_{i-1}\big] \qquad (2.7.7b)$$

where $\tilde{G}_i \equiv G(\tilde{f}_i)$, $\alpha \neq 0$.

The second step, (2.7.7b), can also be written in a form more appropriate for numerical calculation:

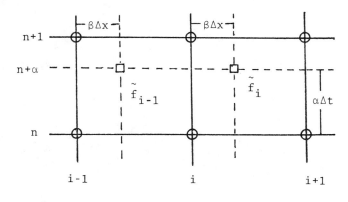

Fig. 2.7.1 Location of point where predictor values $\tilde{f}_i$ are calculated.

$$
f_i^{n+1} = \frac{1}{2\alpha^2}\{(1-\beta)(1-\alpha-\beta)f_{i-1}^n + [2\beta(1-\beta)-\alpha(1-2\alpha)]f_i^n
$$

$$
- \beta(\alpha-\beta)f_{i+1}^n - (1-\alpha-\beta)\tilde{f}_{i-1} + (\alpha-\beta)\tilde{f}_i\}
$$

$$
- \frac{\Delta t}{2\,\alpha\,\Delta x}(\tilde{G}_i - \tilde{G}_{i-1}) \tag{2.7.7b$'$}
$$

because it does not need the storage or recalculation of G_i^n. Note that for schemes (2.7.7a) and (2.7.7b) the corresponding function H considered in (2.7.3) is such that $j = 1$ and

$$
H(a, b) = \frac{1}{2\alpha}[(\alpha-\beta)G(a) - (1-\alpha-\beta)G(b)]
$$

$$
+ \frac{1}{2\alpha}G\left\{\beta a + (1-\beta)b - \alpha\frac{\Delta t}{\Delta x}[G(a)-G(b)]\right\}
$$

In the linear case where $A = dG/df = $ const, it is easy to verify that all the S_β^α schemes are identical with the Lax–Wendroff scheme (2.7.5), where $A_{i+1/2}^n = A_{i-1/2}^n = A$ so that the condition of stability is (2.7.6).

All the finite-difference schemes having the four properties above belong to the class of S_β^α schemes. If $\alpha = \beta = \frac{1}{2}$, the predictor is computed at the center point of the cell: This is the two-step version of the Lax–Wendroff scheme introduced by Richtmyer (1962). If $\alpha = 1$, $\beta = \frac{1}{2}$, we obtain the scheme proposed by Rubin and Burstein (1967). More generally, the schemes $S_{1/2}^\alpha$ have been considered by McGuire and Morris (1973).

The schemes corresponding to $\alpha = 1$ and $\beta = 0$ or $\beta = 1$ are the two versions of the MacCormack (1969) scheme. In this case, the predictor $\tilde{f}_i$ is computed at the same point as the final value f_i^{n+1}. If $\beta = 0$, $\tilde{f}_i$ is defined at $x_i = i\,\Delta x$ and calculated with a forward spatial difference and f_i^{n+1} with a backward difference. If $\beta = 1$, $\tilde{f}_i$ is defined at $x_i = (i+1)\Delta x$ and calculated with a backward difference while the computation of f_i^{n+1} makes use of a forward difference.

The class of schemes S_0^α and S_1^α were considered by Warming, Kutler, and Lomax (1973). As numerical experiments (Lerat and Peyret, 1973a, b, 1974b, 1975) show, the various S_β^α schemes give nonidentical results, in particular, for the case of shock waves. Some lead to shock profiles strongly oscillating, others to profiles more regular and in some cases totally monotonic. These differences in behavior of the numerical solution appear in the nonlinear case only because all the schemes are identical when $A(f) = $ const.

A simple and very useful way, although heuristic in nature, to compare the properties of the various schemes is to make an analysis of the "equivalent equation." For a numerical scheme that approximates the differential equation (2.7.2) with an error of order p and $\sigma = \Delta t/\Delta x$ of order 1, we call an *equivalent equation of order q* the differential equation that is approximated by the same scheme but with an error of order q, where $q > p$. In other terms, the scheme is expected to give results nearer to the solution of the equivalent equation that to the solution of the original equation.

This equation is constructed first by employing Taylor expansions in the manner used for the determination of the truncation error. Then the time derivatives in the error terms are eliminated by expressing them through repeated differentiation of the expanded equation (not the original equation). For details see Warming and Hyett (1974) and Lerat and Peyret (1975).

In the case of schemes (2.7.7a) and (2.7.7b), the equivalent equation of third order is written*

$$\frac{\partial f}{\partial t} + \frac{\partial G(f)}{\partial x} = \frac{\Delta x^2}{6} \frac{\partial}{\partial x} \left\{ (\sigma^2 A^2 - 1) \frac{\partial^2 G}{\partial x^2} - \frac{3}{2\alpha} \frac{dA}{df} \frac{\partial}{\partial x} (\beta f - \alpha \sigma G) \right. \\ \left. \times \frac{\partial}{\partial x} [(\beta - 1)f - \alpha \sigma G] + 2\sigma^2 \frac{dA}{df} \left(\frac{\partial G}{\partial x} \right)^2 \right\}$$

(2.7.8)

where $\sigma \Delta x = \Delta t$. Now, if we consider the numerical results given by the scheme as an approximation of the exact solution of the differential equation (2.7.8) with the initial condition of the original problem, a study of the nature of Eq. (2.7.8) would lead to information about the behavior of the numerical solution. Such a study will permit one to compare the properties of the various schemes S_β^α. For example, assuming $G(f) = f^2/2$, the equivalent equation (2.7.8) becomes

$$\frac{\partial f}{\partial t} + \frac{\partial}{\partial x} \left(\frac{f^2}{2} \right) = \Delta x^2 \left[E_1 \frac{\partial^3 f}{\partial x^3} - E_2 \frac{\partial f}{\partial x} \frac{\partial^2 f}{\partial x^2} + E_3 \left(\frac{\partial f}{\partial x} \right)^3 \right]$$

(2.7.9)

with

$$E_1 = \frac{\eta}{6\sigma} (\eta^2 - 1), \qquad \eta = \sigma f$$

$$E_2 = -\frac{1}{2} \left[(3 - \alpha) \eta^2 + (2\beta - 1) \eta + \frac{\beta(1 - \beta)}{\alpha} - 1 \right]$$

$$E_3 = \frac{\sigma}{4} [2(2 - \alpha) \eta + 2\beta - 1]$$

Now, we can make a heuristic study of the nature of the partial differential equation (2.7.9) by considering the effect of the odd and even higher derivatives. The term $T_1 = E_1 \partial^3 f / \partial x^3$ is of a dispersive nature and leads to dispersive oscillations in the computed solution. If $G(f) = Af$, $A = $ const, this term T_1 is the only one that would remain in the equivalent equation. It is the same for all the S_β^α schemes. The term $T_2 = -E_2 (\partial f / \partial x)(\partial^2 f / \partial x^2)$ can be defined as

---

*We refer to Lerat and Peyret (1974b, 1975) for general hyperbolic systems and application to Gas Dynamics.

a *dissipative* term if $E_2(f) \, \partial f/\partial x < 0$ and *antidissipative* if $E_2(f) \, \partial f/\partial x > 0$. Hence, according to the sign of $E_2 \, \partial f/\partial x$, this term can have a positive benefit by damping the possible oscillations created by the dispersion terms, or a bad effect by creating new oscillations. The third term, $E_3 \, (\partial f/\partial x)^3$, is a differential order lower than the two previous terms and its effect is generally negligible: This fact has been verified by several numerical experiments.

In the case of a compression or shock wave ($\partial f/\partial x < 0$), the scheme is dissipative if $E_2 > 0$. On the other hand, in the case of a rarefaction ($\partial f/\partial x > 0$), the scheme is dissipative if $E_2 < 0$. Therefore, it is not possible to have a scheme with good dissipative properties in all possible events. However, an optimal scheme can be defined by the conditions

$$E_2 \geq 0 \quad \text{for any } \eta \text{ so that } -1 \leq \eta \leq 1 \tag{i}$$

$$\underset{-1 \leq \eta \leq 1}{\mathrm{Max}} \; (E_2) \text{ is minimal} \tag{ii}$$

Condition (i) means that the optimal scheme is dissipative in any compression or shock wave and, due to condition (ii), the effect due to the positiveness of E_2 is the weakest possible. The unique set of (α, β) which satisfies (i) and (ii) is

$$\alpha = 1 + \frac{\sqrt{5}}{2} \approx 2.118, \qquad \beta = \tfrac{1}{2}$$

Such an optimal scheme gives compression and shock profiles with very small spurious oscillations, or even without oscillations, when the dispersive error T_1 is sufficiently small (e.g., if $\sigma = \Delta t/\Delta x$ is chosen so that $\sigma \times \max |f| = 1$). Moreover, due to condition (ii), the shock is smeared over a minimal number of discretization points. Numerical experiments show that, for various schemes of the class S_β^α, the amplitude of the oscillations remains bounded even if the scheme is not everywhere dissipative at the order considered in Eq. (2.7.9). For these schemes, it seems that the weak instability created by the antidissipative term T_2 is balanced by the effect of a term

$$T_4 = \frac{\Delta x^3}{8} \sigma f^2 (\sigma^2 f^2 - 1) \frac{\partial^4 f}{\partial x^4}$$

which would appear in the equivalent equation of fourth order. This conjecture is strengthened by a comparison between the equivalent equations corresponding to the Lax–Wendroff scheme (2.7.5) (where $G = f^2/2$) and the leapfrog scheme

$$f_i^{n+1} = f_i^{n-1} - \tfrac{1}{4}\sigma[(f_{i+1}^n)^2 - (f_{i-1}^n)^2]$$

where $\sigma = \Delta t/\Delta x$. In both cases Eq. (2.7.9) is the same with

$$E_1 = \frac{\eta}{6\sigma}(\eta^2 - 1), \quad E_2 = -\tfrac{1}{2}(3\eta^2 - 1), \quad E_3 = \sigma\eta$$

but, for the leapfrog scheme, the equivalent equation at fourth order does not involve the term T_4 as it exists for the Lax–Wendroff scheme which is stable. This fact can explain the reason why the leapfrog scheme is unstable in the nonlinear case. See also Richtmyer and Morton (1967) and Kreiss and Oliger (1973) for other approaches to nonlinear instability analysis.

Finally, it should be noted that the presence of the dissipative term T_4 can also explain the ability of the stable schemes S_β^α to select the weak solution satisfying the entropy condition. A deeper study of the properties of finite-difference schemes based upon the analysis of the equivalent equation* can be found, for example, in works by Hirt (1968), Yanenko and Shokin (1969), Warming and Hyett (1974), Harten et al. (1976), Lerat and Peyret (1975), Peyret (1977), and Lerat (1981).

In applications, it is often necessary to add an *artificial viscosity* term to some of the schemes described above. Such a term can be necessary (i) to select the solution satisfying the entropy condition, (ii) to prevent nonlinear instabilities, and (iii) to damp spurious oscillations of the numerical solution.

The concept of an artificial viscosity was introduced by Von Neumann and Richtmyer (1950), developed by Lax and Wendroff (1960), and since has been used in various forms in many works. In the simple case of the scalar equation (2.7.2) with $G = f^2/2$, the Lax–Wendroff artificial viscosity consists of adding to Eq. (2.7.5) or to Eq. (2.7.7b) the term $(\Delta t/\Delta x)D_i$ with

$$D_i = \tfrac{1}{2}\chi[\,|f_{i+1}^n - f_i^n|(f_{i+1}^n - f_i^n) - |f_i^n - f_{i-1}^n|(f_i^n - f_{i-1}^n)]$$

where χ is a positive constant of order unity. The condition of stability of the modified scheme is then (see Richtmyer and Morton, 1967)

$$\frac{\Delta t}{\Delta x}|A| \le \left(1 + \frac{\chi^2}{4}\right)^{1/2} - \frac{\chi}{2}$$

which is more strict than the usual CFL condition.

In the present case, the artificial viscosity term D_i corresponds to a discretization of

$$\frac{1}{2}\chi\,\Delta x^3 \frac{\partial}{\partial x}\left(\left|\frac{\partial f}{\partial x}\right|\frac{\partial f}{\partial x}\right)$$

which is a viscosity term of the Von Neumann–Richtmyer type. For the schemes S_β^α with a term D_i, the equivalent equation (2.7.9) has the same form except that the E_2 must be replaced by $\overline{E}_2$, where

$$\overline{E}_2 = \begin{cases} E_2 + \chi & \text{if } \dfrac{\partial f}{\partial x} < 0 \\[2ex] E_2 - \chi & \text{if } \dfrac{\partial f}{\partial x} > 0 \end{cases}$$

---

*also called ''modified equation''.

The term

$$\bar{T}_2 = -\bar{E}_2 \frac{\partial f}{\partial x} \frac{\partial^2 f}{\partial x^2}$$

always has a dissipative nature if $\chi > \text{Max} \, |E_2|$ for $0 \leqslant \eta \leqslant 1$.
Other ways of damping are discussed in the course of this book.

2.7.2 Viscous methods

The two-step explicit schemes which discretize a viscous equation like (2.7.1) with a truncation error of second order in time as well as in space are at least five-point schemes and some are even seven-point schemes. As a consequence, a systematic construction of such schemes leads to a general class of schemes with a large number of parameters. However, if we require that these schemes not be significantly asymmetrical with respect to the point $x_1 = i \, \Delta x$, the number of parameters can be reduced. If, moreover, we consider only the schemes for which the second step does not involve the values f_{i+2}^n and f_{i-2}^n, we obtain a generalization of schemes (2.7.7a) and (2.7.7b):

$$\tilde{f}_i = S_i^{(1)} + \nu\alpha \frac{\Delta t}{\Delta x^2}$$

$$\times \left[\gamma(f_{i+2}^n - 2f_{i+1}^n + f_i^n) + (1 - \gamma)(f_{i+1}^n - 2f_i^n + f_{i-1}^n) \right]$$

$$\tag{2.7.10a}$$

$$f_i^{n+1} = S_i^{(2)} + \frac{\nu \, \Delta t}{2\alpha \, \Delta x^2} \left[(2\alpha - 1)(f_{i+1}^n - 2f_i^n + f_{i-1}^n) + (1 - \beta) \right.$$

$$\left. \times (\tilde{f}_{i+1}^n - 2\tilde{f}_i + \tilde{f}_{i-1}) + \beta(\tilde{f}_i - 2\tilde{f}_{i-1} + \tilde{f}_{i-2}) \right]$$

$$\tag{2.7.10b}$$

In these equations $S_i^{(1)}$ and $S_i^{(2)}$ represent, respectively, the right-hand sides of Eqs. (2.7.7a) and (2.7.7b); i.e., the *nonviscous* part of the scheme. Schemes (2.7.10a) and (2.7.10b) depend on the three arbitrary parameters α, β, and γ, and they will be denoted by $S_{\beta,\gamma}^\alpha$. Among these schemes, we find the two-step Lax–Wendroff schemes ($\beta = \frac{1}{2}$) and the MacCormack scheme ($\beta = 0$ and $\beta = 1$) if γ is chosen so that $\gamma = \beta$.

The five-point schemes are characterized by the simultaneous conditions: $\gamma(1 - \beta) = 0$ and $(1 - \gamma)\beta = 0$. These conditions are satisfied by the schemes $S_{0,0}^\alpha$, $S_{1,1}^\alpha$ which are the direct extensions of the nonviscous schemes S_0^α and S_1^α studied by Warming, Kutler, and Lomax (1973).

Apart from the last schemes, all the others are seven-point schemes which are not convenient for boundary-value problems because they lead to difficulties near the boundaries. This point will be considered later.

Due to this difficulty and because only the steady solution of Eq. (2.7.1) is often looked for, the first extension of the two-step Lax–Wendroff scheme $S_{1/2}^{1/2}$

to the viscous case was a five-point scheme of first order (Δt, Δx^2) during the transient stage and second order (Δx^2) at steady state (Thommen, 1966). Thus, if we look for the general five-point schemes with this property of accuracy, we find the schemes $\overline{S}^\alpha_{\beta,\gamma}$:

$$\tilde{f}_i = \quad S^{(1)}_i + \nu\alpha\,\frac{\Delta t}{\Delta x^2}\,[\gamma(f^n_{i+2} - 2f^n_{i+1} + f^n_i)$$

$$+ (1 - \gamma)(f^n_{i+1} - 2f^n_i + f^n_{i-1})] \tag{2.7.11a}$$

$$f^{n+1}_i = S^{(2)}_i + \nu\,\frac{\Delta t}{\Delta x^2}\,[f^n_{i+1} - 2f^n_i + f^n_{i-1}] \tag{2.7.11b}$$

The scheme introduced by Thommen (1966) is $\overline{S}^{1/2}_{1/2,1/2}$. The schemes $\overline{S}^{1+\sqrt{5}/2}_{1/2,1/2}$ and $\overline{S}^1_{0,0}$ or $\overline{S}^1_{1,1}$ are particularly interesting since they have the properties of the corresponding inviscid schemes and the second step, (2.7.11b), does not require the calculation of a viscous term with $\tilde{f}_i$. This clearly leads to a reduction in the computer time and an increase in efficiency over some of the other schemes.

Due to the complexity of the amplification factor associated with schemes (2.7.10) and (2.7.11), exact criterion of stability are not known. However, by a numerical study of the amplification factor, it is possible to establish the domain of stability. Figures 2.7.2a and 2.7.2b show this domain in the plane

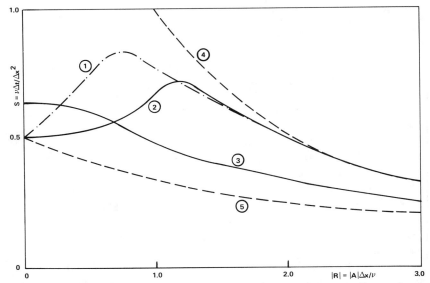

Fig. 2.7.2 (a) Curves of stability for schemes (2.7.10) (stable below curve): ① $\alpha = \beta = \gamma = \frac{1}{2}$ (Richtmyer type); ② $\alpha = 1$, $\beta = \gamma = 0$ or 1 (MacCormack); ③ $\alpha = 1 + \sqrt{5}/2$, $\beta = \frac{1}{2}$, $\gamma = \frac{1}{2}$ (Lerat–Peyret); ④ CFL condition $S = 1/|R|$; ⑤ condition (2.6.24) $S = 1/(2+|R|)$.

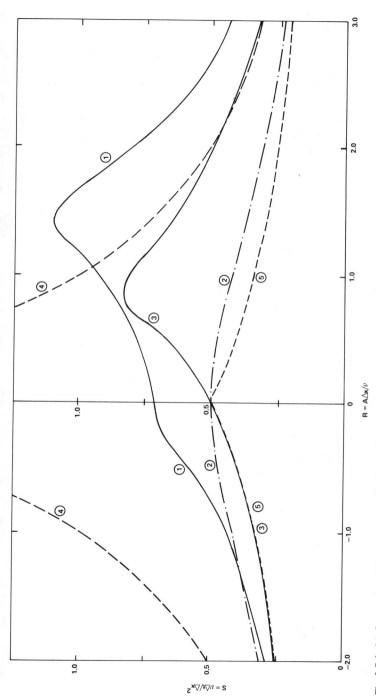

Fig. 2.7.2 (b) Curves of stability for schemes (2.7.10) and (2.7.11) (stable below curve): ① scheme (2.7.10), $\alpha = 1$, $\beta = \frac{1}{2}$, $\gamma = 0$; ② scheme (2.7.11), $\alpha = \beta = \gamma = \frac{1}{2}$ (Thommen); ③ scheme (2.7.11), α, β arbitrary, $\gamma = 0$; ④ CFL condition, $S = 1/|R|$; ⑤ condition (2.6.24), $S = 1/(2+|R|)$.

$R = A \, \Delta x / \nu$, $S = \nu \, \Delta t / \Delta x^2$ for some particular schemes.

Note that both schemes (2.7.10) and (2.7.11) reduce to the S_β^α schemes if $\nu = 0$, so that the CFL criterion is a necessary condition for stability. In the case where $A = 0$, it can be found that the condition $2\nu \, \Delta t \leq \Delta x^2$ is sufficient to insure stability provided $\gamma = \beta$ and $\alpha \geq \frac{1}{2}$. In the case of Thommen's scheme $(\alpha = \beta = \gamma = \frac{1}{2})$ the exact condition of stability is $\Delta t (A^2 \, \Delta t + 2\nu) \leq \Delta x^2$. The condition $\Delta t \leq \Delta x^2 / (2 \, \nu + |A| \, \Delta x)$ can be used, however, as a practical criterion (Figs. 2.7.2a and 2.7.2b).

Now, we consider the parabolic equation

$$\frac{\partial f}{\partial t} + \frac{\partial}{\partial x}\left[F(f) - \nu(f) \, \frac{\partial f}{\partial x} \right] = 0 \tag{2.7.12}$$

where $\nu(f)$ is a positive function of f. This is the case of the compressible Navier–Stokes equations where generally the viscosity depends on the temperature. If the function $G(f)$ defined by

$$G(f) \equiv F(f) - \nu(f) \, \frac{\partial f}{\partial x} \tag{2.7.13a}$$

is introduced, the resulting equation (2.7.12) takes the form

$$\frac{\partial f}{\partial t} + \frac{\partial G}{\partial x} = 0 \tag{2.7.13b}$$

so that the associated second-order numerical schemes are formally given by Eqs. (2.7.7a) and (2.7.7b).

In these schemes $G_i^n = G(f_i^n)$ and $\tilde{G}_i = G(\tilde{f}_i)$ are defined by

$$G_i^n \equiv F_i^n - \frac{M_i^n}{\Delta x}\left[(f_i^n - f_{i-1}^n) + \gamma(f_{i+1}^n - 2f_i^n + f_{i-1}^n) \right] \tag{2.7.14a}$$

$$\tilde{G}_i \equiv \tilde{F}_i - \frac{\tilde{M}_i}{\Delta x}\left[(\tilde{f}_i - \tilde{f}_{i-1}) + [\alpha + (1 - 2\alpha)\gamma](\tilde{f}_{i+1} - 2\tilde{f}_i + \tilde{f}_{i-1}) \right] \tag{2.7.14b}$$

where M_i^n and $\tilde{M}_i$ are approximations of $\nu(f)$:

$$M_i^n = \sum_{j=-1}^{1} m_j \, \nu(f_{i+j}^n), \qquad \tilde{M}_i = \sum_{j=-1}^{1} \overline{m}_j \, \nu(\tilde{f}_{i+j}) \tag{2.7.15}$$

The constants m_j and $\overline{m}_j$ must satisfy the conditions

$$\sum_{j=-1}^{1} m_j = \sum_{j=-1}^{1} \overline{m}_j = 1 \tag{2.7.16a}$$

and

$$(2\alpha - 1) \sum_{j=-1}^{1} j m_j + \sum_{j=-1}^{1} j \overline{m}_j = 0 \tag{2.7.16b}$$

Note the alternate choices of M_i^n and $\tilde{M}_i$ are of the type

$$M_i^n = \nu \left(\sum_{j=-1}^{1} m_j f_{i+j}^n \right)$$

which is equivalent to the first definition to order Δt^2 and identical if ν is a linear function of f.

Schemes (2.7.14) are five-point schemes if and only if $\alpha = 1$, β is arbitrary, and $\gamma = 0$ or $\gamma = 1$. Among them the most interesting are the MacCormack schemes with $\alpha = 1$, $\beta = \gamma = 0$, $m_1 = \overline{m}_{-1} = 0$, $m_{-1} = \overline{m}_1 = m_0 = \overline{m}_0 = \frac{1}{2}$ or $\alpha = 1$, $\beta = \gamma = 1$, $m_{-1} = \overline{m}_1 = 0$, $m_1 = \overline{m}_{-1} = m_0 = \overline{m}_0 = \frac{1}{2}$ because, in this case, the values are located symmetrically with respect to $x_i = i \, \Delta x$ and at points where the final values are computed. This fact avoids some of the difficulties encountered in the use of the schemes near a boundary.

Note that if $\nu = \text{const}$, schemes (2.7.10) and (2.7.14) are identical under the following conditions:

or

(i) $\quad \alpha = \frac{1}{2}, \qquad \beta = \frac{1}{2}, \qquad \gamma$ arbitrary;

(ii) $\quad \alpha \neq \frac{1}{2}, \qquad \beta = \alpha$ and $\gamma = 1$ or $\beta = 1 - \alpha$ and $\gamma = 0$.

It is interesting to point out that the second step, (2.7.7b) with (2.7.14b), can be written in a form analogous to Eq. (2.7.7b'), while schemes (2.7.10) possess the same property only if the above conditions (i) or (ii) are satisfied. Finally, five-point schemes with first-order accuracy in time during the transient stage and second-order accuracy at steady state analogous to (2.7.11) can be obtained for Eq. (2.7.12). The predictor step defining the provisional value $\tilde{f}_i$ is again given by Eq. (2.7.7a) with G_i^n given by (2.7.14a), (2.7.15), (2.7.16a), and the second step defining the final value f_i^{n+1} is

$$f_i^{n+1} = f_i^n - \frac{\Delta t}{2\alpha \, \Delta x}$$

$$\times \left[(\alpha - \beta) F_{i+1}^n + (2\beta - 1) F_i^n + (1 - \alpha - \beta) F_{i-1}^n + \tilde{F}_i - \tilde{F}_{i-1} \right]$$

$$+ \frac{\Delta t}{2 \, \Delta x^2} \left[(\nu_{i+1}^n + \nu_i^n)(f_{i+1}^n - f_i^n) - (\nu_i^n + \nu_{i-1}^n)(f_i^n - f_{i-1}^n) \right]$$

$$(2.7.17)$$

Note that the mean value $\frac{1}{2}(\nu_{i+1} + \nu_i)$ can be replaced by $\nu[(f_{i+1} + f_i)/2]$.

2.7.3 Boundary conditions

In using general multistep methods, the question of how to obtain the values of the intermediate quantities on the boundary arises. The problem is delicate for the case of splitting methods (Yanenko, 1971) in which the finite-difference equation at each step is not consistent with the original equation. Also, in the alternating direction implicit methods, in which each step is consistent with the original equation, an increase of the truncation error with respect to time appears for points adjacent to a boundary (Fairweather and Mitchell, 1967), if the exact time-varying boundary value is used to determine the value of the provisional value on the boundary.

The same phenomenon appears in some of the predictor–corrector methods which have just been described. Let us consider, for example, the behavior of the MacCormack scheme at the first point $i = 1$ near the boundary $x = 0$ on which the value of the function is known, i.e., $f(0,t) = \phi(t)$. The computation of f_1^{n+1} necessitates the knowledge of $\tilde{f}_0$. If the boundary condition is used, i.e., if we take $\tilde{f}_0 = \phi((n + 1) \, \Delta t)$, the truncation error at point $i = 1$ becomes of the order $\nu \, \Delta t / \Delta x$ and this loss of consistency is present in the transient stage as well as in the steady state. The reason is that the final accuracy of the scheme results from a combination of the predictor and the corrector steps. This combination is effective for inner points which do not involve, in the corrector step, predicted values at boundary points. The problems arise mainly for schemes in which provisional values and final values are defined at the same points, since one is tempted to use the boundary value as the predictor value. For other schemes, in which $\beta \neq 0$ or $\beta \neq 1$, like the centered ($\beta = \frac{1}{2}$) Thommen-type schemes, the provisional value $\tilde{f}_0$ is not defined on the boundary and it is usually computed by a modified finite-difference equation. An example is the use of (2.7.11a) with $\gamma = 1$ but care must be taken to compute $\tilde{f}_1$ involved in the calculation of f_1^{n+1} by the same modified scheme. Methods in which the two steps correspond to the same finite-difference operator, such as the Brailovskaya (1965) or the Allen and Cheng (1970) schemes, do not present this loss of accuracy as pointed out by Cheng (1975). In the MacCormack scheme, due to the stability condition which implies $\nu \, \Delta t / \Delta x \approx \Delta x$, the truncation error at point $i = 1$ becomes of the order Δx. In this case, it is plausible that the error between the numerical and the exact solution remains $O(\Delta x^2)$ everywhere. This conjecture, which has been verified numerically, is based on the analogous results obtained by Gustaffson (1975) for a linear hyperbolic equation.

It is possible to obtain a more precise result even though restricted to the linear steady case. We consider the computation of the steady solution of Eq. (2.7.1) in the range $0 < x < 1$, associated with the boundary conditions $f(0) = 1$, $f(1) = 0$, obtained as the limit, when $t \to \infty$, to the numerical solution given by the MacCormack scheme with provisional values at boundaries taken equal to the exact boundary values. The finite-difference problem can be solved explicitly as explained in Section 2.3. Then, by assuming $\nu \, \Delta t / \Delta x \approx \Delta x$, $A \, \Delta x / \nu \ll 1$ when $\Delta x \ll 1$, it is found that the error of the numerical solution with respect to the exact solution is $O(\Delta x^2)$ everywhere, even at a point near a boundary where the truncation error is $O(\Delta x)$. This result is in agreement with the general theory established by Kreiss (1972) concerning the rate of convergence of the numerical solution of boundary-value problems associated with ordinary differential equations. Similar results for a Dirichlet–Poisson Problem were found by Bramble and Hubbard (1962).

A remedy to recover the unconditionally second-order truncation error, for the MacCormack scheme, is to compute a provisional value $\tilde{f}_0$ using either the current scheme (2.7.10a) modified by a second-order noncentered approximation of the second derivative or scheme (2.7.10a) where $\beta = 0$, $\gamma = 1$. This last technique necessitates computing the final value f_1^{n+1} and the corresponding provisional values $\tilde{f}_1$ and $\tilde{f}_2$ with the same scheme $\beta = 0$, $\gamma = 1$. Note that such a method is impossible to apply with Eq. (2.7.12) and the associated second-order schemes (2.7.7a), (2.7.7b), and (2.7.14) except if $\tilde{G}_i$ is no longer defined by (2.7.14b) but rather by

$$\tilde{G}_i = \tilde{F}_i - \frac{\nu(\tilde{f}_{i+1}) + \nu(\tilde{f}_i)}{2} \frac{\tilde{f}_{i+1} - \tilde{f}_i}{\Delta x}$$

which leads to a first-order accuracy in space.

Another way that has been found to be efficient is based upon the remark that the two variants $\beta = \gamma = 0$ and $\beta = \gamma = 1$ have opposite leading part truncation error. So, one simultaneously computes two values $f_1^{n+1,+}$ and $f_1^{n+1,-}$ given, respectively, by each of the two variants and one defines the final value by the average $f_1^{n+1} = (f_1^{n+1,+} + f_1^{n+1,-})/2$. Note that two predictor values corresponding to the two schemes must be computed at $i = 2$.

In conclusion, the degree of arbitrariness existing in the classes of schemes described above allows one to treat points near a boundary by using special approaches to reduce truncation error. However, care must be taken to avoid loss in stability of the boundary scheme.

2.7.4 Implicit methods

The limitation on the time step induced by the stability conditions associated with explicit schemes is often too restrictive in applications, and consequently implicit schemes are being used more and more to overcome this limitation. In the case of a one-dimensional linear equation such as Eq. (2.6.1), the two-level scheme

$$\frac{(f_i^{n+1} - f_i^n)}{\Delta t} + (A \Delta_x^0 - \nu \Delta_{xx}) [\theta f_i^{n+1} + (1 - \theta) f_i^n] = 0 \qquad (2.7.18)$$

where $0 \le \theta \le 1$ with a leading truncation error

$$\frac{\Delta t}{2} \frac{\partial}{\partial t} \left[\frac{\partial f}{\partial t} - 2 \theta \left(A \frac{\partial f}{\partial x} - \nu \frac{\partial^2 f}{\partial x^2} \right) \right]$$

and stability conditions [from (2.6.18)]

$$2 (2 \theta - 1) S + 1 \ge 0, \qquad (2 \theta - 1) T^2 + 2 S \ge 0$$

where $S = \nu \Delta t / \Delta x^2$ and $T = A \Delta t / \Delta x$, is particularly interesting in the case where $\theta = 1/2$ (Crank–Nicolson scheme) because it is second-order accurate and unconditionally stable. At each time step the computational effort is a solution of a linear algebraic system with a tridiagonal matrix.

Another efficient scheme, second-order accurate and unconditionally stable, is the following three-level scheme:

$$\frac{1}{2\,\Delta t}\,(3\,f_i^{n+1} - 4\,f_i^n + f_i^{n-1}) + (A\,\Delta_x^0 - \nu\Delta_{xx})\,f_i^{n+1} = 0 \qquad (2.7.19)$$

As pointed out by Richtmyer and Morton (1967) such a scheme has the property of damping small-wavelength harmonics better than the Crank–Nicolson scheme. This property can be beneficial in some applications.

Both schemes can be written in a form (called the "Δ form"; see Warming and Beam, 1978) which is particularly useful for multidimensional extensions. By introducing $\Lambda_x = A\,\Delta_x^0 - \nu\Delta_{xx}$ and the identity operator I, scheme (2.7.18) can be written

$$[I + \theta\Delta t\,\Lambda_x]\,\Delta f_i^{n+1} = -\Delta t\,\Lambda_x f_i^n$$

$$f_i^{n+1} = f_i^n + \Delta f_i^{n+1}$$

and scheme (2.7.19) becomes

$$\left[I + \frac{2\,\Delta t}{3}\,\Lambda_x\right]\Delta f_i^{n+1} = -\frac{2}{3}\,\Delta t\,\Lambda_x f_i^n + \frac{1}{3}\,\Delta f_i^n$$

$$f_i^{n+1} = f_i^n + \Delta f_i^{n+1}$$

Other implicit schemes for linear one-dimensional equations can be found in the literature but no attempt will be made here to describe further results.

In the case of nonlinear equations in which we are more interested because of application to fluid-mechanics problems, the nonlinearity, if it is treated fully implicitly, necessitates the use of an iterative procedure. When possible, the use of such procedures must be avoided.

If the viscous term is dominant with regard to the nonlinear term, the implicitness can be associated principally with the viscous term. For this case the nonlinear equation (2.7.1) can be approximated with the leapfrog Crank–Nicolson scheme to obtain

$$\frac{1}{2\,\Delta t}\,(f_i^{n+1} - f_i^{n-1}) + \Delta_x^0 G_i^n - \frac{\nu}{2}\,\Delta_{xx}(f_i^{n+1} + f_i^{n-1}) = 0 \qquad (2.7.20)$$

The accuracy is of second order in time as well as in space, and the stability is the CFL condition $|T| = |A(f)|\,\Delta t/\Delta x \le 1$ where $A = dG/df$. However, the scheme is known to be unstable in the nonlinear case if $\nu = 0$. In this case, there exists a decoupling between two meshes. This phenomenon introduces oscillations in the numerical solution which can also exist when ν is present but not large enough to prevent instability. The appearance of such oscillations is generally avoided by averaging the numerical solution periodically every few (15–25) time steps. A possible technique to average the numerical solution, at time $N\,\Delta t$, for example, is the following (Orszag and Tang, 1979):

1. Compute a provisional value, $\tilde{f}_i^{N+1}$, at $N+1$ using Eq. (2.7.20).
2. Define a new value $\bar{f}_i^N = \frac{1}{4}(\tilde{f}_i^{N+1} + 2 f_i^N + f_i^{N-1})$.
3. Compute the new value f_i^{N+1} using Eq. (2.7.20) with f_i^N replaced by $\bar{f}_i^N$.

Such a procedure conserves the second-order accuracy.

Another scheme that can be used with success if the viscosity ν is not too small is the Adams–Bashforth Crank–Nicolson scheme:

$$\frac{1}{\Delta t}(f_i^{n+1} - f_i^n) + \frac{\Delta_x^0}{2}(3 G_i^n - G_i^{n-1}) - \frac{\nu}{2}\Delta_{xx}(f_i^{n+1} + f_i^n) = 0 \quad (2.7.21)$$

The accuracy is of second order, but if $\nu = 0$ the scheme is found to be unstable if the strict Von Neumann criterion is required. The instability is known to be very weak. In any case, it is recommended that such a scheme be used when ν is not zero. The analytical criterion of stability for (2.7.21) is not known. A study based upon the Miller theorem (see the Appendix A) associated with a numerical calculation gives the curves of stability shown in Fig. 2.7.3.

Schemes (2.7.20) and (2.7.21) are not unconditionally stable. Unconditional stability can be obtained if the nonlinear term $\partial G/\partial x$ is evaluated at time $(n+1)\Delta t$ and then linearized in a convenient manner. Let us consider the nonlinear versions of (2.7.18) and (2.7.19) which can be written

$$\frac{1}{2\,\Delta t}[(1 + \epsilon)(f_i^{n+1} - f_i^n) + (1 - \epsilon)(f_i^n - f_i^{n-1})]$$

$$+ \theta\left[\left(\frac{\partial G}{\partial x}\right)_i^{n+1} - \nu\Delta_{xx}f_i^{n+1}\right] \quad (2.7.22)$$

$$+ (1 - \theta)\left[\left(\frac{\partial G}{\partial x}\right)_i^n - \nu\,\Delta_{xx}f_i^n\right] = 0$$

The truncation error with respect to time is $O[(2\theta - \epsilon)\Delta t, \Delta t^2]$ so that the scheme is second-order accurate in time if $2\theta - \epsilon = O(\Delta t)$. Scheme (2.7.18) corresponds to $\epsilon = 1$ and scheme (2.7.19) to $\theta = 1$, $\epsilon = 2$.

Various linearizations are possible:

(i)   If the nonconservative form $\partial G/\partial x = A(f)\,\partial f/\partial x$ is used, the simple formula

$$\left(\frac{\partial G}{\partial x}\right)_i^{n+1} = A_i^n\left(\frac{\partial f}{\partial x}\right)_i^{n+1} + O(\Delta t) \cong A_i^n\,\Delta_x^0 f_i^{n+1}$$

can be considered, but the second-order accuracy in time is destroyed. This second-order accuracy can be preserved by using the linearization proposed by Lindemuth and Killeen (1973):

$$\left(\frac{\partial G}{\partial x}\right)_i^{n+1} = \left(A\,\frac{\partial f}{\partial x}\right)_i^{n+1}$$

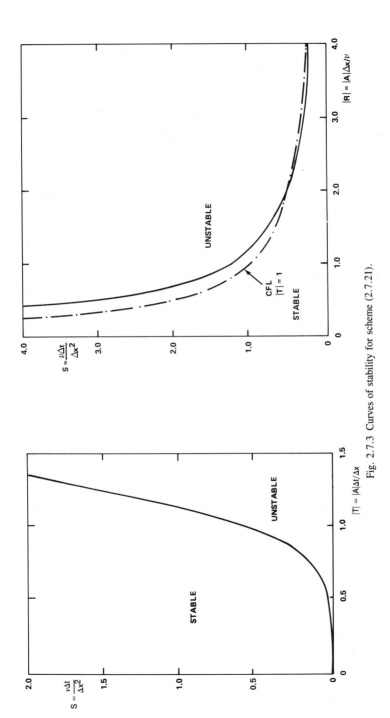

Fig. 2.7.3 Curves of stability for scheme (2.7.21).

$$= \left(A \frac{\partial f}{\partial x}\right)_i^n + \Delta t \left[\frac{\partial}{\partial t}\left(A \frac{\partial f}{\partial x}\right)\right]_i^n + O(\Delta t^2)$$

Then,

$$\left[\frac{\partial}{\partial t}\left(A \frac{\partial f}{\partial x}\right)\right]_i^n = \left[\frac{dA}{df}\frac{\partial f}{\partial t}\frac{\partial f}{\partial x} + A \frac{\partial^2 f}{\partial x \partial t}\right]_i^n$$

and, approximating the time derivative with a forward difference and the space derivative with a centered difference, we obtain

$$\left(\frac{\partial G}{\partial x}\right)_i^{n+1} = \left(A \frac{\partial f}{\partial x}\right)_i^{n+1} \cong A_i^n \Delta_x^0 f_i^{n+1} + \left(\frac{dA}{df}\right)_i^n (f_i^{n+1} - f_i^n)\Delta_x^0 f_i^n$$

$$(2.7.23)$$

(ii)   If $A(f) = Cf$, (C = const), the linearization

$$\left(\frac{\partial G}{\partial x}\right)_i^{n+1} = A_i^n \left(\frac{\partial f}{\partial x}\right)_i^{n+1} + A_i^{n+1}\left(\frac{\partial f}{\partial x}\right)_i^n - A_i^n\left(\frac{\partial f}{\partial x}\right)_i^n + O(\Delta t^2)$$

$$\cong A_i^n \Delta_x^0 f_i^{n+1} + A_i^{n+1} \Delta_x^0 f_i^n - A_i^n \Delta_x^0 f_i^n$$

leads to second-order accuracy.

(iii)   For the conservative form Briley and McDonald (1973, 1975) as well as Beam and Warming (1976, 1978) have proposed the following linearization:

$$\left(\frac{\partial G}{\partial x}\right)_i^{n+1} = \left(\frac{\partial G}{\partial x}\right)_i^n + \Delta t \left(\frac{\partial^2 G}{\partial x \partial t}\right)_i^n + O(\Delta t^2) \qquad (2.7.24\text{a})$$

then

$$\left(\frac{\partial^2 G}{\partial x \partial t}\right)_i^n = \left[\frac{\partial}{\partial x}\left(A \frac{\partial f}{\partial t}\right)\right]_i^n \cong \Delta_x^0\left(A_i^n \frac{f_i^{n+1} - f_i^n}{\Delta t}\right) \qquad (2.7.24\text{b})$$

and, finally

$$\left(\frac{\partial G}{\partial x}\right)_i^{n+1} \cong \Delta_x^0 G_i^n + \Delta_x^0 [A_i^n (f_i^{n+1} - f_i^n)] \qquad (2.7.24\text{c})$$

Now if (2.7.23) or (2.7.24c) are introduced into Eq. (2.7.22), the resulting schemes with $\theta = \frac{1}{2}$, $\epsilon = 1$ or $\theta = 1$, $\epsilon = 2$ are second-order accurate and linearly unconditionally stable. However, the application of the technique of factorization to the solution of the tridiagonal algebraic system associated with the schemes may impose some limitation on the mesh Reynolds number or on the time step (conditions (2.2.7a)).   For example, these conditions for scheme (2.7.22) with $\theta = \frac{1}{2}$, $\epsilon = 1$, and $\nu = 0$ give

$$(|A_{i+1}^n| + |A_{i-1}^n|)\frac{\Delta t}{\Delta x} < 4$$

that is, a condition two times less restrictive than the usual CFL condition. Although such a condition can be avoided if another technique of solution is used for the algebraic system, it can happen that diagonal dominance can also be necessary to ensure that the exact solution of the finite-difference equation is not oscillatory.

(iv)   An efficient scheme that leads to a strictly diagonally dominant matrix without restrictions has been considered by Lerat (1979, 1981) for inviscid equations of conservation laws. The method is easily extended to the viscous equation if only first-order accuracy in time is required. The scheme can be constructed from the Taylor expansion (2.7.24a), but in (2.7.24b) the derivative $\partial f / \partial t$ is expressed in terms of the equation itself, i.e.:

$$\left(\frac{\partial^2 G}{\partial x\, \partial t}\right) = \frac{\partial}{\partial x}\left(A\,\frac{\partial f}{\partial t}\right) = \frac{\partial}{\partial x}\left[A\left(-\frac{\partial G}{\partial x} + \nu\,\frac{\partial^2 f}{\partial x^2}\right)\right]$$

Now by using $\partial G /\partial x = A\, \partial f /\partial x$, we obtain

$$\left(\frac{\partial^2 G}{\partial x\, \partial t}\right) = -\frac{\partial}{\partial x}\left[\frac{1}{2}\left(A\,\frac{\partial G}{\partial x} + A^2\,\frac{\partial f}{\partial x}\right) - \nu A\,\frac{\partial^2 f}{\partial x^2}\right] \qquad (2.7.25)$$

Then, in order to avoid the introduction of third-order differences, the viscous term is neglected in the above expression and, finally, the scheme is written

$$\frac{1}{\Delta t}\,(f_i^{n+1} - f_i^n) + \Delta_x^0 G_i^n - \frac{\Delta t}{4}\,\Delta_x^1[A_i^n\,\Delta_x^1 G_i^n + (A_i^n)^2\,\Delta_x^1 f_i^{n+1}]$$

$$- \frac{\nu}{2}\,\Delta_{xx}(f_i^{n+1} + f_i^n) = 0$$

where

$$\Delta_x^1 f_i = \frac{f_{i+1/2} - f_{i-1/2}}{\Delta x}$$

and

$$A_{i+1/2} = A\left(\frac{f_{i+1} + f_i}{2}\right)$$

The truncation error in time is $O(\nu\,\Delta t)$ and it does not disappear at steady state. Second-order accuracy can be recovered if the viscous term in (2.7.25) is conserved and evaluated at level n, but the unconditional stability is lost. Note that in Lerat's work (1979, 1981), the above scheme with $\nu = 0$ is obtained from a study of a general class of implicit schemes. Finally, it is important to note that if the function $G(f)$ possesses the homogeneous property $G(\lambda f) = \lambda G(f)$ (as for the gas-dynamics equations for perfect gases; see Lerat and Peyret, 1974a; Beam and Warming, 1976; Warming and Beam, 1978) so that $G = Af$, then the linearization (2.7.24c) simplifies to

$$\left(\frac{\partial G}{\partial x}\right)^{n+1} = \frac{\partial}{\partial x}\left[G^n + \Delta t\left(\frac{\partial G}{\partial t}\right)^n + O(\Delta t^2)\right] = \frac{\partial}{\partial x}(A^n f^{n+1}) + O(\Delta t^2)$$

2.8   Multidimensional Equation

2.8.1   Explicit schemes for the advection–diffusion equation

Let us consider first the two-dimensional advection–diffusion equation:

$$\frac{\partial f}{\partial t} + A\,\frac{\partial f}{\partial x} + B\,\frac{\partial f}{\partial y} - \nu\,\nabla^2 f = 0 \tag{2.8.1}$$

where A, B, and $\nu\,(>0)$ are assumed to be constant.

The direct extension of the explicit scheme to solve Eq. (2.8.1) is written

$$\frac{1}{\Delta t}(f_{i,j}^{n+1} - f_{i,j}^n) + A\,\Delta_x^0 f_{i,j}^n + B\,\Delta_y^0 f_{i,j}^n - \nu(\Delta_{xx} + \Delta_{yy})f_{i,j}^n = 0 \tag{2.8.2}$$

where $f_{i,j}^n \cong f(i\,\Delta x, j\,\Delta y, n\,\Delta t)$ with i, j, n integers, and

$$\Delta_x^0 f_{i,j} = \frac{1}{2\,\Delta x}(f_{i+1,j} - f_{i-1,j})$$

$$\Delta_y^0 f_{i,j} = \frac{1}{2\,\Delta y}(f_{i,j+1} - f_{i,j-1})$$

$$\Delta_{xx} f_{i,j} = \frac{1}{\Delta x^2}(f_{i+1,j} - 2f_{i,j} + f_{i-1,j})$$

$$\Delta_{yy} f_{i,j} = \frac{1}{\Delta y^2}(f_{i,j+1} - 2f_{i,j} + f_{i,j-1})$$

Scheme (2.8.2) is first-order accurate in time and second-order in space. The conditions of stability (Hindmarsh et al., 1984) are*:

$$(A^2 + B^2)\,\Delta t \leqslant 2\,\nu, \qquad \frac{\nu\,\Delta t}{\Delta x^2} \leqslant \frac{1}{4} \qquad \text{where } \Delta y = \Delta x$$

The upwind scheme analogous to (2.6.22) can be written, for Eq. (2.8.1), in the form

$$\frac{1}{\Delta t}(f_{i,j}^{n+1} - f_{i,j}^n) + \frac{A}{2}[(1 - \epsilon_A)\Delta_x^+ f_{i,j}^n + (1 + \epsilon_A)\Delta_x^- f_{i,j}^n]$$

$$+ \frac{B}{2}[(1 - \epsilon_B)\Delta_y^+ f_{i,j}^n + (1 + \epsilon_B)\Delta_y^- f_{i,j}^n] \tag{2.8.3a}$$

$$- \nu(\Delta_{xx} + \Delta_{yy})f_{i,j}^n = 0$$

---

*The first condition replaces the necessary (but not sufficient) condition $(|A| + |B|)^2\,\Delta t \leqslant 4\,\nu$ proposed in the previous editions of the book.

where $\epsilon_A = \text{sign}(A)$, $\epsilon_B = \text{sign}(B)$, and

$$\Delta_x^+ f_{i,j} = \frac{1}{\Delta x}(f_{i+1,j} - f_{i,j}), \qquad \Delta_x^- f_{i,j} = \frac{1}{\Delta x}(f_{i,j} - f_{i-1,j})$$

$$\Delta_y^+ f_{i,j} = \frac{1}{\Delta y}(f_{i,j+1} - f_{i,j}), \qquad \Delta_y^- f_{i,j} = \frac{1}{\Delta y}(f_{i,j} - f_{i,j-1})$$

The criterion of stability is, if we assume $\Delta x = \Delta y$,

$$\Delta t \leq \frac{\Delta x^2}{4\,\nu + (|A| + |B|)\,\Delta x} \tag{2.8.3b}$$

2.8.2 The ADI method

In the case where implicit schemes are preferred because of their properties of stability, it is recommended that one select methods leading to the solution of a tridiagonal algebraic system (simple for a scalar equation, by blocks for a vector equation). The alternating direction implicit (ADI) method introduced by Peaceman and Rachford (1955) allows the construction of very efficient implicit schemes. Many works have been published on the subject, we refer here only to the fundamental paper by Douglas and Gunn (1964) and to the books by Yanenko (1971) and by Mitchell (1969).

We assume that A and B in Eq. (2.8.1) are no longer constant and depend on the variable f. The ADI method is a two-step method which is written

$$\frac{2}{\Delta t}(\tilde{f}_{i,j} - f_{i,j}^n) + (A_1 \Delta_x^0 - \nu \Delta_{xx})\tilde{f}_{i,j} + (B_1 \Delta_y^0 - \nu \Delta_{yy})f_{i,j}^n = 0 \tag{2.8.4a}$$

$$\frac{2}{\Delta t}(f_{i,j}^{n+1} - \tilde{f}_{i,j}) + (A_2 \Delta_x^0 - \nu \Delta_{xx})\tilde{f}_{i,j} + (B_2 \Delta_y^0 - \nu \Delta_{yy})f_{i,j}^{n+1} = 0 \tag{2.8.4b}$$

where A_1, A_2, B_1, and B_2 are approximations of A and B. The predictor value $\tilde{f}_{i,j}$ can be considered as an approximation of the exact solution at time $(n + \frac{1}{2})\Delta t$ if A_1, A_2 and B_1, B_2 are suitable approximations. For example, if only first-order accuracy in time is required, the simple expressions $A_1 = A_2 = A_{i,j}^n$ and $B_1 = B_2 = B_{i,j}^n$, where $A_{i,j}^n = A(f_{i,j}^n)$ and $B_{i,j}^n = B(f_{i,j}^n)$, can be used. The accuracy is of second order in the special case where A and B are constant.

For second-order accuracy in time with A and B nonconstant, the suitable approximations are

$$A_1 = c_0 A_{i,j}^{n+1} + (1 - c_0 - c_1)A_{i,j}^n + c_1 A_{i,j}^{n-1}$$
$$A_2 = (1 - c_0 + c_1 + c_2)A_{i,j}^{n+1} + (c_0 - c_1 - 2\,c_2)A_{i,j}^n + c_2 A_{i,j}^{n-1}$$

and

$$B_1 = \gamma_0 B_{i,j}^{n+1} + (1 - \gamma_0 - \gamma_1)B_{i,j}^n + \gamma_1 B_{i,j}^{n-1}$$
$$B_2 = (1 - \gamma_0 + \gamma_1 + \gamma_2)B_{i,j}^{n+1} + (\gamma_0 + \gamma_1 - 2\gamma_2)B_{i,j}^n + \gamma_2 B_{i,j}^{n-1}$$

where c_0, c_1, c_2 and γ_0, γ_1, γ_2 are arbitrary parameters. In the case where $c_0 = \gamma_0 = 0$, $c_1 + c_2 = \gamma_1 + \gamma_2 = -1$, the values $A_{i,j}^{n+1}$ and $B_{i,j}^{n+1}$ do not appear in Eq. (2.8.4), so that both steps (2.8.4a) and (2.8.4b) can be solved successively. If not, an iterative procedure would be required and that is only efficient if it is included into a global iterative procedure as used for the solution of the incompressible Navier–Stokes equations.

In the case where A and B are constant, the above ADI scheme is found to be unconditionally stable. But it must be recalled that a study of stability does not take into account the boundary conditions, and their presence can destroy the unconditional stability (see Bontoux et al., 1978 for the Navier–Stokes equations). At each step (2.8.4a) or (2.8.4b), tridiagonal algebraic systems must be solved, for example, by means of the factorization method. In this case, the strict diagonal dominance condition is satisfied if $|A_1| \, \Delta x / v \leqslant 2$, $|B_2| \, \Delta y / v \leqslant 2$. If these conditions do not hold, the limitation on the time step

$$\Delta t < \frac{2 \, \Delta x^2}{|A_1| \, \Delta x - 2\,v}, \qquad \Delta t < \frac{2 \, \Delta y^2}{|B_2| \, \Delta y - 2\,v} \tag{2.8.4c}$$

ensures that conditions (2.2.7a) are satisfied.

Let us now consider briefly the problem of boundary conditions, already discussed in Section 2.7.3. Assume that the solution is determined in a rectangular domain whose boundary is $\Gamma = \Gamma_1 \cup \Gamma_2$ (Γ_1 and Γ_2 being the parts of Γ parallel to the y- and x-axis, respectively) with boundary conditions of the Dirichlet type, $f(x,y,t) \, |_\Gamma = \phi(x,y,t)$. The solution of the algebraic system corresponding to (2.8.4a) requires knowledge of $\tilde{f}_{i,j}$ on Γ_1. Although $\tilde{f}_{i,j}$ can be considered an approximation of the exact solution, the use of the boundary value $\tilde{f}_{i,j} \, |_{\Gamma_1} = \phi[x,y,(n+\tfrac{1}{2})\Delta t)]$ leads to a large truncation error near the boundary when ϕ is effectively time dependent (Fairweather and Mitchell, 1967; Mitchell, 1969). The reason was explained in Section 2.7.3. The best way to overcome this difficulty is to define a boundary value such that the combination of both time steps yields second-order accuracy at points adjacent to the boundary as well as inner points. This can be accomplished if the value $\tilde{f}_{i,j} \, |_{\Gamma_1}$ is deduced from the finite-difference equations themselves. Therefore, assuming $A_1 = A_2$ in Eq. (2.8.4), the subtraction of (2.8.4b) from (2.8.4a) yields

$$\tilde{f}_{i,j} = \tfrac{1}{2} (f_{i,j}^{n+1} + f_{i,j}^n) + \tfrac{1}{4} \Delta t[(B_2 \Delta_y^0 - v\Delta_{yy})f_{i,j}^{n+1} - (B_1 \Delta_y^0 - v\Delta_{yy})f_{i,j}^n] \tag{2.8.5}$$

This equation furnishes the value $\tilde{f}_{i,j}$ on Γ_1 because $f_{i,j}^n$ and $f_{i,j}^{n+1}$ are known on this boundary.

The limitation on the time step introduced by the conditions (2.8.4c), although not very restrictive, can be avoided if an upwind noncentered approximation of the first-order derivatives (according to the sign of A_1 and B_2) are used instead of centered approximations. In this case, the second-order accuracy is lost. A possible technique to recover second-order accuracy at steady-state consists of alternating the direction of the noncentered difference at each step. The corresponding scheme is

$$\frac{2}{\Delta t}(\tilde{f}_{i,j} - f_{i,j}^n) + (A_{i,j}^n \Delta_x^* - \nu \Delta_{xx})\tilde{f}_{i,j} + (B_{i,j}^n \Delta_y^{**} - \nu \Delta_{yy})f_{i,j}^n = 0$$

$$(2.8.6a)$$

$$\frac{2}{\Delta t}(f_{i,j}^{n+1} - \tilde{f}_{i,j}) + (A_{i,j}^n \Delta_x^{**} - \nu \Delta_{xx})\tilde{f}_{i,j} + (B_{i,j}^n \Delta_y^* - \nu \Delta_{yy})f_{i,j}^{n+1} = 0$$

$$(2.8.6b)$$

with

$$\Delta_x^* = \tfrac{1}{2}[(1 - \epsilon_A)\Delta_x^+ + (1 + \epsilon_A)\Delta_x^-]$$
$$\Delta_x^{**} = \tfrac{1}{2}[(1 + \epsilon_A)\Delta_x^+ + (1 - \epsilon_A)\Delta_x^-] \qquad \epsilon_A = \text{sign}(A_{i,j}^n) \qquad (2.8.6c)$$

and analogous definitions of Δ_y^* and Δ_y^{**} with $\Delta_x^\pm$, replaced by $\Delta_y^\pm$ and ϵ_A by $\epsilon_B = \text{sign }(B_{i,j}^n)$.

During the transient stage, the scheme is only first-order accurate in time and at steady state the truncation error is $O(\Delta t \, \Delta x, \, \Delta t \, \Delta y, \, \Delta x^2, \, \dots)$. Note that the unconditional stability is preserved. Such a technique has been proposed by Peyret (1971) for the solution of the incompressible Navier–Stokes equations in the primitive variables formulation and has been applied to the stream-function vorticity equations by Daube and Ta Phuoc Loc (1978).

2.8.3 Explicit schemes for a nonlinear equation in conservative form

We present here explicit schemes of the predictor–corrector type which are direct extensions of the one-dimensional case considered in Section 2.7.1. The possible extensions are numerous. We refer to works by Lerat (1981) and Laval (1981a, b) for a complete description of such schemes in the nonviscous case, and to Peyret and Viviand (1975) for some viscous extensions. In this section, we begin by describing the second-order accurate scheme proposed by MacCormack (1969), then we present a generalization of the scheme proposed by Thommen (1966) which is first order in time during the transient stage and second order at steady state.

Let us consider the nonlinear equation

$$\frac{\partial f}{\partial t} + \frac{\partial}{\partial x}\left[F(f) - \nu(f)\frac{\partial f}{\partial x}\right] + \frac{\partial}{\partial y}\left[H(f) - \nu(f)\frac{\partial f}{\partial y}\right] = 0 \qquad (2.8.7)$$

which represents a model of the two-dimensional Navier–Stokes equations for compressible flow with a nonconstant viscosity $\nu(f)$ [except there is an absence in Eq. (2.8.7) of mixed derivatives].

The MacCormack (1969) scheme applied to Eq. (2.8.7) is

$$\tilde{f}_{i,j} = f_{i,j}^n - \Delta t \, \Delta_x^+ (F_{i,j}^n - M_{i,j}^n \, \Delta_x^- f_{i,j}^n) - \Delta t \, \Delta_y^+ (H_{i,j}^n - N_{i,j}^n \, \Delta_y^- f_{i,j}^n)$$

(2.8.8a)

$$f_{i,j}^{n+1} = \tfrac{1}{2} (f_{i,j}^n + \tilde{f}_{i,j}) - \tfrac{1}{2}\Delta t \, \Delta_x^- (\tilde{F}_{i,j} - \tilde{M}_{i,j} \, \Delta_x^+ \tilde{f}_{i,j})$$
$$- \tfrac{1}{2}\Delta t \, \Delta_y^- (\tilde{H}_{i,j} - \tilde{N}_{i,j} \, \Delta_y^+ \tilde{f}_{i,j})$$

(2.8.8b)

where the predicted value $\tilde{f}_{i,j}$ is defined at the same points as the final value $f_{i,j}^{n+1}$. In the above equations we have $\tilde{F}_{i,j} = F(\tilde{f}_{i,j})$, $\tilde{H}_{i,j} = H(\tilde{f}_{i,j})$, and

$$M_{i,j}^n = m_0 \nu_{i,j}^n + (1 - m_0)\nu_{i-1,j}^n,$$
$$N_{i,j}^n = n_0 \nu_{i,j}^n + (1 - n_0)\nu_{i,j-1}^n, \qquad \nu_{i,j}^n = \nu(f_{i,j}^n)$$
$$\tilde{M}_{i,j} = (1 - m_0)\tilde{\nu}_{i+1,j} + m_0 \tilde{\nu}_{i,j},$$
$$\tilde{N}_{i,j} = (1 - n_0)\tilde{\nu}_{i,j+1} + n_0 \tilde{\nu}_{i,j}, \qquad \tilde{\nu}_{i,j} = \nu(\tilde{f}_{i,j})$$

(2.8.8c)

where m_0 and n_0 are arbitrary constants. MacCormack (1969) suggested that $m_0 = n_0 = \tfrac{1}{2}$; another possibility is $m_0 = n_0 = 1$. The effect of such a choice appears only in the truncation error. An alternate possibility is to define the above approximation of viscosity by expressions of the type $M_{i,j}^n = \nu[m_0 f_{i,j}^n + (1 - m_0)f_{i-1}^n]$, which are identical with the previous ones if ν is a linear function of f.

Scheme (2.8.8) is only one of four possible variants of the MacCormack scheme. The other variants are deduced from (2.8.8) by exchanging, respectively, Δ_x^+ by Δ_x^- and Δ_y^+ by Δ_y^-. Although no theoretical study of the best variants has been made for the viscous case as it has been for in the nonviscous one-dimensional case (Lerat and Peyret, 1974a, b) it is likely that the direction of propagation of the shock wave (if $\nu = 0$ or $\nu << 1$) will play an important role in the choice of the variant.

No complete theoretical study of the stability of scheme (2.8.8) is known. In the inviscid case a necessary condition of stability can be deduced from the requirement that the numerical domain of dependence must contain the exact domain. As a result one requires

$$\Delta t \leq \left(\frac{|A|}{\Delta x} + \frac{|B|}{\Delta y} \right)^{-1}$$

where $A = dF/df$ and $B = dH/df$. When viscosity is present, this condition can be replaced by (2.8.3b) with $\Delta x = \Delta y$ assumed.

As was mentioned for the one-dimensional case, the MacCormack scheme has the advantage of being compact and simplifying for the treatment of boundary conditions. However, it can happen that its nonsymmetrical character is a disadvantage when flow around a body is computed. In this case, centered schemes can be useful. Besides, when only the steady solution of Eq. (2.8.7) is sought or when a first-order accuracy in time is considered as sufficient, schemes of the Thommen type (Thommen, 1966) [Eq. (2.7.11)] can be used. In the two-dimensional case, the originality of the method proposed by Thommen, as a variant of the two-step Lax–Wendroff scheme (Richtmyer, 1962), is to consider predictors $\tilde{f}^x_{i+1/2,j}$ and $\tilde{f}^y_{i,j+1/2}$ defined at two different points and appearing in the x difference and the y difference, respectively, of the nonviscous part of the finite-difference equation at the second step. In the second step, the viscous terms are evaluated with the values $f^n_{i,j}$. Such an evaluation leads to a compact nine-point scheme, which is first-order accurate in time during the transient stage and second-order accurate (with no dependence on Δt) at steady state.

The Thommen scheme can be generalized by considering the predictor values $\tilde{f}^x_{i,j}$ and $\tilde{f}^y_{i,j}$ defined, respectively, at $x = (i + \beta^x)\Delta x$, $y = j\,\Delta y$, $t = (n + \alpha^x)\Delta t$ and $x = i\,\Delta x$, $y = (j + \beta^y)\Delta y$, $t = (n + \alpha^y)\Delta t$ (see Fig. 2.8.1) as Lerat (1981) did for the nonviscous equation and in the same manner as the S^α_β schemes (2.7.7) were constructed. The general scheme is then

$$
\begin{aligned}
\tilde{f}^x_{i,j} = {}& (1 - \beta^x)f^n_{i,j} + \beta^x f^n_{i+1,j} - \alpha^x\,\Delta t\,\Delta^+_x F^n_{i,j} \\
& - \alpha^x\,\Delta t\{\gamma^x_0[\lambda^x\Delta^+_y + (1 - \lambda^x)\Delta^-_y]H^n_{i+1,j} \\
& + (1 - \gamma^x_0)[\mu^x\Delta^+_y + (1 - \mu^x)\Delta^-_y]H^n_{i,j}\} \\
& + \alpha^x\,\Delta t\{\epsilon^x\Delta^-_x(M^n_{i+1,j}\,\Delta^+_x f^n_{i+1,j}) + (1 - \epsilon^x)\Delta^+_x(M^n_{i,j}\,\Delta^-_x f^n_{i,j})\} \\
& + \alpha^x\,\Delta t[\gamma^x_1\Delta^-_y(\overline{N}^n_{i+1,j}\,\Delta^+_y f^n_{i+1,j}) + (1 - \gamma^x_1)\Delta^-_y(\overline{N}^n_{i,j}\Delta^+_y f^n_{i,j})]
\end{aligned}
$$

$$(2.8.9a)$$

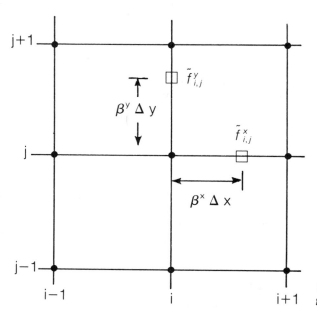

Fig. 2.8.1 Grid system to generalize Thommen's scheme.

$$\tilde{f}^y_{i,j} = (1 - \beta^y)f^n_{i,j} + \beta^y f^n_{i,j+1} - \alpha^y \, \Delta t \, \Delta^+_y H^n_{i,j}$$
$$- \alpha^y \, \Delta t\{\gamma^y_0[\lambda^y\Delta^+_x + (1 - \lambda^y)\Delta^-_x]F^n_{i,j+1} + (1 - \gamma^y_0)$$
$$\times [\mu^y\Delta^+_x + (1 - \mu^y)\Delta^-_x]F^n_{i,j}\}$$
$$+ \alpha^y \, \Delta t[\epsilon^y \, \Delta^-_y \, (N^n_{i,j+1}\Delta^+_y f^n_{i,j+1}) + (1 - \epsilon^y) \, \Delta^+_y \, (N^n_{i,j}\Delta^-_y f^n_{i,j})]$$
$$+ \alpha^y \, \Delta t[\gamma^y_1\Delta^-_x \, (\overline{M}^n_{i,j+1} \, \Delta^+_x f^n_{i,j+1}) + (1 - \gamma^y_1)\Delta^-_x \, (\overline{M}^n_{i,j} \, \Delta^+_x f^n_{i,j})]$$

$$(2.8.9b)$$

$$f^{n+1}_{i,j} = f^n_{i,j} - \frac{\Delta t}{2 \, \alpha^x} \{[(\alpha^x - \beta^x)\Delta^+_x + (\alpha^x + \beta^x - 1)\Delta^-_x]F^n_{i,j} + \Delta^-_x \tilde{F}^x_{i,j}\}$$

$$- \frac{\Delta t}{2 \, \alpha^y} \{[(\alpha^y - \beta^y)\Delta^+_y + (\alpha^y + \beta^y - 1)\Delta^-_y]H^n_{i,j} + \Delta^-_y \tilde{H}^y_{i,j}\}$$

$$+ \Delta t[\Delta^-_x \, (\overline{M}^n_{i,j} \, \Delta^+_x f^n_{i,j}) + \Delta^-_y \, (\overline{N}^n_{i,j} \, \Delta^+_y f^n_{i,j})]$$

$$(2.8.9c)$$

where

$$\tilde{F}^x_{i,j} = F(\tilde{f}^x_{i,j}) \qquad \text{and} \qquad \tilde{H}^y_{i,j} = H(\tilde{f}^y_{i,j})$$

$$M^n_{i,j} = \sum_{k=-1}^{1} m_k v^n_{i+k,j} \qquad \text{with} \sum_k m_k = 1, \qquad m_k = \text{const}$$

$$N^n_{i,j} = \sum_{k=-1}^{1} n_k v^n_{i,j+k} \qquad \text{with} \sum_k n_k = 1, \qquad n_k = \text{const}$$

where

$$\overline{M}^n_{i,j} = \tfrac{1}{2}(v^n_{i+1,j} + v^n_{i,j}) \qquad \text{and} \qquad \overline{N}^n_{i,j} = \tfrac{1}{2}(v^n_{i,j+1} + v^n_{i,j}). \qquad (2.8.9d)$$

The parameters α^x, β^x, γ^x_0, γ^x_1, λ^x, μ^x and the analogous ones with the superscript y are arbitrary. They define the particular scheme of the class. The original Thommen scheme (Thommen, 1966) is obtained when all these parameters are taken equal to $\tfrac{1}{2}$. In the inviscid case, the original scheme has been used for a transonic calculation by Magnus and Yoshihara (1975). Again, for invisicid transonic flow, the scheme for which all the parameters are equal to $\tfrac{1}{2}$, except $\alpha^x = \alpha^y = 1 + \sqrt{5}/2$, has been used by Lerat and Sides (1981). In the inviscid case $v = 0$, the MacCormack schemes are obtained for $\alpha^x = \alpha^y = 1$. The forward variant, identical to the one described by Eq. (2.8.8) with $v = 0$, corresponds to $\beta^x = \gamma^x_0 = \beta^y = \gamma^y_0 = 0$ and $\mu^x = \mu^y = 1$. The other variants mentioned above correspond to the following values of the parameters:

$$\beta^x = \gamma^x_0 = \mu^x = 0, \quad \beta^y = \gamma^y_0 = \lambda^y = 1 \quad x\text{—forward;} \ y\text{—backward}$$

$$\beta^x = \gamma^x_0 = \lambda^x = 1, \quad \beta^y = \gamma^y_0 = \mu^y = 0 \quad x\text{—backward;} \ y\text{—forward}$$

$$\beta^x = \gamma^x_0 = \beta^y = \gamma^y_0 = 1, \quad \lambda^x = \lambda^y = 0 \quad x\text{—backward;} \ y\text{—backward}$$

Note that the scheme defined by $\gamma_0^x = \lambda^x = \mu^x = \gamma_0^y = \lambda_0^y = \mu^y = \frac{1}{2}$, $\alpha^x = \alpha^y = 1$, and β^x, $\beta^y = 0$ or 1 is very close to the MacCormack scheme but has better properties of stability (Lerat, 1981). In the viscous case $\nu \neq 0$, the above choice of parameters defines schemes analogous to the MacCormack scheme, but they are only first-order accurate in time. With regard to stability, we indicate for the special linear cases in which $\gamma_0^x = \lambda^x = \mu^x = \gamma_0^y = \lambda^y = \mu^y = \frac{1}{2}$ that the conditions for the inviscid case are

$$\frac{\Delta t}{\Delta x}\,|A| \leq \frac{1}{\sqrt{8}}, \qquad \frac{\Delta t}{\Delta y}\,|B| \leq \frac{1}{\sqrt{8}}$$

For the viscous case, condition (2.8.3b) is an approximate guide.

2.8.4   Explicit splitting methods

A way to extend the general one-dimensional viscous schemes to multi-dimensional problems is to construct explicit splitting methods (Strang, 1968; Gourlay and Morris, 1968, 1970; McGuire and Morris, 1972; Laval 1980, 1981a, 1981b for the inviscid case; MacCormack 1971 for the viscous case) based upon the one-dimensional schemes. We describe here the splitting technique associated with the MacCormack scheme (2.7.7) and (2.7.14) (with $\alpha = 1$, $\beta = \gamma = 0$ or 1) applied to Eq. (2.8.7). The technique can be extended to any of the schemes of the general class introduced previously.

Difference operators $L_x(\Delta t)$ and $L_y(\Delta t)$ are defined by applying the one-dimensional scheme successively to the x- and the y-derivative terms in Eq. (2.8.7). Thus the condensed expression

$$f_{i,j}^x = L_x(\Delta t)f_{i,j}^n \tag{2.8.10a}$$

stands for the two-step formula (2.7.7) ($f_{i,j}^x$ being identified with $f_{i,j}^{n+1}$), where $G = F - \nu\,\partial f/\partial x$ is approximated by Eq. (2.7.14a) and (2.7.14b). Similarly

$$f_{i,j}^y = L_y(\Delta t)f_{i,j}^n \tag{2.8.10b}$$

stands for the analogous formulas where the role of x is replaced by y and where $G = H - \nu\,\partial f/\partial y$. Applying the operators L_x and L_y successively leads to the scheme

$$f_{i,j}^{n+1} = L_x(\Delta t)L_y(\Delta t)f_{i,j}^n \tag{2.8.11}$$

In very special cases (in particular the scalar case with $\nu = $ const) such a finite-difference scheme is second-order accurate. In the general case where (2.8.7) is a vector equation (e.g., the compressible Navier–Stokes equations), scheme (2.8.11) is not second-order accurate. The second-order accuracy is recovered if symmetric sequences are considered, such as

$$f_{i,j}^{n+1} = L_y\!\left(\frac{\Delta t}{2}\right) L_x\!\left(\frac{\Delta t}{2}\right) L_x\!\left(\frac{\Delta t}{2}\right) L_y\!\left(\frac{\Delta t}{2}\right) f_{i,j}^n \tag{2.8.12}$$

or

$$f_{i,j}^{n+1} = L_y\left(\frac{\Delta t}{2}\right) L_x(\Delta t) L_y\left(\frac{\Delta t}{2}\right) f_{i,j}^n \qquad (2.8.13)$$

in which the sum of the time steps must be equal to $2\,\Delta t$.

The advantages of a splitting method over the two-dimensional schemes (2.8.8a)–(2.8.8c) or (2.8.9a)–(2.8.9d) are: (i) the associated stability conditions are of one-dimensional type, and therefore less restrictive than for the two-dimensional scheme; and (ii) different time steps can be used with the L_x and L_y operators using the following sequence

$$f_{i,j}^{n+1} = L_y\left(\frac{\Delta t}{2M}\right)^M L_x(\Delta t)\, L_y\left(\frac{\Delta t}{2M}\right)^M f_{i,j}^n \qquad (2.8.14)$$

where M is an integer ≥ 1. This is particularly interesting when the maximum allowable time steps in the two directions—Δt_x and Δt_y— are much different (in general due to much different mesh sizes Δx and Δy) since it allows advancement in time in each direction with the corresponding maximum time step. Thus if $\Delta t_y << \Delta t_x$, scheme (2.8.14) can be used with advantage by choosing $\Delta t = \Delta t_x$ and M equal to the smallest integer such that $M \geq \Delta t_x / 2\, \Delta t_y$. However, difficulties may arise if M is too large (MacCormack, 1971).

The major inconvenience of the splitting techniques lies in the treatment of boundary conditions. Because each step is not consistent with the exact differential equation to be solved, a problem arises in determining the boundary conditions for the intermediate values. Techniques for handling the boundary conditions associated with the Richtmyer (two-step Lax–Wendroff) scheme for the approximation of inviscid problems have been proposed by Gourlay and Morris (1970) and McGuire and Morris (1972). In the general viscous case, the common technique is to use the exact boundary conditions for the intermediate values. As discussed in Section 2.7.3, such a procedure leads to an increase in the magnitude of the truncation error near the boundary (see Yanenko, 1971). However, its practical effect on the accuracy of the numerical solution does not seem to be important.

2.8.5 Generalized ADI methods

Implicit schemes applied to two- or three-dimensional problems can be easily solved if generalized ADI methods (Douglas and Gunn, 1964; Yanenko, 1971; Berezin et al., 1972, 1975a, 1975b; Lindemuth and Killeen, 1973; Briley and McDonald, 1973, 1975, 1977, 1980; Beam and Warming, 1978) are used to reduce the problem to a successive solution of linear algebraic systems with tridiagonal (simple or by blocks) matrices.

Let us describe briefly the method. Assume the implicit scheme to be written

$$(I + \Lambda_x + \Lambda_y + \Lambda_z)\, f_{i,j,k}^{n+1} = Q\, f_{i,j,k}^n \qquad (2.8.15)$$

where Λ_x, Λ_y, Λ_z are difference operators each corresponding, respectively, to derivatives in only one direction and Q is a difference operator with respect to any direction and possibly containing mixed operators. The above operators depend also on $f_{i,j,k}^n$ when the original equation is nonlinear.

Schemes like (2.8.15) can be constructed by performing a straightforward multidimensional extension of the implicit schemes described in Section 2.7.4. Moreover, it can approximate a scalar equation as well as a vector equation. Equation (2.8.15) is first written in the Δ form:

$$(I + \Lambda_x + \Lambda_y + \Lambda_z)(f_{i,j,k}^{n+1} - f_{i,j,k}^n) = R f_{i,j,k}^n \qquad (2.8.16)$$

with $R = Q - (I + \Lambda_x + \Lambda_y + \Lambda_z)$. Form (2.8.16) makes apparent the incremental quantity $f_{i,j,k}^{n+1} - f_{i,j,k}^n$. Equation (2.8.16) is solved by the generalized ADI method:

$$(I + \Lambda_x)\psi_{i,j,k}^* = R f_{i,j,k}^n \qquad (2.8.17a)$$

$$(I + \Lambda_y)\psi_{i,j,k}^{**} = \psi_{i,j,k}^* \qquad (2.8.17b)$$

$$(I + \Lambda_z)\psi_{i,j,k}^{***} = \psi_{i,j,k}^{**} \qquad (2.8.17c)$$

$$f_{i,j,k}^{n+1} = f_{i,j,k}^n + \psi_{i,j,k}^{***} \qquad (2.8.17d)$$

Each step leads to the solution of tridiagonal systems and useful comments on such systems have been made by Briley and McDonald (1980) and by Neron (1981). By eliminating the intermediate values, we find the finite-difference equation effectively solved by the above algorithm. The final equation is

$$(I + \Lambda_x)(I + \Lambda_y)(I + \Lambda_z)(f_{i,j,k}^{n+1} - f_{i,j,k}^n) = R f_{i,j,k}^n \qquad (2.8.18)$$

which differs from Eq. (2.8.16) by a factor $O(\Delta t^2)$ since each operator Λ_x, Λ_y, and Λ_z is $O(\Delta t)$. Note that the scheme (2.8.18) has not necessarily the same stability property as (2.8.16).

The splitting technique (2.8.17) eliminates any difficulty associated with the boundary conditions for the intermediate values in the case of Dirichlet conditions. Therefore, let us assume that $f(x, y, z, t)$ is known on the boundary Γ:

$$f(x, y, z, t)|_\Gamma = \phi(x, y, z, t)$$

Let Γ_1, Γ_2, and Γ_3 be the parts of the boundary parallel to planes (y, z), (x, z), and (x, y) respectively. Equation (2.8.17d) gives the value $\psi_{i,j,k}^{***}$ on Γ_3 needed to solve (2.8.17c); i.e.,

$$\psi_{i,j,k}^{***}|_{\Gamma_3} = (\phi_{i,j,k}^{n+1} - \phi_{i,j,k}^n)|_{\Gamma_3}$$

and, in the same way, Eqs. (2.8.17c) and (2.8.17b) give

$$\psi_{i,j,k}^{**}|_{\Gamma_2} = (I + \Lambda_z)(\phi_{i,j,k}^{n+1} - \phi_{i,j,k}^n)|_{\Gamma_2}$$

$$\psi_{i,j,k}^*|_{\Gamma_1} = (I + \Lambda_y)(I + \Lambda_z)(\phi_{i,j,k}^{n+1} - \phi_{i,j,k}^n)|_{\Gamma_1}$$

In particular, if ϕ is independent of time, each intermediate value is zero on the boundary.

We remark that for a two-dimensional equation the Peaceman–Rachford scheme, (2.8.4) with $A_1 = A_2$ and $B_1 = B_2$, can be included in the general procedure (2.8.17) with $\Lambda_z = 0$ by introducing the change of variable $\tilde{f}_{i,j} = f_{i,j}^n + \frac{1}{2}\psi_{i,j}^*$. The boundary condition (2.8.5) is then identical to the above determination. Finally, it can be noted that at steady state $f_{i,j,k}^{n+1} - f_{i,j,k}^n = 0$ so that $Rf_{i,j,k}^n = 0$ represents a steady finite-difference equation associated with the original equation. Therefore, $Rf_{i,j,k}^n$ can be considered, in a general way, as the result of an explicit scheme $f_{i,j,k}^{n+1} - f_{i,j,k}^n = Rf_{i,j,k}^n$ approximating the original equation. By exploiting this remark, the right-hand side of Eq. (2.8.16) could be replaced by the result of a multistep explicit scheme (Hollanders and Peyret, 1981; Lerat, 1981).

References

Adam, Y. *Comput. Math. Appl.* **1**, 393–406 (1975).

Adam, Y. *J. Comput. Phys.* **24**, 10–22 (1977).

Allen, J. S., and Cheng, S. I. *Phys. Fluids* **19**, 37–52 (1970).

Anucina, N. N., *Sov. Math.* **5**, 60–64 (1964).

Beam, R. M., and Warming, R. F. *J. Comput. Phys.* **22**, 87–110 (1976).

Beam, R. M., and Warming, R. F. *AIAA J.* **16**, 393–402 (1978).

Berger, A. E., Solomon, J. M., Ciment, M., Leventhal, S. H., and Weinberg, B. C. *Math. Comput.* **35**, 695–731 (1980).

Berezin, Yu. A., Kovenja, V. M., and Yanenko, N. N. *Numerical Mathematics in Continuum Mechanics* (in Russian), Vol. 3, pp. 3–18, A. N. SSSR Siberian Computer Center, Novosibirsk (1972).

Berezin, Yu. A., Kovenja, V. M., and Yanenko, N. N., *Lecture Notes in Physics,* Vol. 35, pp. 85–90, Springer-Verlag, New York (1975a).

Berezin, Yu. A., Kovenja, V. M., and Yaneko, N. N. *Comput. Fluids* **3**, 271–281 (1975b).

Bontoux, P., Forestier, B., and Roux, B. *J. Mec. Appl.* **2**, 291–316 (1978).

Brailovskaya, I. Yu. *Sov. Phys.–Dokl.* **10**, 107–110 (1965).

Bramble, J. H., and Hubbard, B. E. *Numer. Math.* **4**, 312–327 (1962).

Briley, W. R., and McDonald, H. An Implicit Numerical Method for the Multidimensional Compressible Navier–Stokes Equations, United Aircraft Research Laboratory Rep. M911363-6 (Nov. 1973).

Briley, W. R., and McDonald, H. *Lecture Notes in Physics,* Vol. 35, pp. 105–110, Springer–Verlag, New York (1975).

Briley, W. R., and McDonald, H. *J. Comput. Phys.* **22**, 372–397 (1977).

Briley, W. R., and McDonald, H. *J. Comput. Phys.* **34**, 54–73 (1980).

Cheng, S. I. A Critical Review of Numerical Solution of Navier–Stokes Equations. In *Progress in Numerical Fluid Dynamics, Lecture Notes in Physics,* Vol. 41, pp. 78–225 Springer–Verlag, New York (1975).

Cheng, S. I., and Shubin, G. *J. Comput. Phys.* **28**, 315–326 (1978).

Ciment, M., Leventhal, S. H., and Weinberg, B. C., *J. Comput. Phys.* **28**, 135–166 (1978).

Collatz, L. *The Numerical Treatment of Differential Equations,* Springer–Verlag, New York (1966).

Collatz, L. Hermitian Methods for Initial-Value Problems in Partial Differential Equations. In *Topics in Numerical Analysis,* Proceedings of the Royal Irish Academy Conference on Numerical Analysis, J. J. H. Miller, Ed., pp. 41–61, Academic Press, New York, (1972).

Courant, R., Friedrichs, K. O., and Lewy, M. *Math. Ann.* **100**, 32–76 (1928). (English translation) *IBM J. Res. Devel.* **11**, 215–234 (1967).

Courant, R., Isaacson, E., and Rees, M. *Commun. Pure Appl. Math.* **5**, 243–255 (1952).

Daube, O., and Ta Phuoc Loc, *J. Mec.* **17**, 651–678 (1978).

Dennis, S. C. R., and Chang, G. -Z. *Phys. Fluids* **12**, Suppl. II, II.83–II.93 (1969).

Douglas, J., and Gunn, J. E. *Numer. Math.* **6**, 428–453 (1964).

Elsaesser, E. Etude de Méthodes Hermitiennes pour la Résolution des Equations de Navier–Stokes pour un Fluide Incompressible en Régime Stationnaire, Thèse Doct. 3^{ieme} Cycle, Université Pierre et Marie Curie, Paris (1980).

Elsaesser, E., and Peyret, R. Méthodes Hermitiennes pour la Résolution Numérique des Equations de Navier–Stokes. In *Méthodes Numériques dans les Sciences de l'Ingénieur,* E. Absi and R. Glowinski, Eds., pp. 249–258, Dunod, Paris (1979).

Fairweather, G., and Mitchell, A. R. *SIAM J. Numer. Anal.* **4**, 163–170 (1967).

Garabedian, P. R., *Math. Tables Aids Comput.* **10**, 183–185 (1956).

Ghia, K. N., Shin, C. T., and Ghia, U. Use of Spline Approximation for Higher-Order Accurate Solutions of Navier–Stokes Equations in Primitive Variables. In Proc. 4th AIAA Computational Fluid Dynamics Conference, pp.284–291, Williamsburg, VA. (1979).

Godunov, S. K., and Ryabenski, V. S. *The Theory of Difference Schemes,* North-Holland, Amsterdam (1964).

Gourlay, A. R., and Morris, J. L. *Math. Comput.* **22**, 715–720 (1968).

Gourlay, A. R., and Morris, J. L. *J. Comput. Phys.* **5**, 229–243 (1970).

Gustaffson, B. *Math. Comput.* **29**, 396–406 (1975).

Harten, A., Hyman, J. M., and Lax, P. D. *Commun. Pure Appl. Math.* **29**, 297–322 (1976).

Hindmarsh, A. C., Gresho, P. M. and Griffiths, D. F. *Int. J. Numer. Methods Fluids* **4**, 853–897 (1984).

Hirsh, R. S. *J. Comput. Phys.* **19**, 90–109 (1975).

Hirt, C. W., *J. Comput. Phys.* **2**, 339–355 (1968).

Hollanders, H., and Peyret, R. *Rech. Aérosp.* No. 1981-4, 287–294 (1981).

Isaacson, E. and Keller, H. B. *Analysis of Numerical Methods,* Wiley and Sons, New York (1966).

Khosla, P. M., and Rubin, S. G. *Comput. Fluids* **2**, 207–209 (1974).

Krause, E., Hirschel, E. H., and Kordulla, W. *Comput. Fluids* **4**, 77–92 (1976).

Kreiss, H. O. *Math. Comput.* **26**, 605–624 (1972).

Kreiss, H. O., and Oliger, J. *Methods for the Approximate Solution of Time-Dependent Problems.* GARP Publ. series No. 10, Global Atmospheric Research Program (1973).

Laval, P. *C. R. Acad. Sci. B,* **204**, 239–242 (1980).

Laval, P. *C. R. Acad. Sci. B,* **221**, 125–128 (1981a).

Laval, P. Schémas Explicites de Désintégration du Second Ordre pour la Résolution des Problèmes Hyperboliques Non Linéaires: Théorie et Applications aux Ecoulements Transsoniques. ONERA, Note Technique No. 1981–10 (1981b).

Lax, P. D. *Commun. Pure Appl. Math.* **7**, 159–193 (1954).

Lax, P. D. *Commun. Pure Appl. Math.* **10**, 537–566 (1957).

Lax, P. D., and Wendroff, B. *Commun. Pure Appl. Math.* **13**, 217–237 (1960).

Lecointe, Y., and Piquet, J. On the Numerical Solution of Some Types of Unsteady Incompressible Viscous Flow. Second International Conference on Numerical Methods in Laminar and Turbulent Flow, July 13–16, 1981, Venice, Italy.

Lerat, A. *C. R. Acad. Sci. A* **288**, 1033–1036 (1979).

Lerat, A. Sur le Calcul des Solutions Faibles des Systèmes Hyperboliques de Lois de Conservation à l'Aide de Schémas aux Différences. ONERA Publ. No. 1981–1 (1981).

Lerat, A., and Peyret, R. *C. R. Acad. Sci. A* **276**, 759–762 (1973a).

Lerat, A., and Peyret, R. *C. R. Acad. Sci. A* **277**, 363–366 (1973b).

Lerat, A. and Peyret, R., *Comput. Fluids* **2**, 35–52 (1974a).

Lerat, A. and Peyret, R. *Lecture Notes in Physics,* Vol. 35, pp. 251–256, Springer–Verlag, New York (1974b).

Lerat, A., and Peyret, R. *Rech. Aerosp.* **1975-2**, 61–79 (1975).

Lerat, A., and Sidès, J. Proceedings of the Conference on Numerical Methods in Aerodynamic Fluid Dynamics, Univ. Reading, March (1981).

Lindemuth, J., and Killeen, J. *J. Comput Phys.* **13**, 181–208 (1973).

MacCormack, R. W. The Effect of Viscosity in Hypervelocity Impact Cratering. AIAA paper No. 69–354 (1969).

MacCormack, R. W. In *Lecture Notes in Physics,* Vol. 8, pp. 151–163, Springer–Verlag, New York (1971).

Magnus, R., and Yoshihara, H. *AIAA J.* **13**, 1622–1628 (1975).

Mehta, U. B. Dynamic Stall of an Oscillating Airfoil. AGARD Conference Proceedings No. 227, Unsteady Aerodynamics, Ottawa, Canada, pp. 23.1–23.32 (1977).

McGuire, G. R., and Morris, J. L. *J. Inst. Math. Applic.* **10**, 150–165 (1972).

McGuire, G. R., and Morris, J. L. *J. Comput. Phys.* **11**, 531–549 (1973).

Mitchell, A. R. *Computational Methods in Partial Differential Equations.* Wiley, New York (1969).

Moretti, G. Proceedings of 1974 Heat Transfer and Fluid Mechanics Institute, Stanford Press, Stanford (1974).

Néron, M. Etude et Application d'une Méthode Implicite de Résolution des Equations de Navier–Stokes pour un Fluide Compressible, en Coordonnées Cylindriques. Thèse Doct.-Ingénieur, Université Pierre et Marie Curie, Paris (1981).

Oleinik, O. A. *Am. Math. Soc. Trans.,* Ser. 2, No. 26, pp. 95–172. [Translation of *Usp. Math. Nauk* **12**, 3–73 (1957).]

Orszag, S. A., and Tang, C.-M. *J. Fluid Mech.* **90**, 129–143 (1979).

Peaceman, D. W., and Rachford, H. H. *J. Soc. Indus. Appl. Math.* **3**, 28–41 (1955).

Peyret, R. *C. R. Acad. Sci. A* **272**, 1274–1277 (1971).

Peyret, R. Résolution Numérique des Systèmes Hyperboliques. Application à la Dynamique des Gaz. ONERA, Publication No. 1977-5 (1977).

Peyret, R. *C. R. Acad. Sci. A* **286**, 59–62 (1978a).

Peyret, R. A Hermitian Finite-Difference Method for the Solution of the Navier–Stokes Equations. In *Proceedings of the First International Conference on Numerical Methods in Laminar and Turbulent Flow,* 43–54, Pentech Press, Plymouth, UK (1978b).

Peyret, R. and Viviand, H. Computation of Viscous Compressible Flows Based on the Navier–Stokes Equations. AGARDograph No. 212 (1975).

Raviart, P. A. Méthodes d' Éléments Finis en Mécanique des Fluides. Ecole d'Eté d'Analyse Numérique, EDF, INRIA, CEA (1979).

Richtmyer, R. D. A Survey of Difference Methods for Non-Steady Gas Dynamics. NCAR, Tech. Note 63-2, Boulder, CO (1962).

Richtmyer, R. D. and Morton, K. W. *Difference Methods for Initial Value Problems,* Interscience, New York (1967).

Roache, P. J. *Computational Fluid Dynamics,* Hermosa Publishers, Albuquerque, NM (1972).

Rubin, E. L., and Burstein, S. Z. *J. Comput. Phys.* **2,** 178–196 (1967).

Rubin, S. G., and Khosla, P. K. *J. Comput. Phys.* **24,** 217–246 (1977).

Shokin, Y. I., *Proc. Steklov Inst. Math.* **122,** 67–86 (1973).

Shokin, Y. I. In *Lecture Notes in Physics,* Vol. 59, pp. 410–414, Springer–Verlag, New York (1976).

Steger, J. L. and Warming, R. F. *J. Comput. Phys.* **40,** 263–293 (1981).

Strang, G. *SIAM J. Numerical Analysis,* **5,** 506–517 (1968).

Ta Phuoc Loc, *J. Mec.* **14,** 109–134 (1975).

Ta Phuoc Loc, *J. Fluid Mech.* **100,** 111–128 (1980).

Thommen, H. U., *Z. Angew. Math. Phys.* **17,** 369–384 (1966).

Varga, R. S. *Matrix Iterative Analysis,* Prentice-Hall, Englewood Cliffs, NJ (1962).

Veldman, A. E. P. *Comput. Fluids* **1,** 251–271 (1973).

Von Neumann, J., and Richtmyer, R. D. *J. Appl. Phys.* **21,** 232–257 (1950).

Warming, R. F., and Beam, R. M. *SIAM-AMS Proc.* **11,** 85–129 (1978).

Warming, R. F., and Hyett, B. J. *J. Comput. Phys.* **14,** 159–179 (1974).

Warming, R. F., Kutler, P., and Lomax, H. *AIAA J.* **11,** 189–196 (1973).

Yanenko, N. N. *The Method of Fractional Steps,* Springer–Verlag, New York (1971). [In French, *Méthodes à Pas Fractionnaires,* Armand Colin, Paris (1968).]

Yanenko, N. N., and Shokin, Y. I. *Phys. Fluids* **12,** Suppl. II., II.28–II.33 (1969).

Young, D. M. *Iterative Solution of Large Linear Systems,* Academic Press, New York (1971).

CHAPTER 3

Integral and Spectral Methods

In Chapter 2 the discussion centered on the use of finite-difference approximations to solve the differential equations of fluid flow. Various alternatives to the finite-difference approach are available, including integral approaches such as moment methods, least squares, Galerkin techniques, and Rayleigh–Ritz variational formulations. Each of these approaches can be applied on a subdomain of the flow or over the entire region of interest. The subdomain approach may be a finite-element or spectral method in the classical sense, depending on the functions employed. In addition to these formulations, there are the pseudospectral methods that do not utilize integral forms of the differential equations. There are also specialized cell methods for compressible flows which utilize localized models of the physics for each cell to develop a numerical equivalent of the flow equations. In this chapter we will outline those techniques which have been found to be the most practical. First, we will consider the integral-type approaches.

In general, integral approaches for developing a numerical equivalent of the flow equations can follow two lines of thought. The first is to use variational calculus to formulate a variational equivalent of the problem to be solved (Rayleigh–Ritz). The second is to multiply the conservation equations, which for simplicity we write as $L(f) = 0$, by a weighting function W_i and to integrate the product over an interval of space (weighted residual). The resulting integral relationship $\int L(f)W_i \, ds = 0$ is used as the basis in solving for the flow variables. The procedures vary but they all ultimately proceed by assuming an analytical form of the solution with unknown constants or functions. The assumed solution is then substituted into the integral equation to obtain one or more equations for the unknowns.

In order to employ the variational method, it is necessary to establish an integral over the domain of interest

$$J = \int\int F(x,y,f,f_x,f_y, \ldots) \, dx \, dy$$

so that J can be an extremum. In most physical problems, J would be formed based on energy considerations. The difficulty with this formulation is that, for general viscous fluid motions, variational principles are not established. As a result the approach is convenient for special cases (for example, irrotational potential flow or Stokes flow) and is not easily extended to the complete flow equations.

A more general approach, which encompasses the variational method, is the weighted residual method. This technique forms an integral equation by multiplying the flow equations by a weighting function W_i and then integrating over a prescribed interval. A functional form of the independent variables with unknown coefficients is then assumed and substituted into the integral equation. However, the method must generate the same number of equations as unknowns. This can be accomplished in two ways. The first is to divide the space of interest into N intervals of integration for N unknowns; the second way is to employ N weighting functions for N unknowns. The first approach is essentially a finite-element or integral-relationship approach, and the second is termed a spectral-type approach. This description is simplified but, in principle, finite elements and spectral methods can be viewed as the same approach applied in a slightly different manner. Encompassed in this approach are the subdomain or integral-relation method where $W_i = 0$ or 1, the method of moments where $W_i = x^i$, the Galerkin approach where $W_i = F_i$ with F_i being a function of the set assumed to represent the solution, and the least-squares approach where $W_i = 2\,\partial L(f)/\partial a_i$ with a_i being an unknown variable of the assumed solution. A considerable amount of work has been carried out in each of these areas. Holt (1977) has recently surveyed and reported much of the work on integral methods and moment methods. As a consequence, our attention in this section will focus on the Galerkin approach with some comments on the least-squares method. In order not to burden the reader with an extensive discussion of both finite-element and spectral methods, we have attempted to unify the two areas in the following discussion.

In the discussion the following nomenclature for spectral-type methods has been adopted:

1. A collocation method is one that utilizes only grid points in real space.
2. A spectral method solves a problem only in spectral or transformed space.
3. A pseudospectral method is a combination of collocation and spectral methods in one problem.

3.1 Finite-Element and Spectral-Type Methods

The fundamental step underlying the finite-element and spectral methods for solving fluid-flow problems is the reduction of the original partial differential equations to a set of ordinary differential or algebraic equations which can be solved by straightforward techniques. Generally, for the spectral method the procedure follows the steps used in the classical analytical methods for solving linear partial differential equations by expansions in a set of functions. For example, if one seeks to satisfy the simple model equation

$$\frac{\partial f}{\partial t} + \frac{\partial f}{\partial x} = \nu \frac{\partial^2 f}{\partial x^2} \tag{3.1.1}$$

in the interval $0 \le x \le L$. One typically assumes an expansion of the form

$$f = \sum_{i=0}^{N} a_i(t)\, F_i(x) \tag{3.1.2}$$

where the a_i terms are unknowns and the $F_i(x)$ are chosen functions. One then tries to satisfy the initial and boundary conditions as well as the differential equation. In this approach, things go well if the boundary conditions can be easily satisfied.

However, if this cannot be accomplished, the problem becomes one of translating the boundary conditions from real space into spectral space—i.e., in terms of the a_i's—as well as satisfying the differential equation. This task can be formidable for complicated flow problems. Later, we will discuss how to avoid this problem. However, assume at this point that one can properly eliminate the boundary conditions as a problem. Then, one proceeds by introducing Eq. (3.1.2) into (3.1.1) to determine the a_i terms. The resulting equation is

$$\sum_{i=0}^{N} [a_i' F_i + a_i(F_i' - \nu F_i'')] = 0 \quad \text{where } a_i' = \frac{da_i}{dt} \tag{3.1.3}$$

The next step is to determine relationships for the a_i terms from this equation. This can be accomplished by (i) equating coefficients of each F_i term to zero, (ii) using orthogonality relationships, or (iii) using weighting functions W_i with integration over the region $0 \le x \le L$ to form $N + 1$ equations for $N + 1$ unknowns. First, we discuss cases (i) and (ii). In order to equate coefficients or employ orthogonality one needs to relate $F_i' - \nu F_i''$ to F_i. Typically, this can be accomplished for many polynomials by a relationship of the form

$$F_i' - \nu F_i'' = \sum_{m=0}^{M} \beta_{m,i} F_m \tag{3.1.4}$$

However, this may not be easy to accomplish and can govern the choice of polynomials. Assuming that this relationship can be developed, one can write

$$\sum_{i=0}^{N} \left(a_i' F_i + a_i \sum_{m=0}^{N} \beta_{m,i} F_m \right) = 0 \tag{3.1.5}$$

In this relationship one can equate coefficients of F_i to zero or introduce orthogonality relationships of the form

$$\int_0^L w(x) F_j F_i \, dx = \begin{cases} 0, & i \ne j \\ c_j, & i = j \end{cases} \tag{3.1.6}$$

where c_j is a constant, to obtain relationships for a_i. Each procedure should yield the same result for the a_i relationships. For the simple case where $\beta_{m,i} = 0$ for $m \ne i$—the textbook case—we obtain, by equating coefficients of each F_i term to zero, the result

$$a_i' + \beta_{i,i} a_i = 0 \tag{3.1.7}$$

For the case where $\beta_{m,i} = 0$ for $m \neq i + 1$ and $m \neq i - 1$, we obtain

$$a_i' + a_{i-1}\beta_{i,i-1} + a_{i+1}\beta_{i,i+1} = 0 \qquad (3.1.8)$$

Note that the indices are shifted to equate terms of equal F_i. For applied problems the relationships may be even more complicated, and numerical methods will be required to solve the equations for the a_i terms. In order to establish these concepts clearly, we present an example using Chebyshev polynomials as F_i with $\nu = 0$. For this case one has the relationship

$$F_i' = \sum_{m=0}^{N} \beta_{m,i} F_m \qquad (3.1.9)$$

where the $\beta_{m,i}$ is obtained from the recurrence relationship $2F_m = F_{m+1}'/(m + 1) - F_{m-1}'/(m - 1)$. Equation (3.1.3) may be written then as

$$\sum_{i=0}^{N} \left(a_i' F_i + a_i \sum_{m=0}^{N} \beta_{m,i} F_m \right) = 0 \qquad (3.1.10)$$

One now can equate coefficients of F_i to zero or employ the orthogonality condition of Chebyshev polynomials to obtain the relationships for the a_i terms. In order to employ orthogonality one must normalize the problem either in the range $-1 \leq x \leq 1$ or $0 \leq x \leq 1$. For our case we choose the first interval and introduce $x = 2x/L - 1$ and $t = 2t/L$ into the original equation for normalization. The polynomials of the expansion are then in the new x variable for the interval $-1 \leq x \leq 1$. Equation (3.1.10) then holds for this definition of x. If we now apply the relationships

$$\int_{-1}^{+1} \frac{1}{\sqrt{1-x^2}} F_i F_j \, dx = \begin{cases} 0, & j \neq i \\ \frac{1}{2}\pi, & j = i \neq 0 \\ \pi, & j = i = 0 \end{cases} \qquad (3.1.11)$$

by multiplying (3.1.10) by $F_j/\sqrt{1 - x^2}$ and integrating, one can show that (see Gottlieb and Orszag, 1977)

$$a_i' + \frac{2}{\alpha_i} \sum_{\substack{p=i+1 \\ p+i \, \text{odd}}}^{N} p a_p = 0, \qquad \alpha_0 = 2, \; \alpha_i = 1 \text{ for } i > 0 \qquad (3.1.12)$$

The equations are somewhat complicated and in most cases would require numerical integration to solve a problem. The initial conditions for the a_i terms in such a solution are obtained by expanding the initial values of f in a Chebyshev expansion. The coefficients of each F_i term would be the initial value of a_i.

Up to this point in the analysis, we assumed that all boundary conditions were satisfied by the expansion. However, suppose that this were not the case and that the condition that $f = 1$ at $x = -1$ must be satisfied in our example. At this point one is confronted with N equations for the a_n terms and an additional constraint that

$$1 = \sum_{i=0}^{N} a_i(t) F_i(-1) \tag{3.1.13}$$

This obviously is an ill-posed problem. For Chebyshev polynomials there is a way around the problem, but for other functions the procedure may not be so clear. For Chebyshev polynomials, or any other polynomial, it is important to note that a derivative reduces the order. For example, in the Chebyshev case $F_0 = 1$, $F_1 = x$ and $F_2 = (2 x^2 - 1)$; therefore $F_2' = 4 F_1$ or $F_2'' = 4 F_0$. As a consequence of this, it is clear that the equations obtained by the procedure of equating coefficients of terms of equal F_i or orthogonality will be deficient for a_N if only the first derivative appears and for a_N and a_{N-1} if the second derivative appears. As a result in the example problem one can see how to eliminate the ill-posed condition along with the inaccuracy of the a_N term by simply discarding the equation for a_N and replacing it by Eq. (3.1.13). The problem then is well-posed. We solve for the a_i terms up to a_{N-1} by Eq. (3.1.12) and for a_N by Eq. (3.1.13). This technique is called the *tau* (τ) method and can be employed for second derivatives as well by discarding the equation for a_{N-1} and replacing it by a second boundary condition posed in the manner of Eq. (3.1.13). It is important to note in this type of solution that if an explicit time integration method is used to obtain the a_i terms up to a_{N-1} then the term a_N which is the highest order in the series will be satisfying the boundary condition. This would mean that the highest-order term in the series is dominant and may well lead to an unstable solution after a few time steps. As a result, the user is cautioned about satisfying boundary conditions in this fashion.

The presented example is rather limited in extent, but it serves to guide the reader. For an extension of this example the reader is referred to Gottlieb and Orszag (1977). For a more complex example, the reader is referred to Taylor (1962) who employed the outlined approach using both Bessel functions and Legendre polynomials to develop a_i equations for the solution of low-Reynolds-number mass transfer from a sphere.

An alternative approach to determining the a_i terms directly from the differential equations is to integrate over the interval $a \leq x \leq b$. We then obtain

$$\int_a^b W_j \left(\frac{\partial f}{\partial t} + \frac{\partial f}{\partial x} \right) dx = \nu \int_a^b W_j \frac{\partial^2 f}{\partial x^2} dx, \qquad j = 0, \dots, N \tag{3.1.14}$$

Note that a and b must lie between 0 and L, but they can be chosen arbitrarily in this region. $W_j(x)$ can also be arbitrarily chosen for the region $a \leq x \leq b$. Equation (3.1.14) therefore has many options that can be used to generate equations for determining the a_i terms. Inserting Eq. (3.1.2) into (3.1.14) yields

$$\sum_{i=0}^{N} \int_a^b W_j [a_i' F_i + a_i (F_i' - \nu F_i'')] dx = 0 \qquad j = 0, \dots, N \tag{3.1.15}$$

Equation (3.1.15) is the basis of an integral method. If $b - a = L$, the method fits the description of a spectral scheme and if $b - a < L$, then the method becomes a cell or element technique. For the element approach, the number of terms is typically small for each element and the elements may overlap if one chooses.

For Galerkin finite-element techniques, the functions F_j are usually not very complicated. W_j is usually taken to be F_j. In the finite-element definitions, the expansion (3.1.2) would be rearranged and the F_j terms would be the shape functions; i.e., in finite-element notation,

$$f = \sum_{i=0}^{N} f_i(t)\, \overline{F}_i(x) \tag{3.1.16}$$

where f_i is the value of f at the nodal point i and $\overline{F}_i$ is typically an interpolation coefficient with the properties

$$\overline{F}_i(x_m) = \begin{cases} 1 & i=m \\ 0 & i \neq m \end{cases} \tag{3.1.17}$$

where x_m denotes a point where f is evaluated. Note that this property implies that $\overline{F}_i$ vanishes totally outside of the element. These functions typically are linear or quadratic in form and can be in the Lagrange interpolation coefficient family. However, expansions (3.1.2) and (3.1.16) should be equivalent. Demonstrating this equivalence can be difficult except for simple functions.

Equation (3.1.17) has both good and bad aspects. A negative aspect is that it restricts the available functions while a positive aspect is that it reduces the complexity of the matrices required to solve for the nodal values of f. In applications, one must always examine the trade-off in these two aspects. The reason is that by employing the more general expansions (3.1.2) over a larger region, the representation is known to become more efficient—i.e., it takes less terms to obtain the same accuracy. The trade-off, however, is the effort required to invert the matrices. It is interesting to note that Chebyshev polynomials expansions have the property of rapid convergence for most functional representations and permit the use of fast Fourier transforms to connect coefficients and real spaces. As a result, these functions can be very useful for solving problems.

For the non-Galerkin approaches, such as least-squares, W_j is no longer a shape function and consequently may differ significantly from the Galerkin technique. In fact, the least-squares approach is formulated as an integral of the square of the residual error, i.e.,

$$J = \int_s |L(f)|^2\, ds \tag{3.1.18}$$

One seeks to minimize this function since it represents a measure of the overall error. If the function f is represented as a polynomial or series with a set of unknown coefficients a_i, then one seeks to minimize the integral with respect to each a_i. In a straightforward approach, the derivative of J with respect to each a_i is taken and the result set equal to zero. One then obtains

$$\frac{\partial J}{\partial a_i} = \int_s 2 \frac{\partial L(f)}{\partial a_i} L(f) \, ds = 0 \tag{3.1.19}$$

in which W_i is represented by $2 \, \partial L(f)/\partial a_i$. An alternate to this approach is to utilize optimization theory for profit functions in order to minimize the integral J with respect to the a_i terms. This is the approach used by Bristeau et al. (1978) in their extensive studies of finite elements. Its explanation requires the introduction of optimization theory and as a consequence will not be discussed here. The interested reader is referred to Bristeau et al. (1978, 1979) and Periaux (1979) for the details. The overall approach is fairly complex and an example is presented later in our discussion of incompressible flows to show the nature of the approach and the quality of results.

The example employed to derive the integral equation and display the nature of the various methods is an unsteady case, but the procedure works equally well for steady-state equations. The difference is that the equations for the coefficients in the steady state are usually more difficult to solve since they tend to require complicated matrix inversions, while the unsteady cases can employ explicit time integration to simplify or avoid the matrix inversion problems. However, if implicit time integration is employed the solution effort is about the same. It is important to note that the time integration scheme can be used to relax the solution to the steady state if desired.

In the discussion, we have assumed that it is feasible to incorporate the boundary conditions into the expansion. For some problems, this is not straightforward when spectral expansions of the form (3.1.2) are employed. The finite-element approach expansion (3.1.16), however, always permits inclusion of the boundary point in the calculation because of the nature of the expansion. When the boundary conditions offer problems in spectral space, one may find it preferable to solve the problem in real space. A pseudospectral procedure which permits this is discussed later in this chapter.

So far the discussion has been general and has not provided any specific examples. In order to demonstrate both the spectral and finite-element approaches, we choose simple equations here; later we will display results from flow-field calculations.

3.2 Steady-State Finite-Element Examples

For the first example, we choose the ordinary differential equation

$$\frac{df}{dx} + f = 0 \tag{3.2.1}$$

with the condition that $f = 1$ at $x = 0$. We seek a solution to this equation $0 \leq x \leq 1$ by using the integral form

$$\int_0^1 W_j \left(\frac{df}{dx} + f \right) dx = 0 \tag{3.2.2}$$

We next assume the expansion for f to be

$$f = \sum_{i=1}^N f_i F_i(x) \tag{3.2.3}$$

Upon substitution, we obtain

$$\sum_{i=1}^N f_i \int_0^1 W_j (F_i' + F_i) \, dx = 0 \qquad j = 1, \ldots, N \tag{3.2.4}$$

This equation represents a set of algebraic equations for the constants f_i with coefficients given by

$$b_{j,i} = \int_0^1 W_j (F_i' + F_i) \, dx \tag{3.2.5}$$

Note that in order to have a set of equations which is well posed, it is necessary to have the number of i's equal to the number of unknown f_i's in the problem. Also, the number of unknowns is influenced by the boundary conditions. Up to this point, the spectral and finite-element approaches are not different. We now consider each approach in turn, considering the finite-element approach first.

In the finite-element method, one considers the unknown f_i terms to be nodal values of f to be determined. Their distribution and location are a matter of choice, and in one dimension this is rather straightforward. In two dimensions, one typically chooses points at the corners of either triangular or quadrilateral elements. The typical $F_i(x)$ functions employed are linear or quadratic. For this example a linear element in the range $0 \leq x \leq 1$ is selected so that

$$F_1(x) = 1 - x \tag{3.2.6}$$

$$F_2(x) = x$$

Then,

$$f = f_1(1 - x) + f_2 x \tag{3.2.7}$$

Now, since $f = 1$ at $x = 0$ we have that $f_1 = 1$. Also, the coefficients $b_{j,i}$ for finite elements require that $W_j = F_j$ so that for (3.2.5) we have

$$b_{j,i} = \int_0^1 F_j (F_i' + F_i) dx \tag{3.2.8}$$

and hence

$$b_{1,1} = \int_0^1 (-x)(1-x)\, dx = -\frac{1}{6}, \qquad b_{1,2} = \int_0^1 (1-x)(1+x)\, dx = \frac{2}{3}$$

$$b_{2,1} = \int_0^1 (-x)(x)\, dx = -\frac{1}{3}, \qquad b_{2,2} = \int_0^1 x(1+x)\, dx = \frac{5}{6}$$

Since f_1 has been determined by the boundary condition not all of the equations defined by Eq. (3.2.4) are required. Since only f_2 is unknown, the equation

$$\sum_{i=1}^{2} f_i \int_0^1 F_2(F_i' + F_i)\, dx = 0 \qquad (3.2.9)$$

is required.* Inserting the values of the integrals yields

$$-\tfrac{1}{3}f_1 + \tfrac{5}{6}f_2 = 0 \qquad (3.2.10)$$

Noticing that $f_1 = 1$ we obtain $f_2 = 0.4$. The value obtained from the exact solution $y = e^{-x}$ is $f_2 = 0.368$.

Note that in this procedure the principal effort is evaluating the $b_{j,i}$ terms in coefficients of the matrix. This is generally the problem in applying finite elements. For two-dimensional problems, the procedure follows the same principles except that the integrals become double integrals over the elements whose typical geometry is a triangle or quadrangle. We next demonstrate the method for a second-derivative case.

Consider the solution of the equation

$$\frac{d^2f}{dx^2} - f = 0 \qquad (3.2.11)$$

with $f=0$ at $x = -1$ and $f=1$ at $x = 1$. For this example consider two elements as shown in Fig. 3.2.1. For this solution we take linear elements and adopt the standard finite-element notation, using N_j for the shape functions

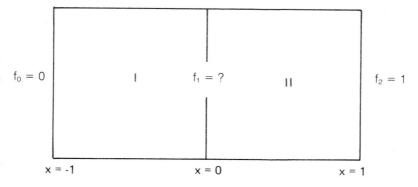

Fig. 3.2.1 One-dimensional finite-element example.

---

*Use the weight F_2 to find f_2.

instead of F_j. Following the weighting function approach, we multiply the original equation by N_j and integrate over the region $-1 \leq x \leq 1$ to obtain

$$\int_{-1}^{+1} N_j \left(\frac{d^2f}{dx^2} - f \right) dx = 0 \qquad (3.2.12)$$

Integrating by parts and noting that $N_j = 0$ at $x = -1$ and $x = +1$ when we limit j to interior points only, we obtain

$$\int_{-1}^{+1} \left(\frac{dN_j}{dx} \frac{df}{dx} + N_j f \right) dx = 0 \qquad (3.2.13)$$

The purpose of this step is to reduce the order of the derivative in the equation. This is a standard finite-element formulation. Next, a form for f must be adopted. We can do this in a general form for the complete region or per element. The general form eventually reduces to element-by-element considerations due to the nodal function forms. For this example we will consider the element-by-element case and later, in a two-dimensional example, the general form. For each element in the example we assume f to have the form*

$$f^{\mathrm{I}} = f_0 N_0(x) + f_1 N_1^{\mathrm{I}}(x)$$
$$f^{\mathrm{II}} = f_1 N_1^{\mathrm{II}}(x) + f_2 N_2(x) \qquad (3.2.14)$$

For the finite-element method the $N_i(x_j)$, by definition, must satisfy

$$N_i(x_j) = \begin{cases} 0, & i \neq j \\ 1, & i = j \end{cases} \qquad (3.2.15)$$

Therefore for the interval $-1 \leq x \leq 0$, we take

$$N_0(x) = -x \quad \text{and} \quad N_1^{\mathrm{I}}(x) = (1 + x)$$

For the interval $0 \leq x \leq 1$, we take

$$N_1^{\mathrm{II}}(x) = (1 - x) \quad \text{and} \quad N_2(x) = x$$

Note that the three points of interest are $x = -1, 0$, and 1 and these functions satisfy (3.2.15) at these points. Now if we insert these functions into (3.2.13) and integrate over the correct intervals for each cell, we obtain

$$\int_{-1}^{0} \frac{dN_j}{dx} \left[f_0 \frac{dN_0}{dx} + f_1 \frac{dN_1^{\mathrm{I}}}{dx} \right] dx + \int_{0}^{1} \frac{dN_j}{dx} \left[f_1 \frac{dN_1^{\mathrm{II}}}{dx} + f_2 \frac{dN_2}{dx} \right] dx$$
$$+ \int_{-1}^{0} N_j [f_0 N_0 + f_1 N_1^{\mathrm{I}}] \, dx + \int_{0}^{1} N_j [f_1 N_1^{\mathrm{II}} + f_2 N_2] \, dx = 0 \qquad (3.2.16)$$

For this problem we know $f_0 = 0$ and $f_2 = 1$. Also we have only one unknown point f_1 so $j = 1$ for the problem. As a result we can write

---

*Note we start the expansion at $i = 0$ for this solution.

$$\int_{-1}^{0} 1[0 + f_1]dx + \int_{0}^{1} (-1)[f_1(-1) + 1]\, dx$$

$$+ \int_{-1}^{0} (1 + x)[0 + f_1(1 + x)]\, dx \qquad (3.2.17)$$

$$+ \int_{0}^{1} (1 - x)[f_1(1 - x) + x]\, dx = 0$$

or

$$f_1 + f_1 - 1 + f_1 \int_{-1}^{0} (1 + x)^2\, dx + f_1 \int_{0}^{1} (1 - x)^2\, dx$$

$$+ \int_{0}^{1} x(1 - x)\, dx = 0 \qquad (3.2.18)$$

Noting that

$$\int_{-1}^{0} (1 + x)^2\, dx = \frac{1}{3}, \qquad \int_{0}^{1} (1 - x)^2 = \frac{1}{3}, \quad \text{and} \quad \int_{0}^{1} x(1 - x)\, dx = \frac{1}{6}$$

we obtain $f_1 = 5/16 = 0.313$. The correct value is given by the exact solution of (3.2.11), $f = Ae^{-x} + Be^{+x}$, where $A = e^{-1}/(e^{-2} - e^{2})$, $B = -e^{2}A$, and at $x = 0$, $f = 1/(e^{-1} + e^{+1}) = 0.324$. These results display the finite-element approach applied element-by-element. Next, we consider the general technique for a two-dimensional problem.

For two dimensions, the application of the finite-element approach becomes tedious, and, in practical applications, one must develop a number of back-ground calculational routines. In order to demonstrate some of these points, consider the solution of the problem shown in Fig. 3.2.2. First assume

$$u = \sum_{i} N_i(x, y)u_i \qquad (3.2.19)$$

For this problem, assume the N_i terms are linear functions and that the four rectangular elements shown in Fig. 3.2.3 are employed. Only the midpoint is unknown since all the other points lie on the boundaries. In order to compute this midpoint by finite elements, one first multiplies the original equation by N_j and integrates over the complete domain to obtain

$$\int_{-1}^{+1} \int_{-1}^{+1} \left(\frac{\partial^2 u}{\partial x^2} + \frac{\partial^2 u}{\partial y^2} \right) N_j\, dx\, dy = 0 \qquad (3.2.20)$$

Integrating by parts and noting that $N_j = 0$ on the boundary, one obtains

$$\int_{-1}^{+1} \int_{-1}^{+1} \left(\frac{\partial N_j}{\partial x} \frac{\partial u}{\partial x} + \frac{\partial N_j}{\partial y} \frac{\partial u}{\partial y} \right) dx\, dy = 0 \qquad (3.2.21)$$

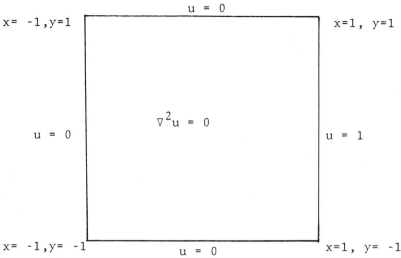

Fig. 3.2.2 Two-dimensional finite-element example.

If expansion (3.2.19) is inserted into this expression, the result is

$$\sum_i \int_{-1}^{+1} \int_{-1}^{+1} \left(\frac{\partial N_j}{\partial x} \frac{\partial N_i}{\partial x} + \frac{\partial N_j}{\partial y} \frac{\partial N_i}{\partial y} \right) dx \, dy \;\; u_i = 0 \tag{3.2.22}$$

Defining

$$c_{j,i} = \int_{-1}^{+1} \int_{-1}^{+1} \left(\frac{\partial N_j}{\partial x} \frac{\partial N_i}{\partial x} + \frac{\partial N_j}{\partial y} \frac{\partial N_i}{\partial y} \right) dx \, dy \tag{3.2.23}$$

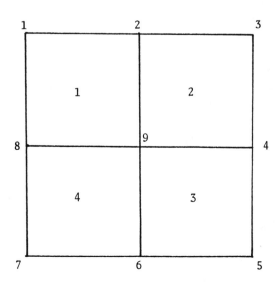

Fig. 3.2.3 Two-dimensional finite-element nodal points.

one then has

$$\sum_i c_{j,i} u_i = 0 \tag{3.2.24}$$

where u_i for $i = 1$–8 are known and $u = 0$ for $i = 1, 2, 6, 7$, and 8. As a result, this expression reduces to

$$c_{j,3} u_3 + c_{j,4} u_4 + c_{j,5} u_5 + c_{j,9} u_9 = 0 \tag{3.2.25}$$

where u_3, u_4, and u_5 are known. This raises a further question: Should u_3 and u_5 values have a magnitude of 1.0, 0, or 0.5? This we will determine later.

Since the unknown for Eq. (3.2.25) is u_9, we then take $j = 9$ so that it is necessary to calculate $c_{9,3}$, $c_{9,4}$, and $c_{9,5}$, and $c_{9,9}$. This is where the details enter and the problem becomes repetitive in nature. First, it is necessary to establish the proper form of the N_i terms in the problem. Note, however, that the procedure to follow is not problem dependent, only cell-geometry dependent. For a linear shape function, N_i, it has been established (see Zienkiewicz, 1971) that the appropriate form is

$$N_i = \tfrac{1}{4}(1 + \xi\xi_i)(1 + \eta\eta_i) \tag{3.2.26}$$

where ξ_i and η_i are the appropriate scale factors. ξ and η in these expressions vary in an element as shown in Fig. 3.2.4. Note that the expression for N_i will vary with each element. For the current problem, due to symmetry we are concerned only with elements 2 and 3 and points 3, 4, 5, and 9. For $i = 9$, the shape functions are

$$N_9(\text{element } 3) = \tfrac{1}{4}(1 - \xi)(1 + \eta), \qquad \xi = -1 + 2\,x, \ \eta = 1 + 2\,y$$

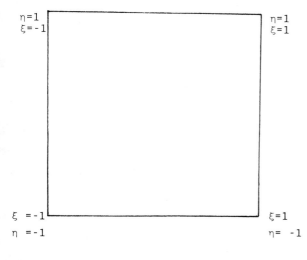

$\eta = 1$
$\xi = -1$

$\eta = 1$
$\xi = 1$

$\xi = -1$
$\eta = -1$

$\xi = 1$
$\eta = -1$

Fig. 3.2.4 Normalized finite-element coordinates.

N_9(element 2) $= \frac{1}{4}(1 - \xi)(1 - \eta)$, $\xi = -1 + 2x$, $\eta = -1 + 2y$

or, in terms of x and y

$$N_9(3) = (1 - x)(1 + y)$$

$$N_9(2) = (1 - x)(1 - y) \qquad\qquad (3.2.27)$$

Similarly, one can show that

$$N_3(2) = xy, \qquad\qquad N_4(3) = x(1 + y)$$

$$N_4(2) = x(1 - y), \qquad N_5(3) = -xy. \qquad\qquad (3.2.28)$$

From these expressions, one then computes the $c_{j,i}$ terms.

It is left to the reader to show that

$$c_{9,3} = -\tfrac{1}{3}, \qquad c_{9,4} = -\tfrac{1}{3}, \qquad c_{9,5} = -\tfrac{1}{3}, \qquad c_{9,9} = \tfrac{8}{3}$$

When these expressions are inserted into (3.2.25), one obtains

$$u_9 = \tfrac{1}{8}(u_3 + u_4 + u_5) \qquad\qquad (3.2.29)$$

Now, if one sets $u_3 = u_5 = 0.5$ and $u_4 = 1$, the result for u_9 is

$$u_9 = \tfrac{1}{4} \qquad\qquad (3.2.30)$$

which is the correct value as can be verified by the simple finite-difference calculation. The uncertainty in selecting the values of the corner points u_3 and u_5 can offer difficulty in applying any numerical method that requires evaluation at such points. As a result, the user should beware.

This finite-element example gives some insight into the details required to set up a finite-element calculation. First, it is generally necessary to employ numerical procedures for computing the $c_{j,i}$ coefficients. These procedures should be constructed in the (ξ, η) system so that they can easily be applied to different geometry, using arbitrary triangles or quadrangles in the (x, y) plane transformed into normalized regular elements in (ξ, η). Also, a mesh generation procedure has to be employed. Besides, in the general case Eq. (3.2.24) is an algebraic system for the nodal values u_i which must be suitably ordered to simplify the solution of the system. As a result, the user should consider carefully the problem and the effort required to solve it before undertaking a new finite-element solution.

3.3 Steady-State Spectral Method Examples

The spectral method will now be utilized for the same example (Eq. 3.2.1) as the finite-element approach. Consider that f can be expanded in a series of Chebyshev polynomials whose properties are outlined in Fox and Parker (1968) and Rivlin (1974). For this case,*

---

*Note the solution expansion begins at $i = 0$ for this expansion.

$$f = f_0 T_0^* + f_1 T_1^* + f_2 T_2^* + \cdots \qquad (3.3.1)$$

where T_i^* is the Chebyshev polynomial that applies in the region $0 \leq x \leq 1$. From the definitions in Fox and Parker (1968), we obtain

$$T_0^* = 1, \qquad T_1^* = 2x - 1, \qquad T_2^* = 8x^2 - 8x + 1 \qquad (3.3.2)$$

For only two terms,

$$f = f_0 + f_1(2x - 1) \qquad (3.3.3)$$

Due to the boundary condition,

$$f_0 = 1 + f_1 \qquad (3.3.4)$$

so

$$f = (1 + f_1) + f_1(2x - 1) = (1 + f_i)T_0^* + f_1 T_1^* \qquad (3.3.5)$$

Substituting this expression into Eq. (3.2.2) yields

$$(1 + f_1) \int_0^1 W_j(T_0^{*\prime} + T_0^*) \, dx + f_1 \int_0^1 W_j(T_1^{*\prime} + T_1^*) \, dx = 0 \qquad (3.3.6)$$

For this equation, W_j must be chosen. If one examines the classical theory of expanding a function in terms of Chebyshev polynomials, one finds that the appropriate form of W_j is

$$W_j = \frac{T_j^*}{\sqrt{x(1 - x)}}$$

Noting the property

$$\int_0^1 \frac{T_i^* T_j^*}{\sqrt{x(1 - x)}} \, dx = \begin{cases} \pi, & i = j = 0 \\ \frac{1}{2}\pi, & i = j \neq 0 \\ 0, & i \neq j \end{cases} \qquad (3.3.7)$$

then from Eq. (3.3.6) where $T_0^{*\prime} = 0$, $T_1^{*\prime} = 2T_0^*$ we find

$$(1 + f_1)\pi + f_1 2\pi = 0 \qquad \text{for } j = 0$$

$$(1 + f_1)0 + f_1 \frac{\pi}{2} = 0 \qquad \text{for } j = 1$$

Clearly, the second equation is not valid since f_1 would vanish and, therefore, it should not be enforced. This is due to the error created by truncating the series as was discussed earlier. From the equation for $j = 0$, we obtain

$$f_1 = -\frac{1}{3}, \qquad f = \frac{2}{3} - \frac{1}{3}(2x - 1) \qquad (3.3.8)$$

For $x = 1$, $f = 0.333$ compared to the exact solution $f_{\text{exact}} = 0.368$

This procedure could be carried to higher order. In that case it is convenient to use directly the formula (Gottlieb and Orszag, 1977) giving the expansion of the derivative, so that

$$f(x) = \sum_{i=0}^{\infty} f_i T_i^*(x), \qquad f'(x) = \sum_{i=0}^{\infty} f_i^x T_i^*(x), \tag{3.3.9a}$$

where

$$\alpha_i f_i^x = 4 \sum_{\substack{p=i+1 \\ p+i \, \text{odd}}}^{\infty} p f_p, \qquad \alpha_0 = 2, \; \alpha_i = 1 \text{ for } i > 0 \tag{3.3.9b}$$

Next, consider the case of a second-derivative equation as in the finite-element example,

$$\frac{d^2 f}{dx^2} - f = 0 \tag{3.3.10}$$

with $f(-1) = 0$ and $f(1) = 1$. We solve this problem with a three-term Chebyshev expansion

$$f = f_0 T_0 + f_1 T_1 + f_2 T_2 \tag{3.3.11}$$

in the interval $-1 \leq x \leq 1$. Inserting this expansion into Eq. (3.3.10) and noting that for the interval $-1 \leq x \leq 1$, $T_0 = 1$, $T_1 = x$, and $T_2 = (2 x^2 - 1)$, we obtain

$$4 f_2 T_0 - (f_0 T_0 + f_1 T_1 + f_2 T_2) = 0 \tag{3.3.12}$$

If we employ the orthogonality condition

$$\int_{-1}^{+1} \frac{T_i T_j}{\sqrt{1 - x^2}} \, dx = \begin{cases} \pi, & \text{for } i = j = 0 \\ \frac{1}{2}\pi, & \text{for } i = j \neq 0 \\ 0, & \text{for } i \neq j \end{cases} \tag{3.3.13}$$

we obtain

$$4 f_2 - f_0 = 0 \tag{3.3.14a}$$

$$f_1 = 0 \tag{3.3.14b}$$

$$f_2 = 0 \tag{3.3.14c}$$

Clearly this result will yield only a trivial solution. The reason for this is that the equations of orthogonality for T_1 and T_2, which yield conditions (3.3.14b) and (3.3.14c), are incomplete due to the series truncation at the term T_2. If we added T_3, then Eq. (3.3.14b) would not have been trivial. However, another equation would have been introduced for f_3 that was trivial. As a result of this process, we see that for Chebyshev polynomials the last two equations should be discarded because they are incomplete. These can be replaced by the boundary conditions as pointed out in the earlier discussion. Following this process yields

$$4 f_2 - f_0 = 0 \qquad \text{(differential equation)} \tag{3.3.15a}$$

$$f_0 - f_1 + f_2 = 0 \quad \text{(boundary condition)} \qquad (3.3.15b)$$
$$f_0 + f_1 + f_2 = 1 \quad \text{(boundary condition)} \qquad (3.3.15c)$$

The solution of this set yields

$$f_0 = \tfrac{2}{5}, \qquad f_1 = \tfrac{1}{2}, \qquad f_2 = \tfrac{1}{10}$$

The solution obtained using the Chebyshev expansion is plotted against the exact solution in Fig. 3.3.1. Considering that only three terms were used the agreement is not too bad. For usual applications using higher-order expansions, a formula of type (3.3.9) for the second derivative must be employed (see Section 8.1).

The spectral procedure in two dimensions follows the same procedure as in one dimension except for more detail. If we consider the two-dimensional problem used for finite elements, $\nabla^2 u = 0$, the spectral approach using eigenfunctions expansion can be demonstrated. We attempt the solution of this problem by assuming that u has a solution of the form

$$u = \sum_{n=0}^{N} \sum_{m=0}^{M} u_{m,n} F_n(x) G_m(y) \qquad (3.3.16)$$

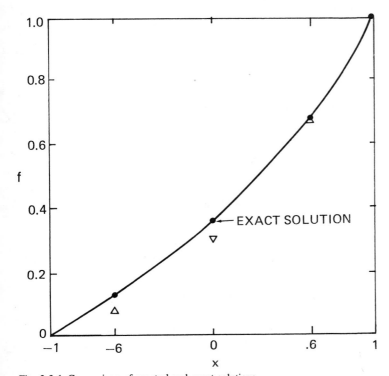

Fig. 3.3.1 Comparison of spectral and exact solutions.

The first step is to choose the forms of F_n and G_m. The selections are dependent on the equation for u and the nature of the boundary conditions. The optimum choice is to utilize functions that satisfy the governing equations and that permit use of optimum inversion techniques for determining the $u_{m,n}$ values.

For the equation $\nabla^2 u = 0$ and the geometry of a box, the natural functions to employ are sines and cosines. As a result assume first that

$$G_m(y) = \sin \frac{m\pi(y + 1)}{2} \tag{3.3.17}$$

This set of functions automatically satisfies the boundary condition $u = 0$ at $y = \pm 1$. Next one must select the $F_n(x)$; there are a number of choices possible. The most appropriate, however, would be a function that together with the $G_m(y)$ will satisfy $\nabla^2 u = 0$. If we seek such a solution we obtain

$$\nabla^2(F_n G_m) = \left[F_n'' - \left(\frac{m\pi}{2} \right)^2 F_n \right] G_m = 0 \tag{3.3.18}$$

This equation is satisfied by

$$F_n = b_n \exp\left[-\frac{m\pi(x + 1)}{2} \right] + c_n \exp\left[\frac{m\pi(x + 1)}{2} \right] \tag{3.3.19}$$

In order that the condition $u = 0$ at $x = -1$ be satisfied, it is necessary to have

$$b_n = -c_n \tag{3.3.20}$$

We next observe that since the function F does not have any need for the n index we can write (3.3.16) in the form

$$u = \sum_{m=1}^{M} u_m \left\{ \exp\left[-\frac{m\pi(x + 1)}{2} \right] - \exp\left[\frac{m\pi(x + 1)}{2} \right] \right\} \sin \frac{m\pi(y + 1)}{2} \tag{3.3.21}$$

This equation must satisfy the condition that $u = 1$ at $x = 1$ so that

$$1 = \sum_{m=1}^{M} u_m [\exp(-m\pi) - \exp(m\pi)] \sin \frac{m\pi(y + 1)}{2} \tag{3.3.22}$$

This equation can be inverted to find the u_m terms numerically by a finite fast Fourier transform. For this simple case, the classical rules of Fourier series lead to

$$u_m[\exp(-m\pi) - \exp(m\pi)] = \int_{-1}^{+1} \sin \frac{m\pi(y + 1)}{2} \, dy \tag{3.3.23}$$

$$= \frac{2}{m\pi} (1 - \cos m\pi)$$

Noting that $\sinh x = (e^x - e^{-x})/2$, then u is approximated by

$$u = \sum_{m=1}^{M} \frac{2}{m\pi} \left[\frac{1 - \cos m\pi}{\sinh m\pi} \right] \sinh \frac{m\pi(x + 1)}{2} \sin \frac{m\pi(y + 1)}{2} \quad (3.3.24)$$

Note that in obtaining the solution one could have selected other expansion functions, but the solution of the resulting equations for the coefficients would probably require numerical inversion of large matrices. Also, note that in this example the discontinuity in boundary conditions produces spurious oscillations (Gibbs phenomenon). Singularities are delicate to handle using spectral-type methods as discussed in Section 3.6.

For the finite-element and spectral examples, one can observe that for a spectral approximation in a single element, or a combination of single elements with simple functions per element, there is not a significant difference between the finite-element and spectral approach. Note the extension of the concept of a spectral scheme in order to include the possibility of using spectral approximations in a combination of elements with appropriate matching conditions (for example, continuity of the function and of some derivatives). The principal difference between both approaches occurs when one wishes to reduce the number of elements (usually to a single one) and increase the number of terms in the spectral expansion. This approach can frequently gain accuracy and reduce computer time if an efficient matrix-inversion scheme is utilized.

The discussion thus far has considered steady-state examples with no time dependence, and therefore one is typically faced with matrix inversions to solve the problems. However, if one employs a time-dependent formulation, either the transient problem or the steady-state problem can be solved by integrating out to long times. Since this approach is useful an example will be discussed next.

3.4 Time-Dependent Finite-Element Examples

In the application of spectral and finite-element methods possibly the least complicated approach is to employ a transient solution to determine a steady state. This approach is in principle the same as that described for finite differences. One begins with some equation of the form

$$\frac{\partial f}{\partial t} + G(f, f_x, f_y, f_z, x, y, z, t) = 0$$

and attempts to find $f(t + \Delta t)$ from $f(t)$. G in this equation usually consists of a set of derivatives and functions which must be approximated in either the finite-element or spectral sense.

In order to demonstrate this approach for finite elements, consider the simple one-dimensional case described earlier, but with the addition of an unsteady term. The equation is

$$\frac{\partial f}{\partial t} + \frac{\partial f}{\partial x} + f = 0 \tag{3.4.1}$$

with the condition that $f = 1$ at $x = 0$. For an initial condition we assume $f = 0$ at $t = 0$ everywhere.

For solution of this equation, first multiply by a weighting function $W_j(x)$ and integrate with respect to x to obtain

$$\frac{\partial}{\partial t} \int_{x_1}^{x_2} W_j f \, dx + \int_{x_1}^{x_2} W_j \left(\frac{\partial f}{\partial x} + f \right) dx = 0 \tag{3.4.2}$$

As in the steady-state case assume $W_j = F_j$ and $F_1 = 1 - x$ and $F_2 = x$, so that in a first approximation*

$$f = f_1(t) F_1(x) + f_2(t) F_2(x) \tag{3.4.3}$$

Since $f = 1$ at $x = 0$ for $t > 0$, then $f_1 = 1$ for $t > 0$. With f_1 known we then have the equation

$$\frac{\partial}{\partial t} \int_{x_1}^{x_2} F_2(F_1 + F_2 f_2) \, dx + \int_{x_1}^{x_2} F_2(F_1' + f_2 F_2' + F_1 + f_2 F_2) \, dx = 0 \tag{3.4.4}$$

From the steady-state example if we choose $x_1 = 0$ and $x_2 = 1$, we know that

$$\int_0^1 F_2(F_1' + F_1) \, dx = -\frac{1}{3} \quad \text{and} \quad \int_0^1 F_2(F_2' + F_2) \, dx = \frac{5}{6} \tag{3.4.5}$$

One can also easily show that

$$\int_0^1 F_1 F_2 \, dx = \frac{1}{6} \quad \text{and} \quad \int_0^1 F_2 F_2 \, dx = \frac{1}{3} \tag{3.4.6}$$

Therefore,

$$\frac{d}{dt} \left(\frac{1}{6} + \frac{f_2}{3} \right) + \left(-\frac{1}{3} + \frac{5}{6} f_2 \right) = 0 \tag{3.4.7}$$

or

$$\frac{df_2}{dt} + \frac{5}{2} f_2 - 1 = 0 \tag{3.4.8}$$

This equation can be integrated numerically or analytically to obtain a solution for $f_2(t)$. Analytically, one obtains the result

$$f_2 = \tfrac{2}{5}[1 - e^{-5t/2}] \tag{3.4.9}$$

which vanishes at $t = 0$ and approaches the steady-state value of 0.4 for large

---

*We start the expansion with $i = 1$.

t. In this example it is important to note once again that a boundary condition replaced one of the equations for the unknowns—i.e., f_1. For each problem one should be aware that this behavior can and usually does occur.

3.5 Time-Dependent Spectral Method Examples

One can solve Eq. (3.4.1) spectrally by employing a Chebyshev polynomial expansion of two terms as in the steady-state case such that*

$$f = f_1(t) + f_2(t)\,(2x - 1) \tag{3.5.1}$$

When this expresion is inserted into Eq. (3.4.2) along with

$$W_j = \frac{T^*_{j-1}}{\sqrt{x(x-1)}} \quad \text{and} \quad F_i = T^*_{i-1} \tag{3.5.2}$$

and one sets $x_1 = 0$ and $x_2 = 1$ one obtains

$$\frac{d}{dt}(f_1 \pi) + f_1 \pi + 2\pi f_2 = 0$$

$$\frac{d}{dt}\left(f_2 \frac{\pi}{2}\right) + f_2 \frac{\pi}{2} = 0 \tag{3.5.3}$$

The boundary condition $f = 1$ at $x = 0$ requires that

$$f_1 = 1 + f_2 \tag{3.5.4}$$

Clearly, the problem is overspecified and one of these equations must be discarded. It is clear that the boundary condition cannot be dismissed, so it is necessary to eliminate one of the differential equations. The best equation to eliminate because of accuracy is the second. The reason was described earlier in the general discussion (Section 3.1). Following this procedure and combining (3.5.4) with (3.5.3) we obtain

$$\frac{df_1}{dt} + 3 f_1 - 2 = 0 \tag{3.5.5}$$

This equation has a solution that vanishes at $t = 0$ of the form

$$f_1 = \tfrac{2}{3}(1 - e^{-3t}) \tag{3.5.6}$$

and consequently

$$f_2 = -(\tfrac{1}{3} + \tfrac{2}{3}e^{-3t}) \tag{3.5.7}$$

As *t* becomes large these results reduce to the steady-state results obtained

---

*Note that the indices here start at $i = 1$.

previously. Note, however, that f_2 does not satisfy an initial condition of $f_2 = 0$ at $t = 0$. This is an inconsistency that arises when one employs the approach usually termed the τ method. This can be troublesome in applications (see Taylor and Murdock, 1980). If for example, one chooses to solve the equation

$$\frac{\partial f}{\partial t} + \frac{\partial f}{\partial x} = 0 \tag{3.5.8}$$

with the conditions $f = 0$ at $t = 0$ and $f = 1$ at $x = 0$ for $t > 0$, by the Chebyshev spectral approach with explicit numerical integration, the solution after the first time step is as shown in Fig. 3.5.1. As a result, one should as mentioned earlier, proceed with caution in using the τ method with explicit time integration.*

The one-dimensional time integration examples displayed give the essential features of including transient terms in spectral and finite-element techniques. The extension to multidimensions just adds more terms and will not contribute particularly to the understanding. As a consequence, these details are not included here.

3.6 Pseudospectral Methods

In addition to finite-element, integral relationships, and spectral method applications by the Galerkin approach, there are other spectral- and cell-type methods that have proven to be useful in fluid-flow computations. In this section, the methods found to be most useful are outlined. One of the most powerful techniques for solving problems is the use of collocation in combination with spectral expansion methods. This technique has evolved with the name pseudospectral method. The advantage of this approach is that it eliminates boundary-condition problems associated with using the spectral expansion approach. To demonstrate the method, consider the equation

$$\frac{\partial f}{\partial t} + \frac{\partial f}{\partial x} = \nu \frac{\partial^2 f}{\partial x^2} \tag{3.6.1}$$

Also assume that f can be represented by

$$f = \sum_{i=0}^{N} f_i(t) F_i(x) \tag{3.6.2}$$

Instead of inserting (3.6.2) into (3.6.1) and deriving equations for the f_i terms, one proceeds by first integrating (3.6.1) with respect to time to obtain

$$f(t + \Delta t, x) - f(t, x) = \int_t^{t+\Delta t} \left(\nu \frac{\partial^2 f}{\partial x^2} - \frac{\partial f}{\partial x} \right) dt \tag{3.6.3}$$

---

*Collocation solution is also displayed to show the improved behavior when a real space calculation is used. See Taylor and Murdock (1980).

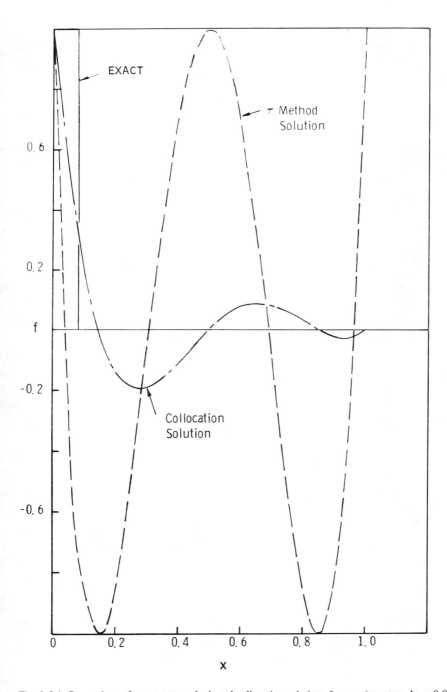

Fig. 3.5.1 Comparison of exact, tau method, and collocation solutions for one time step, $\Delta t = 0.08$.

This equation now relates real-space values of f to the spectral space values if expansion (3.6.2) is substituted only on the right-hand side of Eq. (3.6.3). The right-hand side, however, is a quadrature that can be evaluated by a variety of quadrature formulas. Typically, the formula takes the form

$$\int_t^{t+\Delta t} G\, dt = \sum_{m=0}^M b_m G(t_m, x)\, \Delta t \quad \text{with} \quad M \le n + 1 = \frac{t + \Delta t}{\Delta t} \quad (3.6.4)$$

This equation, when combined with (3.6.3), can be used to solve for $f(t, x)$. The procedure depends strongly on whether the quadrature (3.6.4) is implicit or explicit. For the times where $f(t, x)$ is known, one can develop an expansion of f in the form (3.6.2). Then it is possible to compute

$$G(t_m, x) = \left(\nu \frac{\partial^2 f}{\partial x^2} - \frac{\partial f}{\partial x} \right)_{t_m} \qquad (3.6.5)$$

for each $M < n + 1$ from these expansions.

The next step is to compute $f(t + \Delta t, x)$ from the formula

$$f(t + \Delta t, x) = f(t, x) + \sum_{m=0}^M b_m G(t_m, x)\Delta t \qquad (3.6.6)$$

This is straightforward if $b_{n+1} = 0$ because the equation is explicit in time. However, if $b_{n+1} \ne 0$, one must devise a way to solve this equation. The most practical way would be a predictor-corrector approach since it would work even if the function $G(t, x)$ is quite complicated. For this approach, one would employ an explicit predictor to obtain a predicted value $f_P(t + \Delta t, x)$ and then correct the answer by utilizing $f_P(t + \Delta t, x)$ to compute $G(t + \Delta t, x)$ in the time-quadrature approximation. The result would then yield a new corrected value of $f(t + \Delta t, x)$.

An alternate approach may be employed in handling the viscous terms since they are linear. One first must separate the linear and nonlinear terms in the expression for $G(t, x)$. One then can apply an explicit time integration to the nonlinear term and an implicit time integration to the linear viscous term. For the time integration of the viscous term one approximates values of $\partial^2 f/\partial x^2$ at t by a spectral expansion and $\partial^2 f/\partial x^2$ at $t + \Delta t$ by a grid-point approximation to the complete inversion of the spectral expansion. Morchoisne (1981) has developed a variation of this type of procedure and successfully applied it to a variety of flow calculations. Morchoisne employs a standard central difference approximation for predicting the viscous term; other possible approximations are an area for future study. It is important to note also that the concept of finite-difference predictor and spectral corrector can also be successfully applied to the solution of the Poisson equation as numerical experiments by Hirsh et al. (1982) have shown. Such an approach was introduced by Orszag (1980).

Another area where additional definition is required is in the area of possible errors that may arise in the computation of the nonlinear terms. The errors are termed aliasing errors and result from the multiplication of two spectral expansions, such as $f \, \partial f / \partial x$. From a general view it is known that these errors are not dominant, but some wave-propagation computation results display small oscillations that may be either physical or aliasing errors. The user of the spectral approach should take care to evaluate any small oscillations in order to determine if they are physical or numerical in nature.

In practical applications of the pseudospectral approach, it has been found that the Adams–Bashforth explicit time integration scheme works reasonably well for the solution of inviscid problems by Chebyshev expansions. In the application, however, there is some uncertainty regarding the optimum time step that should be employed in such calculations. Gottlieb and Orszag (1977) suggest the condition $\Delta t \leq 8/N^2$ for stability (N is the number of spectral terms). Based on the physics of the problem, the spectral method can be viewed as a one-element expansion over the interval $-1 \leq x \leq 1$. As a consequence, it contains the information that permits any point in the interval to "communicate" with another point. As a result, it seems that the time step is governed roughly by the time it would take a wave to propagate over the complete interval $-1 \leq x \leq 1$. In a scaled form this would imply that $\Delta t < 2$ is the maximum limit. An experimental study by Myers, Taylor, and Murdock (1981) of the pseudospectral approach has indicated that this limit can be approached and still have stable results. The optimum for accuracy and stability is not known at this time, however.*

For viscous flows the time step will be limited by the viscous diffusion time, $\Delta t \sim L^2/\nu$, unless an implicit time integration is used for these terms. If an implicit scheme is used, the inviscid time-step will be controlling. Since the viscous time step is usually small compared to the inviscid value, it is suggested that an implicit viscous time integration be pursued.

The outlined pseudospectral approach has many advantages. First, it can be made as accurate spatially as one desires by adding more terms to the expansion for $f(t, x)$. Second, it permits introduction of physical boundary points without difficulty. Third, it does not require the extensive quadratures of finite-element methods. Fourth, it permits the choice of a wide variety of functions for the expansion. And lastly, it is easily applied to compressible flows whereas a Galerkin spectral method is not. The functions can be splines, Chebyshev polynomials, sines, cosines, or any other function sets desired. However, if possible, it is important to choose functions that permit the use of rapid techniques such as fast Fourier transforms (FFT) or conjugate-gradient-type methods in transforming between real and spectral space. Taylor et al. (1981) have demonstrated that the conjugate-gradient method can be competitive with the FFT in a recent study. Further studies indicate that matrix preconditioning can make straightforward matrix inversion competitive as well for time integration problems.

---

*The gradients near the boundaries have a strong influence on the time step that will work best.

The constraints for the use of the FFT may be a problem because of the required unequally spaced mesh points as occurs when the Chebyshev poly-nominals are used. However, if one chooses to use equally spaced points and Chebyshev polynomials, one will encounter accuracy problems principally in the solution for the coefficients of the series by matrix inversion. The reason is not totally clear, but Lanczos (1956) points out that the proper way to develop an equal-spaced mesh Chebyshev polynomial expansion is to first develop a sine series and transform it using functional identities which relate sines to a series of Bessel functions and Chebyshev polynomials. This approach, however, does not seem to yield any advantage when it is implemented.

Another troublesome area where caution should be exercised in applying the pseudospectral method involves problems with discontinuities or large gradients. In such cases it is possible to obtain substantial Gibbs-type phenomena (oscillations) near the gradient. As a result, it may be necessary to apply a filter to the calculation to damp this type of noise.

Orszag and Gottlieb (1980) as well as Haidvogel et al. (1980) have employed a spectral-filter approach in which the first step is to solve for the coefficients of the spectral expansion and then to multiply the resulting coefficient, f_i, by a factor g_i, which has the form

$$g_i = \frac{1 - \exp[-(N^2 - i^2)/N_0^2]}{1 - \exp[-(N^2/N_0^2)]}$$

where i is the index of the term, N is the maximum number of terms, and N_0 is an adjustable scale typically taken to be 2 N. Our experience with this type of filter has not always been satisfactory. Numerical experiments have indicated that more satisfactory filters can be found by filtering in real space according to the rule

$$f_j^{new} = f_j^{old} + \sigma(f_{j+1} - 2f_j + f_{j-1})^{old} \qquad (3.6.7)$$

where σ is a damping coefficient and j denotes the spacial location. From the finite-difference discussions presented earlier, it is easy to see that this filter is equivalent to an artificial viscosity. The key to applying this type of filter is the selectivity with which it is applied. Research in this area by Myers et al. (1981) and Taylor, Myers, and Albert (1981) has revealed selection rules that seem to work reasonably well. One is for incompressible flows and the other is for compressible flows. For incompressible flows the rule is as follows: Given the point x_j and the computed value of $f(x_j)$

1. Form $S_{j+1/2} = f(x_{j+1}) - f(x_j)$, etc.
2. If $S_{j+1/2}S_{j+3/2} < 0$ and $S_{j-1/2}S_{j-3/2} < 0$, apply the correction $f^{new}(x_j) = f(x_j) + 0.1(S_{j+1/2} - S_{j-1/2})$.

For compressible flows the best rule known at this time is the flux-correction method of Boris and Book (1976) along with Zalesak (1979). This method uses the following steps:

1. Time integrate the equations of interest to obtain $f(t + \Delta t)$ from $f(t)$ at N points.
2. Diffuse $f(t + \Delta t)$ by the rule

$$\bar{f}_j(t + \Delta t) = f_j(t + \Delta t) + \sigma[f_{j+1}(t + \Delta t) - 2 f_j(t + \Delta t)$$
$$+ f_{j-1}(t + \Delta t)]$$

3. Antidiffuse $\bar{f}(t + \Delta t)$ by the rule

$$f_j(t + \Delta t) = \bar{f}_j(t + \Delta t) - (D_{j+1/2} - D_{j-1/2})$$

where $D_{j+1/2}$ is obtained from the selection rule

$$D_{j+1/2} = S \max[0, \min[S\Delta_{j-1/2}, |D_{j+1/2}^1|, S\Delta_{j+3/2}]]$$

with

$$D_{j+1/2}^1 = \tfrac{1}{8}[f_{j+1}(t + \Delta t) - f_j(t + \Delta t)]$$
$$S = \text{sign } (\Delta_{j+1/2})$$
$$\Delta_{j+1/2} = \bar{f}_{j+1}(t + \Delta t) - \bar{f}_j(t + \Delta t)$$

This procedure is frequently called the flux correction procedure in finite differences and seems to work well in spectral solutions also. An alternate to this procedure would be utilization of an artificial viscosity scheme from finite differences with a rule for selective application. Research and understanding in this area is limited at this time.

Figures 3.6.1, 3.6.2, and 3.6.3 show the densities computed across a rarefaction, contact surfaces, and shock waves obtained by solving one-dimensional compressible flow equations by the pseudospectral method using the filter just described. The calculations were made using a 33-term Chebyshev expansion of each flow variable; the initial conditions for each case are shown on the figures. The results show that the one-dimensional approach looks promising. Figure 3.6.4 shows the density distribution with zero damping for the shock wave.

In closing the discussion on filtering techniques, we mention an approach suggested by Lanczos (1956) for reducing noise in sine-series representations of functions. Lanczos suggests filtering the sine series by truncation and then multiplying each coefficient by the factor

$$\sigma_i = \frac{\sin[i\pi/(N+1)]}{i\pi/(N+1)}$$

where $N + 1$ is the first neglected term in the series and i is the index. In addition, Lanczos has demonstrated that the approach removes the Gibbs phenomena when one attempts to compute a step function. The factor is also suggested for improving convergence of the derivative of the sine series.

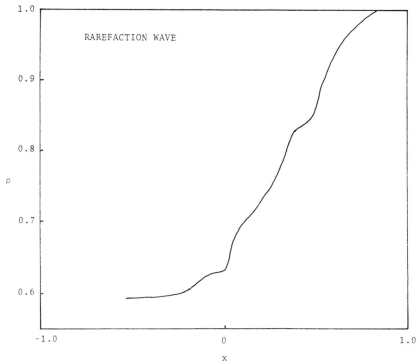

Fig. 3.6.1 Density for pseudospectral calculation through rarefaction wave.

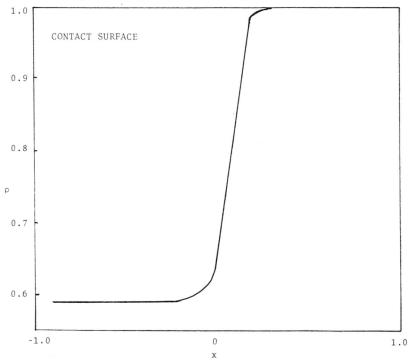

Fig. 3.6.2 Density for pseudospectral calculation through contact surface.

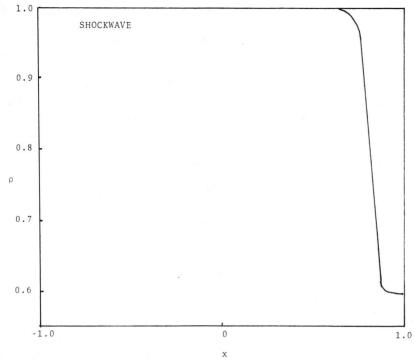

Fig. 3.6.3 Density for pseudospectral calculation through shock wave.

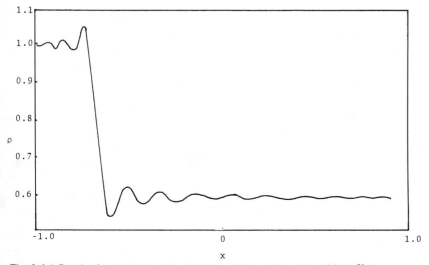

Fig. 3.6.4 Density for pseudospectral calculation through shock wave without filter.

3.7 Finite-Volume or Cell Method

The finite-volume or cell method is based upon an integral form of the equation to be solved. The computational region is divided into elementary volumes within which the integration is carried out. Such a procedure allows one to deal with complicated geometry without considering the equation written in curvilinear coordinates. This also preserves the property of conservation. Only the coordinates of the corners of the volume are really necessary, and curvilinear coordinates—not necessarily orthogonal—can be used to define the set of volumes.

The finite-volume technique is now described for the model equation

$$\frac{\partial f}{\partial t} + \frac{\partial F}{\partial x} + \frac{\partial G}{\partial y} = 0 \tag{3.7.1}$$

where

$$F = F_I(f) - \nu(f) \frac{\partial f}{\partial x}$$

$$\hspace{4cm} (\nu > 0) \tag{3.7.2}$$

$$G = G_I(f) - \nu(f) \frac{\partial f}{\partial y}$$

The integral form of Eq. (3.7.1) is

$$\frac{d}{dt} \int_{\Omega_p} f \, d\sigma + \int_{\Gamma_p} \mathbf{H} \cdot \mathbf{N} \, ds = 0, \qquad \mathbf{H} = (F, G) \tag{3.7.3}$$

where Ω_p is an elementary fixed cell with a boundary Γ_p, and $\mathbf{N}$ is the unit normal to Γ_p. If Cartesian coordinates (x, y) are used in the evaluation of the boundary integral in (3.7.3) we have

$$\mathbf{H} \cdot \mathbf{N} \, ds = F \, dy - G \, dx \tag{3.7.4}$$

Assuming a curvilinear mesh (Fig. 3.7.1), the elementary cell Ω_p is the quadrilateral cell $ABCD$. The evaluation of Eq. (3.7.3) in Ω_p is carried out by first defining an averaged value f_p of f in Ω_p. When necessary, f_p is assumed to be located at the center of Ω_p. Equation (3.7.3) then yields

$$\frac{d}{dt} (S_p f_p) + (H_{AB} + H_{BC} + H_{CD} + H_{DA}) = 0 \tag{3.7.5}$$

where S_p is the area of the cell Ω_p, and H_{AB}, H_{BC}, H_{CD}, and H_{DA} are the fluxes through the sides. Equation (3.7.5) is a differential equation which describes the evolution in time of the averaged value f_p. It is necessary also to (i) define the fluxes H_{AB}, H_{BC}, H_{CD}, H_{DA}, and (ii) to introduce a time discretization of (3.7.5). Here we consider only the first point. Taking account of (3.7.4), we have

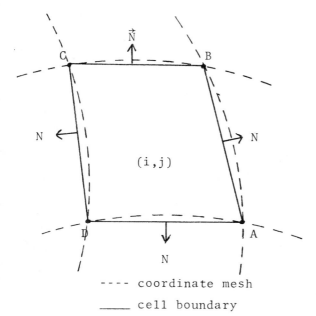

Fig. 3.7.1 Basic element for
finite-volume method.

---- coordinate mesh
_____ cell boundary

$$H_{AB} = F_{AB}\Delta y_{AB} - G_{AB}\Delta x_{AB}$$

$$H_{BC} = F_{BC}\Delta y_{BC} - G_{BC}\Delta x_{BC}$$

$$H_{CD} = F_{CD}\Delta y_{CD} - G_{CD}\Delta x_{CD} \qquad (3.7.6)$$

$$H_{DA} = F_{DA}\Delta y_{DA} - G_{DA}\Delta y_{DA}$$

where F_{AB}, G_{AB}, F_{BC}, G_{BC}, etc. are, respectively, the mean value of F, G on sides AB, BC, etc., and where

$$\Delta x_{AB} = x_B - x_A, \qquad \Delta y_{AB} = y_B - y_A, \text{ etc.} \qquad (3.7.7)$$

The coordinates of a point such as A are (x_A, y_A).

Next, we consider the inviscid equation ($\nu = 0$) deduced from Eq. (3.7.1); that is, there is no derivative of f into the flux $\mathbf{H} \cdot \mathbf{N}$. If we denote by $f_{i,j}$ the mean value of f in the elementary cell (see Fig. 3.7.2), the straightforward way to define the fluxes is

$$H_{AB} = \tfrac{1}{2}(F_{i+1,j} + F_{i,j})\Delta y_{AB} - \tfrac{1}{2}(G_{i+1,j} + G_{i,j})\Delta x_{AB} \qquad (3.7.8)$$

with a analogous expressions for the other fluxes. In the case where a uniform Cartesian mesh is considered, these formulas yield the usual central differencing.

In the same manner, it is possible to define a noncentered scheme by assuming, for instance, the mean of F on AB is defined by $\alpha_1 F_{i,j} +$

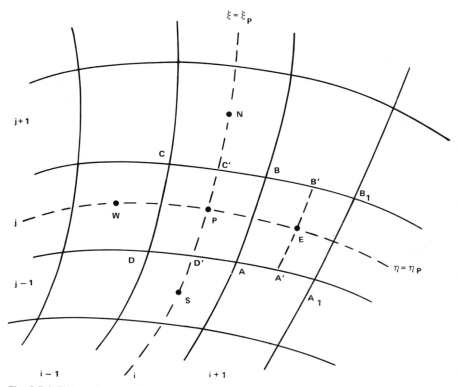

Fig. 3.7.2 Finite-volume mesh system.

$(1 - \alpha_1)F_{i+1,j}$ instead of $\frac{1}{2}(F_{i+1,j} + F_{i,j})$ and the mean value on CD by $(1 - \alpha_1)F_{i,j} + \alpha_1 F_{i-1,j}$ instead of $\frac{1}{2}(F_{i,j} + F_{i-1,j})$. Such a definition with $\alpha_1 = 0$ or 1 is used in the finite-volume methods of MacCormack and Paullay (1972), Rizzi and Inouye (1973), and Deiwert (1975). In these methods, the time discretization is based upon the splitting technique associated with a predictor-corrector scheme as described in Chapter 2.

When viscosity is present, as it is in Eq. (3.7.1), the flux on each side of the cell involves first-order derivatives, and it becomes necessary to define mean values for these derivatives. Let us consider the flux H_{AB} which is now written

$$H_{AB} = \left[\frac{1}{2}(F_{i+1,j} + F_{i,j}) - \frac{1}{2}(\nu_{i+1,j} + \nu_{i,j})\left(\frac{\partial f}{\partial x}\right)_{AB} \right] \Delta y_{AB}$$

$$- \left[\frac{1}{2}(G_{i+1,j} + G_{i,j}) - \frac{1}{2}(\nu_{i+1,j} + \nu_{i,j})\left(\frac{\partial f}{\partial y}\right)_{AB} \right] \Delta x_{AB}$$

and let us describe a simple way to define the averaged derivatives $(\partial f/\partial x)_{AB}$

and $(\partial f/\partial y)_{AB}$. They are considered as mean values on the cell Ω' bounded by $\Gamma' = (A', B', C', D')$.

$$\left(\frac{\partial f}{\partial x}\right)_{AB} = \frac{1}{S'}\int_{\Omega'}\frac{\partial f}{\partial x}\,d\sigma = \frac{1}{S'}\int_{\Gamma'}f\,dy$$

$$\left(\frac{\partial f}{\partial y}\right)_{AB} = \frac{1}{S'}\int_{\Omega'}\frac{\partial f}{\partial y}\,d\sigma = -\frac{1}{S'}\int_{\Gamma'}f\,dx$$

where S' is the area of Ω'. Now, it remains to evaluate the curvilinear integrals on Γ'. Let

$$\int_{\Gamma'}f\,dy = f_E\,\Delta y_{A'B'} + f_B\,\Delta y_{B'C'} + f_P\,\Delta y_{C'D'} + f_A\,\Delta y_{D'A'}$$

$$\int_{\Gamma'}f\,dx = f_E\,\Delta x_{A'B'} + f_B\,\Delta x_{B'C'} + f_P\,\Delta x_{C'D'} + f_A\,\Delta x_{D'A'}$$

where

$$f_E = f_{i+1,j}\,, \qquad f_P = f_{i,j}$$
$$f_A = \tfrac{1}{4}(f_{i+1,j} + f_{i,j} + f_{i,j-1} + f_{i+1,j-1})$$
$$f_B = \tfrac{1}{4}(f_{i+1,j+1} + f_{i,j+1} + f_{i,j} + f_{i+1,j})$$

and

$$\Delta y_{A'B'} = \tfrac{1}{2}(\Delta y_{A_1B_1} + \Delta y_{AB})$$
$$\Delta y_{C'D'} = \tfrac{1}{2}(\Delta y_{BA} + \Delta y_{CD})$$
$$\Delta y_{B'C'} = \tfrac{1}{2}(\Delta y_{B_1B} + \Delta y_{BC}) \tag{3.7.9}$$
$$\Delta y_{D'A'} = \tfrac{1}{2}(\Delta y_{DA} + \Delta y_{AA_1})$$

with analogous definitions for $\Delta x_{A'B'}$, etc. The area $S' = \tfrac{1}{2}(S_{i+1,j} + S_{i,j})$, where $S_{i,j}$ refers to the area of the cell (i,j). The interest of the above approximations is that only the coordinates of the vertices of the cells are involved.

Another approximation could be

$$\Delta y_{A'B'} = -\Delta y_{C'D'} = \Delta y_{AB}\,, \qquad \Delta y_{B'C'} = -\Delta y_{D'A'} = -\Delta y_{PE} \tag{3.7.10}$$

with analogous definitions for $\Delta x_{A'B'}$, etc., and $S' = \Delta x_{PE}\,\Delta y_{AB} - \Delta x_{AB}\,\Delta y_{PE}$. Such an approximation introduces the coordinates of the center of the cells defined as the average of the vertices coordinates. Note that it can be introduced in a different way with an equivalent result. More precisely, let us assume the center of the cell (i, j) belongs to a cuvilinear system $\xi(x, y) = $ const, $\eta(x, y) = $ const (see Fig. 3.7.2). The derivatives of $\partial f/\partial x$ and $\partial f/\partial y$ are expressed in terms of derivatives with respect to ξ and η by

$$\frac{\partial f}{\partial x} = \frac{1}{J}\frac{\partial(f, y)}{\partial(\xi, \eta)}, \qquad \frac{\partial f}{\partial y} = \frac{1}{J}\frac{\partial(f, x)}{\partial(\xi, \eta)}, \qquad J = \frac{\partial(x, y)}{\partial(\xi, \eta)} \tag{3.7.11}$$

The various derivatives with respect to ξ and η involved in the above expressions are approximated with finite-difference formulas, for example,

$$\left(\frac{\partial f}{\partial \xi}\right)_{AB} = \frac{f_E - f_P}{\xi_E - \xi_P}, \qquad \left(\frac{\partial f}{\partial \eta}\right)_{AB} = \frac{f_B - f_A}{\eta_B - \eta_A} \qquad (3.7.12)$$

and analogous expressions hold for the derivatives of x and y. Introducing these expressions into Eq. (3.7.11), we obtain averaged derivatives $(\partial f/\partial x)_{AB}$ and $(\partial f/\partial y)_{AB}$ identical to those obtained using Eq. (3.7.10). It is interesting to note that the values of ξ and η at points E, P, A, B, etc., disappear in the final results: The geometry of the mesh appears only as coordinates of the vertices and the centers of the cells.

Both approximations reduce to the standard centered finite differences in the case of a uniform Cartesian mesh.

Obviously, it is possible to approximate the derivatives in (3.7.12) by other finite-difference formulas, for instance, by noncentered differences as used by Deiwert (1975).

Thus far, we have discussed the general approach. Next, we consider two specific volume (cell) methods that have been developed for inviscid compressible flows.

3.7.1 Godunov method

There is a finite-volume technique available in compressible flows that has proven useful in the computation of extremely complicated inviscid flows with strong gradients. This technique is a method proposed by Godunov (1959). The method is very interesting since it is based on the concept of utilizing localized solutions of one-dimensional physical problems to estimate the flow behavior in a multidimensional flow. For discussion purposes, consider a one-dimensional unsteady compressible flow described by the equations

$$\frac{\partial f}{\partial t} + \frac{\partial F}{\partial x} = 0 \qquad (3.7.13)$$

where

$$f = \begin{pmatrix} \rho \\ \rho u \\ \overline{E} \end{pmatrix}, \qquad F = \begin{pmatrix} \rho u \\ p + \rho u^2 \\ u(\overline{E} + p) \end{pmatrix}$$

$$\overline{E} = \frac{p}{\gamma - 1} + \frac{\rho u^2}{2}$$

In these expressions, p denotes the pressure, ρ the density, u the velocity, and γ the adiabatic gas constant.

Next, we integrate Eq. (3.7.13) with respect to x to obtain

$$\frac{\partial}{\partial t} \int_{x_1}^{x_2} f\, dx + F(x_2) - F(x_1) = 0 \tag{3.7.14}$$

We then define average quantities by the expression

$$f_{av} = \int_{x_1}^{x_2} \frac{f\, dx}{(x_2 - x_1)} \tag{3.7.15}$$

and divide the flow into elements of width $\Delta x = x_2 - x_1$. If one then describes the flow in terms of the average quantities, the properties in each element will differ but will be constant within the element. Figure 3.7.3 displays this variation in a qualitative manner at a time t.

The principle of the technique centers on this distribution. The basic postulate is that, given the initial distribution, the time behavior at the boundaries x_1 and x_2 can be computed from a solution of the Riemann problem (Fig. 3.7.4) between the states given by the average cell values $n-1$ to n and n to $n+1$,[*] respectively. As a consequence, the flux across the cell boundaries over the time interval Δt that goes into Eq. (3.7.14) can be estimated from the Riemann problem solutions. The solution of the Riemann problem requires an iterative solution for large differences between initial states. This solution approach is described by Godunov (1959).

For the weak-gradient or acoustic-wave case, the problem can be linearized, and the resulting equations for the solution between the states n and $n+1$ are

$$u_{x_2} = \bar{u}_2 - \frac{p_{n+1} - p_n}{2\sqrt{\gamma \bar{p}_2 \bar{\rho}_2}} \tag{3.7.16}$$

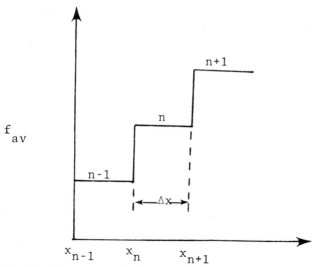

Fig. 3.7.3 First-order distribution of conservation quantities, f.

[*]We have used n here to denote the cell number and not time.

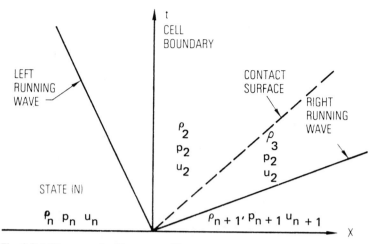

Fig. 3.7.4 Wave map for Riemann problem.

$$\rho_{x_2} = \overline{\rho}_2 - \frac{(\rho u)_{n+1} - (\rho u)_n}{2\sqrt{\gamma \overline{p}_2/\overline{\rho}_2}} \tag{3.7.17}$$

$$p_{x_2} = \overline{p}_2 - \frac{u_{n+1} - u_n}{2}\sqrt{\gamma \overline{p}_2 \overline{\rho}_2} \tag{3.7.18}$$

These equations are used to estimate the fluxes at the boundary if the flow is subsonic where $|u^*| < \sqrt{\gamma p_2/\rho_2}$. This is the case where waves move away from the boundary in each direction. If the flow is supersonic in the direction where

$$|u^*| = \left| \overline{u}_2 - \frac{p_{n+1} - p_n}{2\sqrt{\gamma \overline{p}_2 \overline{\rho}_2}} \right| > \sqrt{\frac{\gamma p_2}{\rho_2}} \tag{3.7.19}$$

one must use the relationships

$$u_{x_2} = \overline{u}_2 - \frac{u_{n+1} - u_n}{2} \tag{3.7.20}$$

$$\rho_{x_2} = \overline{\rho}_2 - \frac{\rho_{n+1} - \rho_n}{2} \tag{3.7.21}$$

$$p_{x_2} = \overline{p}_2 - \frac{p_{n+1} - p_n}{2} \tag{3.7.22}$$

for the boundary fluxes. Note that for this application it is assumed that the flow is supersonic from left to right. If the reverse is true, the indices must be reversed. In these expressions the subscript 2 denotes the average between states n and $n+1$ defined by $\bar{f}_2 = \frac{1}{2}(f_n + f_{n+1})$.

The expression given here for the density is slightly different from that presented by Godunov. The results, however, are consistent. The quantities on the left-hand sides of Eqs (3.7.16)–(3.7.18) and (3.7.20)–(3.7.22) are used to compute the fluxes in Eq. (3.7.14). Equation (3.7.14) is then integrated forward one step by an explicit time integration method. This will yield a new distribution of average cell quantities. The procedure is then repeated for as many time steps as desired.

In the application of this method in two dimensions, one uses splitting and reduces the problem to two one-dimensional problems while applying the same formulas. The method is very stable essentially because it has strong damping due to the gradient terms that appear in Eqs. (3.7.16)–(3.7.18) and (3.7.20)–(3.7.22). These gradient terms in a finite-difference analysis would appear as artificial-viscosity-type terms. For very strong blast-wave-type calculations, experience has shown that the nonlinear Riemann formulas must be used. For practical engineering work, the Godunov method has a lot to offer because of its strong stability. The sophisticated analyst may look down on the scheme, but it should be pointed out that both Taylor et al. (1972) and Sod (1975) have shown it to be as accurate as many second-order finite-difference methods. As a result, the method is highly recommended for inviscid flows with complicated shock patterns.

An interesting extension of the Godunov method is a scheme proposed by Glimm (1965) and developed by Chorin (1976), Sod (1978, 1980a, 1980b), and Colella (1979). A rather complete discussion of the method has been presented by Sod (1980b) and we will therefore give an abbreviated version here. This method does not employ the differential equations in the integral or difference form. It simply applies the Riemann solutions for two one-half time steps and uses random sampling. The concept is as follows.

3.7.2 Glimm method

Consider the one-dimensional problem used to discuss the application of Godunov's method. For that problem, the flow was divided into cells of width Δx. Then the fluxes at the boundaries were obtained by solving the Riemann problem with initial values being the average states on each side of the boundary. Up to the point of solving the Riemann problem, Godunov's and Glimm's methods are identical. At this point, however, they vary. Glimm's approach does not use the flow equations to compute new quantities for a cell; instead, it proceeds to solve a second Riemann problem. However, before doing this it uses a random sampling of the first Riemann solution to obtain the new half-time step state. Since this technique is foreign to most readers, we will attempt to summarize the procedure through graphical means.

Step I:

(a) Initial conditions   (b) Solve Riemann problem

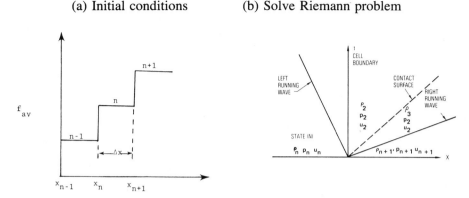

Step II: Randomly sample the state between $x_n - \Delta x/2$ and $x_n + \Delta x/2$ to obtain new average initial conditions at $t + \Delta t/2$

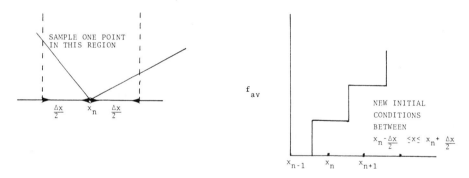

Step III: Repeat Steps I and II using the new initial conditions generated in Step II (note that the x interval is shifted by $\Delta x/2$ for this step).

At the conclusion of Step III, one has generated a new solution for the values of f at the $x_{n+1/2}$ grid points. This procedure is very interesting since it never employs the overall conservation equations—only the local solutions. This is an interesting concept that could possibly be extended to other equations. However, the technique for applying the method to generalized flows with nonrectangular geometries has not been fully developed at this time. For one-dimensional problems the method works extremely well. Figure 3.7.5 shows some typical one-dimensional results obtained by Sod for the solution of the shock-tube problem with initial states $\rho_1 = 1$, $p_1 = 1$, $u_1 = 0$ and $\rho_2 = 0.125$, $p_2 = 0.1$, and $u_2 = 0$. For the calculations $\gamma = 1.4$ and

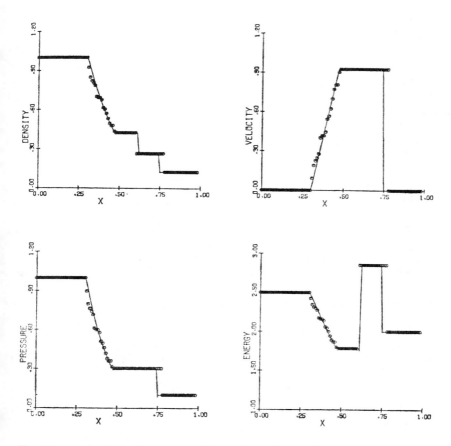

Fig. 3.7.5 Results obtained for solution of shock-tube problem by Glimm's method. (Courtesy of G. Sod.)

$\Delta x = 0.01$. In the calculations Sod pointed out that in the application of the Glimm method a velocity to be applied at the boundary is chosen based on the rule*

$$u_{x_{n+1/2}}^{t+\Delta t/2} = U(x_n + \xi \Delta x, t + \Delta t/2)$$

where $-\frac{1}{2} \leq \xi \leq \frac{1}{2}$ and ξ is chosen randomly. Sod indicates that the best success is obtained if ξ is chosen only once per time step. As one can see the results for the one-dimensional calculation are quite adequate and are better than most finite-difference approaches.

---

*U in this equation is the Riemann solution.

References

Boris, J. P., and Book, D. L. *J. Comput. Phys.* **20,** 397–431 (1976).

Bristeau, M., Glowinski, R., Periaux, J., Perrier, P., Pironneau, O., and Poirier, G. IRIA Laboria Rept. No. 294 (1978).

Bristeau, M., Glowinski, R., Periaux, J., Perrier, P., and Pironneau, O. *Comput. Meth. Appl. Mech. Eng.* **17/18,** 619–657 (1979).

Chorin, A. J. *J. Comput. Phys.* **22,** 517–533 (1976).

Collella, P. An Analysis of the Effect of Operator Splitting and of the Sampling Procedure on the Accuracy of Glimm's Method, PhD Thesis, University of California, Berkeley (1979).

Deiwert, G. S. *AIAA J.* **13,** 1354–1359 (1975).

Fox, L. and Parker, I. B. *Chebyshev Polynomials in Numerical Analysis,* Oxford Press, London (1968).

Glimm, J. *Commun. Pure Appl. Math.* **18,** 697–715 (1965).

Godunov, S. K. *Mat. Sb.* **47,** 271–306 (1959).

Gottlieb, D., and Orszag, S. A. *Numerical Analysis of Spectral Methods,* SIAM, Philadelphia (1977).

Haidvogel, D. B., Robinson, A. R., and Schulman, E. E. *J. Comput. Phys.* **34,** 1–53 (1980).

Hirsh, R. S., Taylor, T. D., Nadworny, M. and Kerr, J. *Proceedings 8th International Conference on Numerical Methods in Fluid Dynamics,* Springer-Verlag, New York (to appear).

Holt, M. *Numerical Methods in Fluid Dynamics,* Springer-Verlag, New York (1977).

Lanczos, C. *Applied Analysis,* Prentice-Hall, Englewood Cliffs, NJ (1956).

McCormack, R. W., and Paullay, A. J. AIAA Paper 72-154, January (1972)

Morchoisne, Y. AIAA Paper No. 81-0109, Aerospace Sciences Mtg., St. Louis (1981).

Myers, R., Taylor, T., and Murdock, J. *J. Comput. Phys.* **43,** 180–188 (1981).

Orszag, S. A. *J. Comput. Phys.* **37,** 70–92 (1980).

Orszag, S. A., and Gottlieb, D. In: *Approximation Methods for Navier–Stokes Problems,* Lecture Notes in Mathematics, Vol. 771, pp. 381–398 Springer-Verlag, New York (1980).

Periaux, J. *Résolution de Quelques Problèmes Non linéaires en Aérodynamique par des Méthodes d'Éléments Finis et de Moindres Carrés Fonctionnels,* Thèse de 3ème cycle, Université Pierre et Marie Curie, Paris (1979).

Rivlin, T. J. *The Chebyshev Polynomials,* Wiley-Interscience, New York (1974).

Sod, G. *J. Comput. Phys.* **27,** 1–31 (1978).

Rizzi, A. W., and Inouye, M. *AIAA J.* **11** (11), 1478–1485 (1973)

Sod, G. A. SAE Tech. Paper No. 800288, Feb. (1980a).

Sod, G. A. Computational Fluid Dynamics with Stochastic Techniques. Princeton University Report MAE No. 1479. Also presented at Von Karman Institute of Fluid Dynamics Lecture, March 24 (1980b).

Taylor, T. Heat, Mass, and Momentum Transfer at Low Reynolds Numbers, Ph.D. Thesis, University of California, Berkeley (1962).

Taylor, T., Ndefo, E., and Masson, B. S. *J. Comput. Phys.* **9,** 99–119 (1972).

Taylor, T., and Murdock, J. In: *Approximation Methods for Navier–Stokes Problems,* Lecture Notes in Mathematics, Vol. 771, pp. 519–539 Springer-Verlag, New York (1980).

Taylor, T., Myers, R. and Albert, J. *Comput. Fluids,* **9,** 469–473 (1981).

Taylor, T., Hirsh, R. S., and Nadworny, M. In *Proceedings of the 4th GAMM Conference on Numerical Methods in Fluid Mechanics,* Paris (1981).

Zalesak, S. T. *J. Comput. Phys.* **31,** 335–362 (1979).

Zienkiewicz, O. C. *The Finite Element Method in Engineering Science,* McGraw-Hill, London (1971).

CHAPTER 4

Relationship Between Numerical Approaches

The previous discussions have presented a variety of numerical methods that can be employed to generate numerical solutions to flow problems. The apparent difference in these methods is not always easy to point out explicitly. However, we can demonstrate some explicit relationships between finite-difference, integral, and spectral methods for simple cases. For the more complicated or higher-order approaches, only qualitative comparisons are possible.

4.1 Finite-Difference Equivalent of Finite-Element Scheme

In Chapter 3, we compared techniques in the same class. In this section, we concentrate on the comparison of techniques of different classes. Consider first a comparison of finite-element and finite-difference techniques. This is best accomplished by deriving a finite-difference method from a finite-element approach. Consider for this case the simple equation

$$\frac{\partial f}{\partial t} + \frac{\partial f}{\partial x} = \frac{\partial^2 f}{\partial x^2} \tag{4.1.1}$$

and assume

$$f = \sum_{j=1}^{N} N_j(x) f_j(t) \tag{4.1.2}$$

where $N_j(x)$ are the one-dimensional shape functions and $f_j(t)$ are the nodal values of f at points j. Following the finite-element approach previously outlined, one obtains the equations

$$\sum_{j=1}^{N} \left(p_{i,j} \frac{df_j}{dt} + q_{i,j} f_j \right) = \sum_{j=1}^{N} r_{i,j} f_j \tag{4.1.3}$$

where

$$p_{i,j} = \int_0^1 N_i N_j \, dx$$

$$q_{i,j} = \int_0^1 \frac{dN_j}{dx} N_i \, dx \tag{4.1.4}$$

$$r_{i,j} = -\int_0^1 \frac{dN_i}{dx} \frac{dN_j}{dx}\, dx$$

For a linear shape function such that for $x_{j-1} \le x \le x_j$

$$N_{j-1} = \frac{x_j - x}{\Delta x}, \qquad N_j = \frac{x - x_{j-1}}{\Delta x}$$

and for $x_j \le x \le x_{j+1}$

$$N_j = \frac{x_{j+1} - x}{\Delta x}, \qquad N_{j+1} = \frac{x - x_j}{\Delta x}$$

one can show that the resulting values for the coefficients of Eq. (4.1.3) are

$$p_{i,j} = \begin{cases} 0, & j \ne i,\ i+1,\ i-1 \\ \frac{1}{6}\Delta x, & j = i-1 \\ \frac{2}{3}\Delta x, & j = i \\ \frac{1}{6}\Delta x, & j = i+1 \end{cases} \tag{4.1.5}$$

$$q_{i,j} = \begin{cases} 0, & j \ne i,\ i+1,\ i-1 \\ -\frac{1}{2}, & j = i-1 \\ 0, & j = i \\ \frac{1}{2}, & j = i+1 \end{cases} \tag{4.1.6}$$

$$r_{i,j} = \begin{cases} 0, & j \ne i,\ i+1,\ i-1 \\ 1/\Delta x, & j = i-1 \\ -2/\Delta x, & j = i \\ 1/\Delta x, & j = i+1 \end{cases} \tag{4.1.7}$$

When these values are inserted into Eq. (4.1.3), the result is

$$\frac{d}{dt}\left[\frac{f_{i+1}}{6} + \frac{2f_i}{3} + \frac{f_{i-1}}{6} \right] + \frac{f_{i+1} - f_{i-1}}{2\,\Delta x} = \frac{f_{i+1} - 2f_i + f_{i-1}}{\Delta x^2} \tag{4.1.8}$$

This equation is the same as the standard second-order finite-difference method presented earlier. The time term differs, however, because it has a three-point spatial average of the variable f. As a result, the difference equation is equivalent to an implicit finite-difference method.

4.2   Finite-Difference Equivalent of Spectral Scheme

A similar equation is found if one applies a Chebyshev expansion to derive a difference equation. In order to derive the equivalent expression assume that

$$f = \sum_{j=0}^{2} a_j(t) T_j(x) \tag{4.2.1}$$

where $T_j(x)$ is a Chebyshev polynomial of degree j and $a_j(t)$ is an unknown function.

In this expansion, we normalize the interval $-1 < x < 1$ and let $f = f_{j-1}$ at $x = -1$, $f = f_j$ at $x = 0$, and $f = f_{j+1}$ at $x = 1$. Noting that $T_0 = 1$, $T_1 = x$, and $T_2 = 2x^2 - 1$, we substitute (4.2.1) into Eq. (4.1.1) to obtain

$$(a_0' T_0 + a_1' T_1 + a_2' T_2) + a_1 T_0 + 4 a_2 T_1 = 4 a_2 T_0 \tag{4.2.2}$$

If we apply the integral relationship

$$\int_{-1}^{+1} \frac{T_i T_j}{\sqrt{1 - x^2}} \, dx = \begin{cases} 0, & i \neq j \\ \frac{1}{2}\pi, & i = j \neq 0 \\ \pi, & i = j = 0 \end{cases}$$

we obtain

$$\begin{aligned} a_0' + a_1 &= 4 a_2 \\ a_1' + 4 a_2 &= 0 \\ a_2' &= 0 \end{aligned} \tag{4.2.3}$$

The last two equations are clearly invalid since they require that $a_2 = $ const and $a_1 \sim $ const $\times\ t$. This occurs because of truncation of the expansion. As a result, they should be discarded. This also becomes apparent if one attempts to derive a difference equation from these three relationships. In order to obtain difference equations, it is necessary to relate the a_0, a_1, and a_2 terms to the values of f at $x = -1, 0$, and 1. This is easily accomplished by evaluating Eq. (4.2.1) at these points and solving the resulting equations to obtain

$$a_0 = \frac{f_{j+1} + 2f_j + f_{j-1}}{4} \tag{4.2.4}$$

$$a_1 = \frac{f_{j+1} - f_{j-1}}{2} = \frac{\Delta^1 f}{2} \tag{4.2.5}$$

$$a_2 = \frac{f_{j+1} - 2f_j + f_{j-1}}{4} = \frac{\Delta^2 f}{4} \tag{4.2.6}$$

Substitution of these into the set of equations (4.2.3) yields

$$\frac{d}{dt}\left(\frac{f_{j+1} + 2f_j + f_{j-1}}{4}\right) + \frac{\Delta^1 f}{2} = \Delta^2 f \tag{4.2.7}$$

$$\frac{d}{dt}\left(\frac{\Delta^1 f}{2}\right) = -\Delta^2 f \tag{4.2.8}$$

$$\frac{d}{dt}(\Delta^2 f) = 0 \tag{4.2.9}$$

Equation (4.2.7) represents the original equation and is equivalent to the finite-difference equation

$$\frac{d}{dt}\left(\frac{f_{j+1} + 2f_j + f_{j-1}}{4}\right) + \frac{f_{j+1} - f_{j-1}}{2\,\Delta x} = \frac{f_{j+1} - 2f_j + f_{j-1}}{\Delta x^2} \tag{4.2.10}$$

This equation (4.2.10) can be interpreted in two ways:

(i) It gives the solution f_j when f_{j-1} and f_{j+1} are boundary conditions. In that case the approximation is quite coarse ($\Delta x = 1$). But, one could derive more accurate finite-difference-type equations by using higher-order expansion polynomials. Note that each of such equations connects N nodal values of f in case on a N-term expansion.

(ii) It is considered as applied to an elementary interval (x_{j-1}, x_{j+1}) of length $2\Delta x$, $\Delta x \ll 1$, normalized in $(-1, 1)$. In that case, conditions on the coefficient a_j in Eq. (4.2.1) are needed to match the solution at the common bound of two elementary intervals. For example, continuity of the function f and of its derivative $\partial f / \partial x$ can be required. That will give supplementary equations connecting the nodal values of the function f. For higher-order expansions, similar matching conditions can be devised.

4.3 Finite-Difference Equivalent of Godunov Method

In addition to the finite-element and spectral equivalents of finite differences, it is also possible to indicate the equivalent finite-difference scheme for the Godunov method. This can be accomplished by examining the flow equations in one dimension. If we write the conservation equations in the form

$$\frac{\partial f}{\partial t} + \frac{\partial F}{\partial x} = 0 \tag{4.3.1}$$

where f is the conserved quantity and F is the flux through an x boundary, then it is possible to relate Godunov's approach to finite differences. In the Godunov scheme, one can observe on close examination [Eqs. (3.7.16)–(3.7.18)] that the value of F at a cell boundary between cells n and $n+1$ is approximated by the functional form

$$F^t_{n+1/2} = \frac{F_n + F_{n+1}}{2} - (h_{n+1} - h_n) \qquad \text{at } t \tag{4.3.2}$$

As a result, when Eq. (4.3.1) is written in difference form, one has

$$\frac{f_n(t + \Delta t) - f_n(t)}{\Delta t} + \frac{F_{n+1/2}^t - F_{n-1/2}^t}{\Delta x} = 0 \qquad (4.3.3)$$

or

$$\frac{f_n(t + \Delta t) - f_n(t)}{\Delta t} + \frac{F_{n+1} - F_{n-1}}{2\Delta x} - \frac{(h_{n+1} - 2h_n + h_{n-1})}{\Delta x} = 0 \qquad (4.3.4)$$

This equation is a finite-difference equivalent of the original equation with an added dissipation term. This dissipation is the stabilizing influence and, from tests of the method, appears to work reasonably well. In fact, it can be shown (Richtmyer and Morton, 1967) that Godunov's method is equivalent to an upwind finite-difference scheme for the equations written in characteristic form.

Reference

Richtmyer, R. D. and Morton, K. W., *Difference Methods for Initial Value Problems*, Interscience, New York (1967).

CHAPTER 5

Specialized Methods

In addition to the finite-difference, finite-element, and spectral techniques, there are computational techniques that do not fall directly into these catego- ries. These techniques can be useful and, therefore, we have included them in this chapter.

These techniques vary in nature from Green's function methods to the method of characteristics. In the discussion, we will outline some of the more practical approaches, but will not attempt to give all the details since each method could be a subject for an independent treatise. The first technique to be considered is an approach used to solve problems that can be described by a Laplace- or Poisson-type differential equation.

5.1 Potential Flow Solution Technique

In the solution of numerous incompressible flow problems it is possible to utilize the classical Green's function techniques in order to introduce an inte- gral formulation of some of the describing equations. Potential flow problems are particularly suited for this approach since it allows the reduction of a three-dimensional problem to a two-dimensional surface integral. This simplification greatly reduces the magnitude of a three-dimensional problem and results because of the introduction of classical analysis tools into a prob- lem. This is important to note since one might anticipate, upon initial exam- ination, that a finite-difference method would be the best approach for solving potential problems. The formulation of a finite-difference equivalent is simple but the difficulty in most practical applications arises due to geometrical re- quirements. These requirements arise either due to complicated geometry on which the boundary condition,*

$$\frac{\partial \phi}{\partial n} = 0$$

must be satisifed or due to a large spatial region in which the number of grid points exceeds practical computer capacity. As a consequence of these fea-

---

*Note we employ n to denote the normal in this section.

tures, the direct solution of the potential equation by finite differences is not the first choice of solution methods. The reason is that for many subsonic flows the solution to a potential problem for a complex geometry can be converted to the solution of an integral equation that can be solved with far less cost and effort. The procedure has been developed by Hess and Smith (1964, 1966), Roberts and Rundle (1972), and Rubbert et al. (1967, 1972).

The principle behind the solution to the potential equation is the well-known approach of summing sources, sinks, and dipoles to form an integral equation that describes the potential. This can be accomplished in two ways. The first is to employ Green's theorem which states that the potential at a point P in space exterior to the surface s is given by the expression

$$\phi(P) = -\frac{1}{4\pi} \iint\limits_{s} \frac{1}{r(P,\,q)} \frac{\partial\phi}{\partial n}(q)\, ds \,+\, \frac{1}{4\pi} \iint\limits_{s} \phi(q)\, \frac{\partial}{\partial n}\left[\frac{1}{r(P,\,q)}\right]_{q} ds$$

$$(5.1.1)$$

where n denotes the normal to the surface s at point q. Figure 5.1.1 gives a clear picture of each variable's meaning.

On the surface Eq. (5.1.1) becomes an integral equation for $\phi(p)$ since $\partial\phi/\partial n_q$ is prescribed. Hess (1972) points out in his studies, however, that this equation has a nonunique solution and is not the best formulation. The formulation yielding the least difficulty is the surface potential given by Kellogg (1953)

$$\phi(p) = \iint\limits_{s} \frac{1}{r(p,\,q)}\, \sigma(q)\, ds \qquad\qquad (5.1.2)$$

where $\sigma(q)$ is the unknown distribution of source strength. Applying the

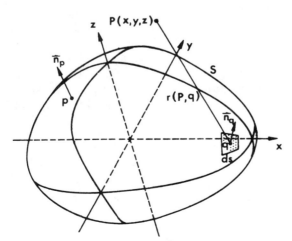

Fig. 5.1.1 Nomenclature for potential formulation.

boundary condition $\partial\phi/\partial n = 0$ on the surfaces and noting a jump in the derivative at the surface (see Kellogg,1953), one obtains

$$2\pi\sigma(p) - \iint\limits_{s} \frac{\partial}{\partial n_p} \left[\frac{1}{r(p,\, q)}\right]\sigma(p)\, ds \;=\; -\mathbf{n}_p\cdot\mathbf{V}_\infty \tag{5.1.3}$$

This is a Fredholm integral equation of the second kind which has a well-behaved solution. Once the value of $\sigma(p)$ has been obtained from this equation, Eq. (5.1.2) can be employed to compute $\phi(p)$ and the velocity field given by $\mathbf{V} = \text{grad } \phi$. Once the velocity field is known, the pressure field can be calculated by the equation

$$p + \tfrac{1}{2}\rho_\infty V^2 = p_\infty + \tfrac{1}{2}\rho_\infty V_\infty^2 \tag{5.1.4}$$

It is important to note that this approach is equally useful for external or interior flows, provided the sign is reversed on the normal for interior flows.

The primary advantages of this approach are that it reduces the dimensionality of the problem by one and can be made as accurate as necessary. In addition, its grid points occur only on the boundaries which are normally finite even though the flow field is infinite. Also, it computes the velocity field as well as the potential with equal accuracy. The disadvantages of the method are that it is inaccurate in regions of extreme body curvature and a high density of points is needed. The degree of concentration of points depends on the accuracy with which the body surface is approximated in the solution of the integral equation, Eq. (5.1.3). This approximation results when the integral equation is reduced to the form

$$2\pi\sigma(p) - \sum_{i}^{N} \sigma_i(p) \iint\limits_{A_i} \frac{\partial}{\partial n_p} \left[\frac{1}{r(p,\, q)}\right] ds \;=\; -\mathbf{n}_p\cdot\mathbf{V}_\infty \tag{5.1.5}$$

where the integral

$$\iint\limits_{A_i} \frac{\partial}{\partial n_p} \left[\frac{1}{r(p,\, q)}\right] ds \tag{5.1.6}$$

must be evaluated for the element A_i. At times the evaluation can be troublesome since it is dependent on the geometry of the surface, and the computation of the derivative with respect to the normal can offer difficulty. The evaluation of the A_i terms is accomplished most easily for the case where the surface is approximated by planar elements. This yields what is frequently called a first-order solution. Figure 5.1.2 displays such an approximation of a body. For surfaces with sharp corners and small radii of curvature, this approximation clearly becomes inaccurate. Improvements beyond the linear area approximations have been introduced by Hess (1973), Roberts and Rundle (1972) and Craggs et al. (1973) who have addressed the nature of the potential solutions near the edges of bodies indicating the importance of such effects on

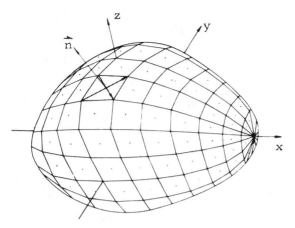

Fig. 5.1.2 Approximation of body by elements.

the solution. The procedure for solving the potential equation consists of the following steps:

1. Approximate the surface of interest by N elements.
2. Choose one point on each element to be the point p at which $\sigma(p)$ is to be evaluated.
3. Evaluate the surface integrals*

$$A_{i,j} = \iint_{A_i} \frac{\partial}{\partial n_p} \left[\frac{1}{r(p,\, q)} \right] ds \tag{5.1.7}$$

for each element. Note that for $p = q$ this integral reduces to 2π so that the constant term in Eq. (5.1.5) is canceled. Much work in evaluating these integrals has been completed by Hess for two- and three-dimensional flows. He points out that those integrals can be evaluated by examining the potential and velocity induced by a surface element of unit source density at a point in space. The procedure to arrive at the result is, unfortunately, a bit lengthy and somewhat tedious. As a consequence, we refer the reader to the work of Hess and Smith (1966) for details of the evaluation.
4. Form the equations

$$\sum_{i=1}^{N} A_{i,j}\sigma_j = -n_j \cdot \mathbf{V}_\infty \tag{5.1.8}$$

5. Solve the algebraic equations of step 4 by a direct Gauss elimination method or by the Gauss–Seidel iteration approach.

In principle, this approach is straightforward except for the details of evaluating the $A_{i,j}$ terms. In general, applications of this method necessitate the

*Here j denotes point p at which the normal is evaluated.

development of computer subroutines for the $A_{i,j}$ computations. Once this is accomplished, the method becomes a large matrix manipulation which can be troublesome.

In addition to the inviscid potential flow, it is also possible to utilize the Green's function approach in the solution of formulation and solution of other problems. This is particularly true of stream-function vorticity problems.

5.2 Green's Functions and Stream-Function Vorticity Formulation

The two-dimensional incompressible flow equations of Chapter 1 can be written in the form

$$\frac{\partial \omega}{\partial t} + u \frac{\partial \omega}{\partial x} + v \frac{\partial \omega}{\partial y} = \frac{1}{\text{Re}} \nabla^2 \omega \qquad (5.2.1)$$

$$\nabla^2 \Psi + \omega = 0 \qquad (5.2.2)$$

$$u = + \frac{\partial \Psi}{\partial y}, \qquad v = - \frac{\partial \Psi}{\partial x} \qquad (5.2.3)$$

where ω is the vorticity, Ψ the stream function, u the velocity in the x direction, and v the velocity in the y direction. The equation for the stream function can be converted into an integral equation by employing results of the potential theory. First the solution of the stream-function equation is given by the integral* (see Batchelor, 1967)

$$\Psi = \frac{-1}{2\pi} \iint \omega' \ln r \, dV' \qquad (5.2.4)$$

where

$$r = \sqrt{(x - x')^2 + (y - y')^2}$$

$$dV' = dx' \, dy'$$

$$\omega' = \omega(x', y')$$

From this expression one obtains the velocities

$$u_p = - \frac{1}{2\pi} \iint \frac{(y - y')\omega'}{r^2} \, dV' \qquad (5.2.5)$$

$$v_p = + \frac{1}{2\pi} \iint \frac{(x - x')\omega'}{r^2} \, dV' \qquad (5.2.6)$$

---

*In this expression $\ln r$ is the Green's function in two dimensions.

These velocities are for solution of the stream-function equation with vorticity. However, it is possible to add solutions u_h and v_h which satisfy $\nabla^2 \Psi = 0$ with the result that

$$u = u_p + u_h \tag{5.2.7}$$

$$v = v_p + v_h \tag{5.2.8}$$

These expressions, when combined with the vorticity transport equation, yield a formulation in terms of u, v, and ω. The boundary conditions on u and v are satisfied by selecting the appropriate u_h and v_h solutions. There are varying views regarding the necessity for the homogeneous solutions to the stream-function equation. Thompson et al. (1973, 1974) argue, for example, that if the vorticity satisfies exact viscous boundary conditions then the homogeneous solutions need only satisfy the free-stream conditions; therefore, $u_h = U_\infty$ and $v_h = V_\infty$. The key to this argument is the ability to determine the vorticity on the boundaries so that the exact boundary conditions and flow equations are satisfied. This turns out to be the principal problem with this approach. Generally, the boundary vorticity is obtained by a Taylor series expansion of the definition (see Section 6.5)

$$\boldsymbol{\omega} = \nabla \times \mathbf{V} \tag{5.2.9}$$

Wu (1975, 1976) has studied this question and suggests that both the first- and second-order Taylor series boundary approximations can violate an overall integral law for vorticity. He further suggests applying Eqs. (5.2.7) and (5.2.8) at the boundary in order to obtain an integral equation for the surface vorticity. The procedure for solving a problem would be as follows:

1. Integrate Eq. (5.2.1) forward in time to obtain new ω values at each interior field point.
2. Form the relationships

$$u = U_\infty - \frac{1}{2\pi} \sum_i \sum_j \frac{(y - y_j)\omega_{i,j}\, \Delta x\, \Delta y}{(x - x_i)^2 + (y - y_j)^2} + u_s \tag{5.2.10}$$

$$v = V_\infty + \frac{1}{2\pi} \sum_i \sum_j \frac{(x - x_i)\omega_{i,j}\, \Delta x\, \Delta y}{(x - x_i)^2 + (y - y_j)^2} + v_s \tag{5.2.11}$$

requiring that the singularity (i, j) be avoided. In these expressions u_s and v_s are surface vorticity potentials with undetermined strengths and the $\omega_{i,j}$ are the values from step 1.
3. Evaluate the relationships of step 2 at the boundary and determine appropriate vorticity potentials to satisfy the boundary conditions.

The outlined procedure is very similar to the potential solution method of Hess and Smith (1966) discussed previously. The procedure for finding u_s and v_s can be equally tedious. As a result, the user may find it necessary to invest a fair amount of effort in computing a fully accurate solution. For some problems the user may wish to pursue the simplified approach of Thompson et al. (1973, 1974). For this case assume u_s and v_s are zero and compute the surface vorticity by the formula

$$\omega_{i-1/2,j} = \mathbf{n}_{i-1/2,j} \times \frac{1}{3\Delta x}[-\mathbf{V}_{i+1,j} + 9\,\mathbf{V}_{i,j}] + 0(\Delta x^2) \qquad (5.2.12)$$

The difference in results obtained by using Green's formula or approximations analogous to (5.2.12) is discussed in Section 6.6.1.

The methods discussed thus far are for unsteady cases which, of course, can be integrated to a steady state. Wu and Wahbah (1976), Wu and Rizk (1979) have also presented an alternate integral formulation for both velocity and vorticity for application to steady and unsteady flows. However, we will not attempt to review the formulation here.

Independent of the approach employed one is also faced with the problem of starting the calculation. If one is not interested in the exact transient of the flow and seeks only the steady-state result, then the most natural course is to assume potential flow.

The advantages of the Green's function formulation are primarily in the integral formulation which tends to lend stability and frequently a decrease in computation time when the region of vorticity is limited in size. Studies to date have not revealed any large advantage over other approaches in two dimensions, but they do indicate a time advantage over finite differences in three dimensions. However, the method still needs further investigation to draw conclusions regarding its principal advantages. For the reader wishing to pursue this approach, additional detail can be found in the work of Thompson et al. (1973, 1974) as well as Wu and Thompson (1973). In addition to the Green's function formulation, there is an additional approach to solving the stream-function vorticity equations. This is the discrete vortex summation method which will now be outlined.

5.3 The Discrete Vortex Method

The discrete vortex element method for solving incompressible two-dimensional flows has been discussed by a variety of authors in the literature—Rosenhead (1931), Birkhoff and Fisher (1959), Chorin (1973), Milinazzo and Saffman (1977), Moore (1976), Hald (1979), Saffman and Baker (1979), and Leonard (1980). The basic concept of the approach is to represent the flow by discrete vortices that move with the fluid. For an inviscid flow, this concept is suggested by the vorticity equation

$$\frac{D\omega}{Dt} = 0 \tag{5.3.1}$$

which physically implies that if one travels with an element of fluid the vorticity remains constant. As a result, one can conceive of following vortex elements through a flow. The trajectories of these elements are given by the definitions of fluid velocity, i.e.,

$$\frac{dx}{dt} = u \tag{5.3.2}$$

$$\frac{dy}{dt} = v \tag{5.3.3}$$

If one knows the velocities, these equations can be integrated. The remaining step is to find an expression for the velocities. If it is assumed that the field is composed of a set of point vortices of strength k_i at the points (x_i, y_i), then the stream function of the field by superposition would take the form

$$\Psi = -\frac{1}{4\pi} \sum_i k_i \ln \left[(x - x_i)^2 + (y - y_i)^2 \right] \tag{5.3.4}$$

since a single point vortex has the solution

$$\Psi_p = -\frac{k_i}{2\pi} \ln r = -\frac{k_i}{4\pi} \ln \left[(x - x_i)^2 + (y - y_i)^2 \right] \tag{5.3.5}$$

The velocity field then is given by

$$\frac{dx_j}{dt} = u = +\frac{\partial \Psi}{\partial y} = -\frac{1}{2\pi} \sum_{\substack{i \\ i \neq j}} \frac{k_i(y_j - y_i)}{r_{i,j}^2} \tag{5.3.6}$$

$$\frac{dy_j}{dt} = v = -\frac{\partial \Psi}{\partial x} = +\frac{1}{2\pi} \sum_{\substack{i \\ i \neq j}} \frac{k_i(x_j - x_i)}{r_{i,j}^2} \tag{5.3.7}$$

where

$$r_{i,j}^2 = (x_i - x_j)^2 + (y_i - y_j)^2$$

These realtionships represent the trajectory equations for each vortex. As a consequence, their integration will yield the position of each vortex and hence the velocity field at each position and time. However, there are two problems with this approach. First, the approach does not satisfy any boundary conditions as written, and second, the integration of the equations tends to blow up due to the singular nature of the right-hand side of the equations for i approaching j. Chorin and Bernard (1973) have indicated that the latter problem can be overcome by utilizing "blobs" in place of point vortices, i.e., use

$$\Psi_p = -\frac{k_i}{2\pi} \ln |r| \qquad \text{for } |r| \geq \delta$$

$$\qquad\qquad\qquad\qquad\qquad\qquad\qquad\qquad (5.3.8)$$

$$= -\frac{k_i}{2\pi} \frac{r}{\delta} \qquad \text{for } |r| < \delta$$

and choose δ so that vortex displacement during a time step is of the same order as the separation of vortices. The selection, however, is up to the analyst.

The discussion thus far has considered only the inviscid case. The point vortex method has also been applied to viscous flow problems (Chorin (1973)). The procedure is to employ the point vortex approach with a random walk added into the trajectory. It is speculated that this approach should represent solutions to the viscous flow problems. At this time, the approach remains controversial as the work of Milinazzo and Saffman (1977) demonstrate.

Application of the general concept of the vortex method to free field flows without solid boundaries is not difficult in principle, but the optimum procedure for including boundaries is not totally clear. For inviscid flows, one can employ images to eliminate flow through the boundary. In the viscous case, however, vorticity is generated at the boundary, and the source strength and location of vortices introduced to satisfy the Navier–Stokes equations is not well defined. Chorin (1978) has discussed this issue and proposed a method that seems to work for boundary layers but encounters difficulty with separated flows.

Based on this discussion, it is apparent that the vortex method is rather novel but remains somewhat controversial in its value for application to viscous flows. Perhaps future research can resolve the question. The extension of the method to three dimensions can be difficult. Rehbach (1977), however, has developed a technique to compute three-dimensional flows which is similar in nature, and the reader is referred to Rehbach's work or Guiraud-Vallée et al. (1978) for the details.

An alternate to the discrete vortex approach has been proposed by Christiansen (1973). The technique combines the vortex tracking concept with a mesh for the stream function. This approach will now be outlined.

5.4 The Cloud-in-Cell Method

The application of the discrete vortex method is clearly troubled by the source singularities and the fact that a large number of vortices are required for reasonable accuracy. An approach that seems to take the best features of the vortex tracking technique and the stream-function vorticity finite-difference methods has been proposed by Roberts and Christiansen (1972). This approach integrates the trajectory equations for each vortex, i.e.,

$$\frac{dx_n}{dt} = u_n \qquad\qquad\qquad\qquad\qquad\qquad (5.4.1)$$

$$\frac{dy_n}{dt} = v_n \tag{5.4.2}$$

The velocities are obtained from the stream function but not by the series summation. A numerical solution of the stream-function equation is constructed using the vorticity averaged over cells. For the solution, a mesh overlays the field of vortex movement. The vorticity at each mesh point of a grid surrounding a vortex (Fig. 5.4.1) is obtained by allocating the vorticity of a point vortex located at the point $x_n = x_i + \overline{\delta x}$, $y_n = y_j + \overline{\delta y}$ according to the normalized weights ($\delta x = \overline{\delta x}/\Delta x$, $\delta y = \overline{\delta y}/\Delta y$)

$$w_{i,j} = (1 - \delta x)(1 - \delta y) = A_1, \qquad w_{i+1,j} = \delta x(1 - \delta y) = A_2$$
$$w_{i,j+1} = (1 - \delta x)\delta y = A_3, \qquad w_{i+1,j+1} = \delta x\, \delta y = A_4 \tag{5.4.3}$$

When the vorticity of each vortex has been attributed to all the grid points, one then has a mesh of points (i,j) with a given vorticity. The stream-function equation,

$$\nabla^2 \Psi + \omega = 0 \tag{5.4.4}$$

is then solved by any desired technique—i.e., the finite-difference, finite-element, or spectral method. The appropriate boundary condition for the velocity must be known in order to completely set the stream function at each boundary of the overall mesh. Once the stream function and hence the velocity is known at each point of the mesh then the velocities for translating the vortex elements are computed by the rule

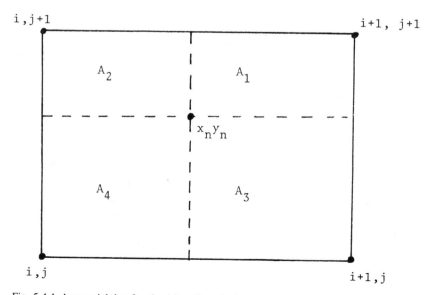

Fig. 5.4.1 Area weighting for cloud-in-cell method.

$$u_n = u(x_n, y_n) = A_1 u_{i,j} + A_2 u_{i+1,j} + A_3 u_{i,j+1} + A_4 u_{i+1,j+1} \qquad (5.4.5)$$

Next, the trajectory equations are integrated for each vortex, i.e.,

$$\frac{dx_n}{dt} = u_n, \qquad \frac{dy_n}{dt} = v_n \qquad (5.4.6)$$

This procedure seems to work well and has been successfully applied by Roberts and Christiansen (1972) and Christiansen (1973), as well as Milinazzo and Saffman (1977). The applications indicate that the cloud-in-cell approach is about 20 times faster for $N = 1000$ (vortex count) than the discrete vortex approach.

5.5 The Method of Characteristics

In the past years the solution of inviscid supersonic or inviscid transient compressible flow problems were frequently developed by a technique termed the method of characteristics. This approach took advantage of the fact that these flows could be characterized by families of intersecting lines along which disturbances propagate. These lines became known as the characteristics of the flow and made up the computational grid. The trajectory of these lines were unknown *a priori* and had to be determined by calculation. As a result, the grid shape was unknown. This approach had a fundamental problem because in regions of steep flow gradients the characteristics tended to collapse on top of each other, and hence the grid for the calculations would degenerate. As a consequence, the method was frequently troublesome to apply. However, it did have the asset of being fast when it worked. In recent years, the direct characteristics method is not often utilized to solve problems because of advances in new methods. The principles, however, are still important and are frequently employed to relate interior flow mesh solutions from all types of methods to boundary values. They are also prevalent in the understanding of both the Godunov and Glimm methods discussed earlier. As a result of this importance, the basic principle of the approach for both the steady and transient cases is discussed.

Consider the transient case for a one-dimensional isentropic compressible flow. For this case the equations take the form

$$\frac{\partial \rho}{\partial t} + \frac{\partial \rho u}{\partial x} = 0 \qquad \text{(mass)} \qquad (5.5.1)$$

$$\frac{\partial \rho u}{\partial t} + \frac{\partial}{\partial x}(p + \rho u^2) = 0 \qquad \text{(momentum)} \qquad (5.5.2)$$

$$\frac{p}{\rho^\gamma} = \text{const} \qquad (5.5.3)$$

For use in these equations the sound speed c is defined by the relationships

$$c^2 = \frac{dp}{d\rho} = \frac{\gamma p}{\rho} \tag{5.5.4}$$

Introducing these relationships, the original equations can be rearranged to obtain

$$\frac{\partial \rho}{\partial t} + u \frac{\partial \rho}{\partial x} + \rho \frac{\partial u}{\partial x} = 0 \tag{5.5.5}$$

$$\frac{\partial u}{\partial t} + u \frac{\partial u}{\partial x} + \frac{c^2}{\rho} \frac{\partial \rho}{\partial x} = 0 \tag{5.5.6}$$

These equations can be employed to derive the characteristics, using the concept of determining where derivatives of the functions ρ and u may be indeterminant in the flow. This is accomplished by forming a set of equations with (5.5.5) and (5.5.6) and the relationships

$$du = \frac{\partial u}{\partial x} dx + \frac{\partial u}{\partial t} dt \tag{5.5.7}$$

$$d\rho = \frac{\partial \rho}{\partial x} dx + \frac{\partial \rho}{\partial t} dt \tag{5.5.8}$$

One then attempts to solve this set for the derivatives of ρ and u. Proceeding with this task and finding the condition when these derivatives are indeterminant, one will obtain the relationships

$$\frac{dx}{dt} = u \pm c \tag{5.5.9}$$

and

$$c \frac{d\rho}{\rho} \pm du = 0 \tag{5.5.10}$$

Equation (5.5.9) is known as the characteristic equation. When integrated it yields a characteristic network since one set of lines $x = \int_t^{t+\Delta t} (u + c)\, dt$ intersects the other set $x = \int_t^{t+\Delta t} (u - c)\, dt$. Figure 5.5.1 shows a simple example. Along the characteristic Eq. (5.5.10) holds; the positive sign holding along $u + c$ and the negative sign along $u - c$. From these equations and the known conditions at points 1 and 2, it is possible to solve for the position of point 3 and the values of ρ and u at that point. Using this procedure from a line $t = 0$ one can advance forward stepwise in time. The mesh may become irregular, however, depending on the nature of the flow and the position of a boundary. It is important to note that the equations can be applied at all points in the flow and provide a means for connecting interior points in a flow to

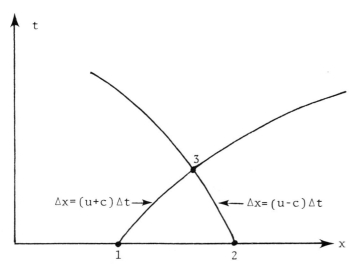

Fig. 5.5.1 Characteristics grid.

boundary points. For steady supersonic flows another set of characteristic equations can be derived by following a similar logic. One utilizes the equations

$$u\frac{\partial u}{\partial x} + v\frac{\partial u}{\partial y} + \frac{1}{\rho}\frac{\partial p}{\partial x} = 0 \qquad \text{(momentum)} \tag{5.5.11}$$

$$u\frac{\partial v}{\partial x} + v\frac{\partial v}{\partial y} + \frac{1}{\rho}\frac{\partial p}{\partial y} = 0 \qquad \text{(momentum)} \tag{5.5.12}$$

$$\frac{\partial \rho u}{\partial x} + \frac{\partial \rho v}{\partial y} = 0 \qquad \text{(mass)} \tag{5.5.13}$$

$$\frac{\partial v}{\partial x} - \frac{\partial u}{\partial y} = 0 \qquad \text{(irrotationality)} \tag{5.5.14}$$

$$\frac{p}{\rho^\gamma} = \text{const}$$

One can show that the equations for the characteristics are

$$\frac{dy}{dx} = \frac{uv \pm c\sqrt{u^2 + v^2 - c^2}}{u^2 - c^2}$$

and the conditions along the characteristics are

$$(u^2 - c^2)\frac{dy}{dx}\,du + (v^2 - c^2)\,dv = 0$$

These equations can easily be transformed to the classical mach line–streamline formulation given in texts such as by Liepmann and Roshko (1957).

The discussion thus far considers only one-dimensional unsteady and two-dimensional steady flows. However, characteristics can be applied to multi-dimensions. There is more than one procedure (see Ferri, 1954; Holt, 1956; Butler, 1960; Holt, 1963; Cline and Hoffman, 1972; Rakich, 1967; Holt (1974)) for this but, for practical purposes, the splitting methods can be applied to reduce the problem to multiple one-dimensional unsteady or multiple two-dimensional marching problems. The details of this procedure are not included here since, in general, the other numerical methods discussed in this text are less difficult and more efficient to apply. This is confirmed in the studies of Rakich and Kutler (1972) who compared the characteristics approach with a finite-difference shock-capturing calculation for a three-dimensional supersonic flow. They found on a point-by-point basis that the shock-capturing calculation was about four times faster.

The principal future use that can be seen for the method of characteristics is basically for extrapolation from computed points in a compressible flow to an unknown boundary point. The principle is simply to extend a characteristic from a point until it intersects with a boundary and then integrate the compatibility relationship along the characteristic. When the boundary condition is combined with these results the flow at the boundary point can usually be determined.

References

Batchelor, G. K. *An Introduction to Fluid Dynamics,* Cambridge University Press, Cambridge (1967).

Birkhoff, G., and Fisher, J. *Circ. Mat. Palermo* **8,** 77–90 (1959).

Butler, D. C. *Proc. R. Soc. London. A* 1281, **255,** 232–252 (1960).

Christiansen, J. P. *J. Comput. Phys.* **13,** 363–379 (1973).

Chorin, A. J. *J. Fluid Mech.* **57,** 785–796 (1973).

Chorin, A. J., and Bernard, P. S. *J. Comput. Phys.* **13,** 423–428 (1973).

Chorin, A. J. *J. Comput. Phys.* **22,** 428–442 (1978).

Cline, M. C., and Hoffman, J. D. AIAA Paper No. 72–190 January (1972).

Craggs, J. W., Mangler, K. W., and Zamir, M. *Aeronaut. Q.* **24,** February (1973).

Ferri, A. *General Theory of High Speed Aerodynamics,* Vol. 6, pp. 583–669, Princeton University Press, Princeton, NJ (1954).

Guiraud-Vallée, D. et al. ONERA Report No. T.P. 1978-1, Chatillon, France (1978). Also *Proceedings of the Conference on Numerical Methods in Fluid Dynamics,* University of Reading, U.K., January (1978).

Hald, O. H. *SIAM J. Numer. Anal.* **16,** 726–755 (1979).

Hess, J. L., and Smith, A. M. O. Douglas Aircraft Co. Rept. No. ES40622, March (1962). Abbreviated version: *J. Ship Res.* **8,** 22–44 (1964).

Hess, J. L., and Smith, A. M. O. *Progress in Aeronautical Science,* Vol. 8, pp. 1–138, Pergamon Press, New York (1966).

Hess, J. L. McDonnell-Douglas Corp. Rept. MDCJ5679-01, Long Beach, October (1972).

Hess, J. L. McDonnell-Douglas Corp. Engr. Paper 6060: Long Beach, California (1973).

Holt, M. *J. Fluid Mech.* **1,** 409–423 (1956).

Holt, M. The Method of Near Characteristics for Unsteady Flow Problems, University of California, Berkeley, Rept. No. AS63-2, June (1963).

Holt, M. Recent Contributions to the Method of Characteristics for Three-Dimensional Problems in Gas Dynamics, University of California, Berkeley, Rept. No. FM-74-2, April (1974).

Kellogg, O. D. *Foundations of Potential Theory,* Dover Press, New York (1953).

Leonard, A. *J. Comput. Phys.* **37,** 289–335 (1980).

Liepmann, H. W., and Roshko, A. *Elements of Gasdynamics,* Wiley & Sons, New York (1957).

Milinazzo, F., and Saffman, P. G. *J. Comput. Phys.* **23,** 380–392 (1977).

Moore, D. W. *Mathematika* **23,** 35–44 (1976).

Rakich, J. V. *AIAA J.* **5,** 1906–1908 (1967).

Rakich, J. V., and Kutler, P. AIAA Paper 72–191 (1972).

Rehbach, C. *Rech. Aéros.* No. 1977–5, 289–298 (1977).

Roberts, A. L., and Rundle, K. BAC (Weybridge) Rept. Aero. MA 19 (1972).

Roberts, K. V., and Christiansen, J. P. *Comput. Phys. Commun.* **3,** 14–32 (1972).

Rosenhead, L. *Proc. R. Soc. Lond. A* **134,** 170–192 (1931).

Rubbert, P. E., et al. Vol. I, USAAVLABS Tech. Rept. No. 67-61A December (1967).

Rubbert, P. E., and Saaris, G. R. AIAA Paper No. 72–188 (1972).

Saffman, P. G., and Baker, G. G. *Ann. Rev. Fluid Mech.* **11,** 95–122 (1979).

Thompson, Jr., J. F., Shanks, S. P., and Wu. J. C. "Numerical Solution of 3-D Navier–Stokes Equations in Integro-Differential Form" In Proceedings of the AIAA Computational Fluid Dynamics Conference, Palm Springs, California, July (1973).

Thompson, Jr., J. F., Shanks, S. P., and Wu, J. C. *AIAA J.* **6,** 787–794 (1974).

Wu, J. C. AIAA Paper No. 75–47 (1975).

Wu, J. C. *AIAA J.* **14,** 1042–1049 (1976).

Wu, J. C., and Rizk, Y. M. *Lecture Notes in Physics,* Vol. 90, pp. 558–564, Springer-Verlag, New York (1979).

Wu, J. C., and Thompson, J. F. *Comput. Fluids* **1,** 197–215 (1973).

Wu, J. C., and Wahbah, M. M. *Lecture Notes in Physics,* Vol. 59, pp. 448–453, Springer-Verlag, New York (1976).

PART II
INCOMPRESSIBLE FLOWS

In this part of the book we address the problem of computing incompressible flows. For such flows there are various possibilities for the formulation of the problem. These include primitive variables, stream-function vorticity, and vorticity velocity. The primitive-variable approach offers the fewest complications in extending two-dimensional calculations to three dimensions. The primary difficulty with this approach is specification of boundary conditions on pressure. However, this problem can be eliminated as will be described in the following chapters.

We also discuss the use of the stream-function vorticity formulation for plane two-dimensional flows. Here the difficulty is primarily associated with determination of vorticity at a boundary. A number of procedures for surmounting this problem are available and some are described in the discussion that follows. An inconvenience of the stream-function vorticity formulation is that the pressure is not directly obtained and consequently additional calculations are required for its determination.

We have not addressed the formulation for three dimensions which employs an extension of the concept of stream-function vorticity. The interested reader is referred to Aziz and Hellums (1967), Mallinson and De Vahl Davis (1973, 1977) for application details and to Hirasaki and Hellums (1970), Richardson and Cornish (1977) for boundary-condition questions.

We note the possibility of utilizing a formulation in two dimensions of a stream-function only equation. We have not investigated this approach and the reader is referred to Bourcier and Francois (1969), Roache and Ellis (1975), Morchoisne (1979), and Cebeci et al. (1981) for the details of this approach.

One additional formulation which is not included should be mentioned. This is the velocity vorticity approach. This approach requires the vorticity equation, the continuity equation, and the equations that define vorticity in terms of velocity gradients. A combination of the continuity equation and the definition of vorticity can be made after differentiation of both equations. This yields elliptic equations for the velocity components. The interested reader is referred to Fasel (1976) and Dennis et al. (1979) for application of the approach.

References

Aziz, K., and Hellums, J. D. *Phys. Fluids* **10**, 314–324 (1967).

Bourcier, M., and François, C. *Rech. Aérosp.* No. 131, 23–33 (1969).

Cebeci, T., Hirsh, R. S., Keller, H. B., and Williams, P. G. *Comput. Meth. Appl. Mech. Engr.* **27**, 13–44 (1981).

Dennis, S. C. R., Ingham, D. B., and Cook, R. N. *J. Comput. Phys.* **33**, 325–339 (1979).

Fasel, M. *J. Fluid Mech.* **78**, 355–383 (1976).

Hirasaki, G. J., and Hellums, J. D. *Q. Appl. Math.* **28**, 293–297 (1970).

Mallinson, G. D., and De Vahl Davis, G. *J. Comput. Phys.* **12**, 435–461 (1973).

Mallinson, G. D., and De Vahl Davis, G. I. *J. Fluid Mech.* **83**, 1–31 (1977).

Morchoisne, Y. *Rech. Aérosp.* No. 1979-5, 293–306 (1979).

Richardson, S. M. and Cornish, A. R. H. *J. Fluid Mech.*, **82**, 309–319 (1977).

Roache, P. J., and Ellis, M. A. *Comput. Fluids* **3**, 305–320 (1975).

CHAPTER 6

Finite-Difference Solution of the Navier–Stokes Equations

6.1 The Navier–Stokes Equations in Primitive Variables

The Navier–Stokes equations without external force can be written in dimensionless form

$$\frac{\partial \mathbf{V}}{\partial t} + \mathbf{A}(\mathbf{V}) + \nabla p = \frac{1}{\mathrm{Re}} \nabla^2 \mathbf{V} \tag{6.1.1a}$$

$$\nabla \cdot \mathbf{V} = 0 \tag{6.1.1b}$$

where $\mathbf{A}(\mathbf{V})$ is expressed by one of the following equations according to the choice of *conservative* or *nonconservative* form of the convective term:

$$\mathbf{A}(\mathbf{V}) = \nabla \cdot (\mathbf{V}\mathbf{V}) \tag{6.1.2a}$$

$$\mathbf{A}(\mathbf{V}) = (\mathbf{V} \cdot \nabla)\mathbf{V} \tag{6.1.2b}$$

In Cartesian coordinates (x, y) the two components $a(u, v)$, $b(u, v)$ of $\mathbf{A}(\mathbf{V}) = a\mathbf{i} + b\mathbf{j}$ where $\mathbf{V} = (u, v)$ are given by, respectively,

$$a(u, v) = \frac{\partial}{\partial x} (u^2) + \frac{\partial}{\partial y} (uv)$$
$$\tag{6.1.3a}$$
$$b(u, v) = \frac{\partial}{\partial x} (uv) + \frac{\partial}{\partial y} (v^2)$$

for (6.1.2a) and by

$$a(u, v) = u \frac{\partial u}{\partial x} + v \frac{\partial u}{\partial y}$$
$$\tag{6.1.3b}$$
$$b(u, v) = u \frac{\partial v}{\partial x} + v \frac{\partial v}{\partial y}$$

for (6.1.2b).

A typical problem associated with Eq. (6.1.1) is the initial boundary-value problem defined as follows: Find the $\mathbf{V}$, p, solution of Eq. (6.1.1) in a bounded domain Ω with the boundary Γ such that $\mathbf{V}$ is given on the boundary Γ by the equation

$$\mathbf{V} = \mathbf{V}_\Gamma \tag{6.1.4}$$

and $\mathbf{V}$ is given at initial time $t = 0$ by

$$\mathbf{V} = \mathbf{V}^0 \tag{6.1.5}$$

The boundary value $\mathbf{V}_\Gamma$ must satisfy the condition

$$\int_\Gamma \mathbf{V}_\Gamma \cdot \mathbf{N} \, ds = 0 \tag{6.1.6}$$

where $\mathbf{N}$ is the normal unit vector to Γ, and the initial condition $\mathbf{V}^0$ must satisfy

$$\nabla \cdot \mathbf{V}^0 = 0 \tag{6.1.7}$$

The main difficulties associated with the solution of the Navier–Stokes equations in velocity–pressure formulation are the following:

(i) The presence of the constraint $\nabla \cdot \mathbf{V} = 0$, which must be satisfied at any time, does not allow the use of a simple explicit method that avoids solution of an algebraic system of equations.

(ii) There is a lack of boundary conditions for the pressure.

In the discussion which follows we will describe various approaches that have been employed in finite-difference solutions to overcome these difficulties. In Chapters 7 and 8 we also discuss procedures employed for finite-element and spectral methods. We begin the discussion by considering the artificial compressibility method for steady flows.

6.2 Steady Navier–Stokes Equations: The Artificial Compressibility Method

The steady Navier–Stokes equations are deduced from (6.1.1) and are written

$$\mathbf{A}(\mathbf{V}) + \nabla p = \frac{1}{\mathrm{Re}} \, \nabla^2 \mathbf{V} \tag{6.2.1a}$$

$$\nabla \cdot \mathbf{V} = 0 \tag{6.2.1b}$$

Typical boundary-value problems associated with these equations concern the solution in a bounded domain with velocity $\mathbf{V}$ on the boundary given by

$$\mathbf{V} = \mathbf{V}_\Gamma \tag{6.2.2}$$

As for the unsteady Navier–Stokes equations, the numerical difficulty lies in the constraint $\nabla \cdot \mathbf{V} = 0$. In the steady case, this difficulty can be surmounted by using the so-called *artificial compressibility method*.

6.2.1 Description of the method

The method has been introduced independently and under slightly different forms by Vladimirova et al. (1965) (see Yanenko, 1971) and by Chorin (1967). The principle of the method is to consider the solution of the steady equations

(6.2.1) as the limit when $t \to \infty$ of the solution of unsteady equations obtained by associating the unsteady momentum equation (6.1.1a) with a perturbed divergence equation in order to get a system of equations of evolution which can be easily solved by standard methods (explicit or not).

The techniques described by Yanenko (1971) and Chorin (1967) vary slightly in concept. We have chosen to display the Chorin method since there is limited information on the success of the Yanenko approach. Taylor and Ndefo (1970) experimented with the approach but were not successful in solving a channel-flow problem and found it necessary to solve the problem by using the Poisson pressure equation. More research is required, however, to establish the limits of the Yanenko approach.

The Chorin method is established by first writing a perturbed continuity equation

$$\frac{\partial p}{\partial t} + c^2\,\mathbf{\nabla}\cdot\mathbf{V} = 0 \tag{6.2.3}$$

where c^2 is an arbitrary constant. This equation has no physical meaning before the steady state $\partial/\partial t = 0$ is reached. So the constraint $\mathbf{\nabla}\cdot\mathbf{V} = 0$ is satisfied at convergence only. The method, which consists of solving Eqs. (6.1.1a) and (6.2.3) can be called a *pseudo-unsteady method* because the time t involved has no physical meaning.

The parameter c^2 in Eq. (6.2.3) must be chosen to ensure convergence, i.e., to ensure the existence of a steady numerical solution of the system (6.1.1a) and (6.2.3) with boundary conditions (6.2.2) and initial conditions

$$\mathbf{V} = \mathbf{V}^0 \quad \text{at } t = 0 \qquad \text{and} \qquad p = p^0 \quad \text{at } t = 0 \tag{6.2.4}$$

where $\mathbf{V}^0$ and p^0 are arbitrary.

It has been shown by Fortin et al. (1971) that the exact solution of the unsteady Stokes problem associated with (6.1.1a), (6.2.3), (6.2.2), and (6.2.4) tends toward the solution of the corresponding steady Stokes problem.

The term "artificial compressibility method" (Chorin, 1967) was coined because Eqs. (6.1.1a) and (6.2.3) can be derived from the Navier–Stokes equations for a compressible fluid whose state law would be

$$p = c^2\rho \quad \text{with} \quad c^2 = \text{const}$$

As a matter of fact, Eq. (6.2.3) becomes Eq. (6.1.1b) if $c^{-2} \to 0$, and it is possible to consider (Temam, 1969a) the solution of system (6.1.1a) and (6.2.3) with $c^{-2} \ll 1$ as an approximation of the unsteady solution of the system (6.1.1). This point of view has been used by Ganoulis and Thirriot (1976) with a finite-difference method. However, possible numerical difficulties can be associated with the use of a very large value of c^2, and hence the artificial compressibility method will likely have the most value in the computation of steady solutions. In this way, it can be considered a procedure to build a special iterative method for solving the steady problem.

6.2.2 Discretization

The spatial discretization makes use of the staggered marker-and-cell (MAC) mesh (Fig. 6.2.1) introduced by Harlow and Welsh (1965), and we consider a very simple explicit discretization in time. The approximation of (6.1.1a) and (6.2.3) is

$$\frac{1}{\Delta t}(u_{i+1/2,j}^{n+1} - u_{i+1/2,j}^n) + a_{i+1/2,j}^n + \Delta_x^1 p_{i+1/2,j}^n = \frac{1}{Re} \nabla_h^2 u_{i+1/2,j}^n \qquad (6.2.5a)$$

$$\frac{1}{\Delta t}(v_{i,j+1/2}^{n+1} - v_{i,j+1/2}^n) + b_{i,j+1/2}^n + \Delta_y^1 p_{i,j+1/2}^n = \frac{1}{Re} \nabla_h^2 v_{i,j+1/2}^n \qquad (6.2.5b)$$

$$\frac{1}{\Delta t}(p_{i,j}^{n+1} - p_{i,j}^n) + c^2(\Delta_x^1 u_{i,j}^{n+1} + \Delta_y^1 v_{i,j}^{n+1}) = 0 \qquad (6.2.5c)$$

where the difference operators Δ_x^1, Δ_y^1, and ∇_h^2 are defined by

$$\Delta_x^1 f_{l,m} = \frac{1}{\Delta x}(f_{l+1/2,m} - f_{l-1/2,m})$$

$$\Delta_y^1 f_{l,m} = \frac{1}{\Delta y}(f_{l,m+1/2} - f_{l,m-1/2})$$

$$\nabla_h^2 f_{l,m} = \Delta_{xx} f_{l,m} + \Delta_{yy} f_{l,m} = (\Delta_x^1 \Delta_x^1 + \Delta_y^1 \Delta_y^1)f_{l,m} \qquad (6.2.6)$$

$$\Delta_{xx} f_{l,m} = \frac{f_{l+1,m} - 2f_{l,m} + f_{l-1,m}}{\Delta x^2}$$

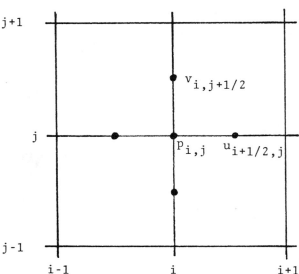

Fig. 6.2.1 The MAC mesh.

$$\Delta_{yy}f_{l,m} = \frac{f_{l,m+1} - 2\,f_{l,m} + f_{l,m-1}}{\Delta y^2}$$

where l, m are integers or not. The term $a_{i+1/2,j}^n$ and $b_{i,j+1/2}^n$ are the approximations of $a(u, v)$ and $b(u, v)$ as defined in (6.1.3). In the case where the nonconservative form (6.1.3b) is used, we have

$$a_{i+1/2,j}^n = u_{i+1/2,j}^n \Delta_x^0 u_{i+1/2,j}^n + \hat{v}_{i+1/2,j}^n \Delta_y^0 u_{i+1/2,j}^n$$

$$b_{i,j+1/2}^n = \hat{u}_{i,j+1/2}^n \Delta_x^0 v_{i,j+1/2}^n + v_{i,j+1/2}^n \Delta_y^0 v_{i,j+1/2}^n \qquad (6.2.7a)$$

where

$$\hat{u}_{i,j+1/2} = \tfrac{1}{4}(u_{i+1/2,j} + u_{i+1/2,j+1} + u_{i-1/2,j+1} + u_{i-1/2,j})$$

$$\hat{v}_{i+1/2,j} = \tfrac{1}{4}(v_{i+1,j+1/2} + v_{i,j+1/2} + v_{i,j-1/2} + v_{i+1,j-1/2})$$

$$\Delta_x^0 f_{l,m} = \frac{1}{2\,\Delta x}(f_{l+1,m} - f_{l-1,m})$$

$$\Delta_y^0 f_{l,m} = \frac{1}{2\,\Delta y}(f_{l,m+1} - f_{l,m-1})$$

In the case where the conservative form (6.1.3a) is used, we have

$$a_{i+1/2,j}^n = \Delta_x^1(u^2)_{i+1/2,j}^n + \Delta_y^1(uv)_{i+1/2,j}^n$$

$$b_{i,j+1/2}^n = \Delta_x^1(uv)_{i,j+1/2}^n + \Delta_y^1(v^2)_{i,j+1/2}^n \qquad (6.2.7b)$$

where

$$(u^2)_{i,j} = \tfrac{1}{4}(u_{i+1/2,j} + u_{i-1/2,j})^2$$

$$(uv)_{i,j} = \tfrac{1}{4}(u_{i+1/2,j+1} + u_{i+1/2,j})(v_{i+1,j+1/2} + v_{i,j+1/2})$$

$$(v^2)_{i,j} = \tfrac{1}{4}(v_{i,j+1/2} + v_{i,j-1/2})^2$$

All these approximations are of second-order accuracy. Moreover, the latter (6.2.7b) have good conservation properties since the momentum and kinetic energy (in the case $Re^{-1} = 0$) are conserved (Zabusky and Deem, 1971).

6.2.3 Convergence toward a steady state

The object of the artifical compressibility method is to obtain a steady solution, characterized by

$$\max\left(\frac{1}{\Delta t}\left|u^{n+1} - u^n\right|, \frac{1}{\Delta t}\left|v^{n+1} - v^n\right|, \frac{1}{c^2\,\Delta t}\left|p^{n+1} - p^n\right|\right) < \epsilon \qquad (6.2.8)$$

where ϵ is a small number. The possibility of obtaining convergence is based upon two facts: (i) the proof mentioned above, concerning the limit when $t \to \infty$ of the solution of the perturbed unsteady Stokes problem toward the solution of the steady problem and (ii) the consistency and stability of scheme

(6.2.5). (This is a heuristic extension of the Lax equivalence theorem.) The consistency follows from the construction of the finite-difference scheme.

In order to establish stability an approximate linear analysis is made. In the linearization process approximations (6.2.7a) and (6.2.7b) are identical, so we consider only (6.2.7a) with the assumption

$$u_{i+1/2,j} = \hat{u}_{i,j+1/2} = u_0 \text{ (const)}$$

$$v_{i,j+1/2} = \hat{v}_{i+1/2,j} = v_0 \text{ (const)}$$

A first study is made by neglecting the pressure term in (6.2.5a) and (6.2.5b) so that the divergence equation is disregarded and each of the two momentum equations is of the advection–diffusion type:

$$f_{i,j}^{n+1} = \left[I - \Delta t \left(u_0 \Delta_x^0 + v_0 \Delta_y^0 - \frac{1}{\mathrm{Re}} \nabla_h^2 \right) \right] f_{i,j}^n \qquad (6.2.9)$$

where I is the identity operator.

The conditions of stability of this finite-difference scheme are (see Section 2.8.1):

$$\tfrac{1}{2}(u_0^2 + v_0^2) \, \Delta t \, \mathrm{Re} \leqslant 1 \qquad (6.2.10)$$

$$\frac{4 \, \Delta t}{\mathrm{Re} \, \Delta x^2} \leqslant 1 \qquad (6.2.11)$$

where $\Delta x = \Delta y$.

A second approximate analysis of stability can be easily accomplished by neglecting the convective term ($u_0 = v_0 = 0$, i.e., the Stokes approximation) and conserving the pressure terms. The scheme is then written

$$u_{i+1/2,j}^{n+1} = \left(I - \frac{\Delta t}{\mathrm{Re}} \nabla_h^2 \right) u_{i+1/2,j}^n - \Delta t \, \Delta_x^1 p_{i+1/2,j}^n \qquad (6.2.12a)$$

$$v_{i,j+1/2}^{n+1} = \left(I - \frac{\Delta t}{\mathrm{Re}} \nabla_h^2 \right) v_{i,j+1/2}^n - \Delta t \, \Delta_y^1 p_{i,j+1/2}^n \qquad (6.2.12b)$$

$$p_{i,j}^{n+1} = p_{i,j}^n - \Delta t \, c^2 (\Delta_x^1 u_{i,j}^{n+1} + \Delta_y^1 v_{i,j}^{n+1}) \qquad (6.2.12c)$$

The spectral radius of the amplification matrix associated with the above scheme is not larger than 1 if the following conditions obtained by the technique developed in Appendix A are satisfied:

$$4 \frac{\Delta t}{\Delta x^2} \left(\frac{1}{\mathrm{Re}} + \frac{\Delta t \, c^2}{2} \right) \leqslant 1, \qquad c^2 > 0 \qquad (6.2.13)$$

This condition includes the classical parabolic criterion (6.2.11): Thus, the conditions of stability are (6.2.10) and (6.2.13). Numerical experience shows that the convergence is better when c^2 is large but slightly smaller than c_{max}^2 given by (6.2.13).

Scheme (6.2.5) can be modified to improve the convergence and simplify programming (Fortin et al., 1971). The modification consists of using the values of the unknown at time $n + 1$ in (6.2.5) as soon as they are computed. (This technique, which is nothing more than a Gauss–Seidel technique, can be more or less implicit according to whether the equations are solved simultaneously or successively.) As an example, Eq. (6.2.5a) along with (6.2.7a) would be replaced by

$$\frac{1}{\Delta t}(u_{i+1/2,j}^{n+1} - u_{i+1/2,j}^n) + \hat{u}_{i+1/2,j}^n\left(\frac{u_{i+3/2,j}^n - u_{i-1/2,j}^{n+1}}{2\,\Delta x}\right)$$

$$+ \hat{v}_{i+1/2,j}^n\left(\frac{u_{i+1/2,j+1}^n - u_{i+1/2,j-1}^{n+1}}{2\,\Delta y}\right) + \frac{1}{\Delta x}(p_{i+1,j}^n - p_{i,j}^n)$$

$$- \frac{1}{\mathrm{Re}}\left(\frac{u_{i+3/2,j}^n - 2u_{i+1/2,j}^n + u_{i-1/2,j}^{n+1}}{\Delta x^2}\right.$$

$$\left. + \frac{u_{i+1/2,j+1}^n - 2u_{i+1/2,j}^n + u_{i+1/2,j-1}^{n+1}}{\Delta y^2}\right) = 0$$

(6.2.14)

in which it is assumed that the computations are made with increasing indices i and j.

Obviously, the resulting scheme is no longer consistent with Eqs. (6.1.1a) and (6.2.3) during the transient stage. More precisely, the finite-difference equation (6.2.14) can be considered as approximating the equation

$$K\frac{\partial u}{\partial t} + u\frac{\partial u}{\partial x} + v\frac{\partial u}{\partial y} + \frac{\partial p}{\partial x} - \frac{1}{\mathrm{Re}}\left(\frac{\partial^2 u}{\partial x^2} + \frac{\partial^2 u}{\partial y^2}\right) = 0 \qquad (6.2.15)$$

with

$$K = 1 - \Delta t\left(\frac{u + v}{2\,\Delta x} + \frac{2}{\mathrm{Re}\,\Delta x^2}\right) \qquad (6.2.16)$$

where $\Delta x = \Delta y$. Equation (6.2.15) is parabolic in the direction $t > 0$ only if $K > 0$. This is a necessary condition for the solution of (6.2.15) to tend toward a limit when $t \to \infty$.

Without performing a complete analysis of the convergence of the modified scheme, it is possible to obtain a rough idea of its convergence. If convective and pressure terms are neglected, the numerical method reduces to the successive relaxation method applied to the solution of a Laplace equation with $\omega = 4\,\Delta t/\mathrm{Re}\,\Delta x^2$ as the relaxation parameter. It is easy to verify that the condition of convergence of the successive relaxation method $0 < \omega < 2$ is nothing more than the condition $K > 0$ for the case $u = v = 0$. We note that the criterion of convergence of the modified semi-implicit scheme is two times less restrictive than criterion (6.2.11).

We note also that $K < 1$, if $4 + (u + v)\mathrm{Re}\,\Delta x > 0$; i.e., the convergence toward the steady solution is faster than the convergence given by a consistent scheme (for the same Δt). This analogy between the iterative procedure and the evolution equations shown briefly here has been discussed in Section 2.2.3.

6.2.4 Treatment of boundary conditions

Let Γ be the boundary of the computational domain and assume that the velocity $\mathbf{V}$ is given on Γ; i.e., $\mathbf{V}_\Gamma = (u_\Gamma, v_\Gamma)$. There is no condition for the pressure p. When the staggered MAC mesh is used, the boundary Γ is located as shown in Fig. 6.2.2. The vertical sides pass through the points where the horizontal component u of the velocity is defined, and the horizontal sides pass through the points where the vertical component v is defined. Therefore, the pressure p is not defined on the boundary and this fact is essential: Equation (6.2.5c) can be applied to the first points adjacent to the boundary without modification. Algorithm (6.2.5a)–(6.2.5c) allows the computation of the pressure without requiring the explicit prescription of boundary conditions for it.

The MAC mesh is well adapted for the pressure but, on the other hand, it presents some disadvantage concerning the velocity. As seen in Fig. 6.2.2 difficulties occur when v_1 or u_1 have to be computed: For example, if v_1 is computed from (6.2.5b), the approximations of $b(u, v)$ and $\partial^2 v / \partial x^2$ involve the value v_0 which is outside the computational domain. Various ways to define the outside value v_0 can be considered.

The first is the so-called *reflection technique* which consists of writing the velocity v_Γ on Γ as the mean value of the two velocities v_0 and v_1, so that

$$v_0 = 2\, v_\Gamma - v_1 \qquad\qquad (6.2.17)$$

i.e., v_0 is defined by a linear extrapolation.

A second technique which is partly identical to the previous one consists of approximating derivatives at point 1 (Fig. 6.2.2) with noncentered first-order differences. So, for the first-order derivative

$$\left.\frac{\partial v}{\partial x}\right|_1 = \frac{1}{2\,\Delta x}\,(v_2 + v_1 - 2\,v_\Gamma) \qquad\qquad (6.2.18a)$$

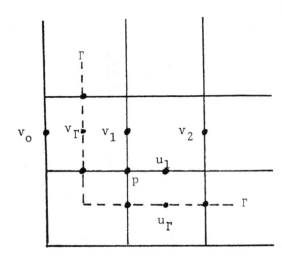

Fig. 6.2.2 The MAC mesh near a boundary Γ.

and for the second-order derivative

$$\left.\frac{\partial^2 v}{\partial x^2}\right|_1 = \frac{4}{3\,\Delta x^2}(v_2 - 3\,v_1 + 2\,v_\Gamma) \qquad (6.2.18b)$$

In fact, from the point of view of programming, at point 1 we can use the general centered approximation and formulas (6.2.18a) and (6.2.18b) to define the outside value v_0. Thus for the first-order derivatives,

$$\left.\frac{\partial v}{\partial x}\right|_1 = \frac{v_2 - v_0}{2\,\Delta x} = \frac{1}{2\,\Delta x}(v_2 + v_1 - 2\,v_\Gamma)$$

This equality gives a value v_0 that is identical to (6.2.17). In the same way, the equation

$$\left.\frac{\partial^2 v}{\partial x^2}\right|_1 = \frac{v_2 - 2\,v_1 + v_0}{\Delta x^2} = \frac{4}{3\,\Delta x^2}(v_2 - 3\,v_1 + 2\,v_\Gamma)$$

yields

$$v_0 = \tfrac{1}{3}(v_2 - 6\,v_1 + 8\,v_\Gamma) \qquad (6.2.19)$$

Finally, if second-order noncentered differences are used we have the formulas

$$v_0 = \tfrac{1}{3}(v_2 - 6\,v_1 + 8\,v_\Gamma) \qquad \text{for } \frac{\partial v}{\partial x} \qquad (6.2.20a)$$

and

$$v_0 = -\tfrac{1}{5}(v_3 - 5\,v_2 + 15\,v_1 - 16\,v_\Gamma) \qquad \text{for } \frac{\partial^2 v}{\partial x^2} \qquad (6.2.20b)$$

It should be noted that if (6.2.17) is used in the approximation of the second-order derivatives at point 1, the error associated with it is $O(1)$. That is to say, at point 1 the resulting difference equation (6.2.5b) is no longer consistent with the differential equation. However, this local loss of consistency does not necessarily lead to a complete loss of accuracy of the numerical solution (Bramble and Hubbard, 1962; Kreiss, 1972).

The following is devoted to a simple study of the effect of the treatment of boundary conditions when a staggered mesh is used. Let us consider the problem

$$\frac{d^2 f}{dx^2} - \delta\frac{df}{dx} = 0, \qquad 0 < x < 1, \qquad \delta = \text{const} \qquad (6.2.21a)$$

$$f(0) = U_0, \qquad f(1) = U_1 \qquad (6.2.21b)$$

whose exact solution is

$$f(x) = \frac{(U_0 - U_1)\exp(\delta x) + U_1 - U_0\exp(\delta)}{1 - \exp(\delta)} \qquad (6.2.22)$$

The interval $[0, 1]$ is discretized so that half-meshes appear near the limits (Fig. 6.2.3); thus $x_i = (i - \frac{1}{2})\Delta x$ for $i = 1, \ldots, N$ and $\Delta x = 1/N$.

Let us assume (6.2.21a) is approximated by

$$\frac{1}{\Delta x^2}(f_{i+1} - 2 f_i + f_{i-1}) - \frac{\delta}{2 \Delta x}(f_{i+1} - f_{i-1}) = 0, \qquad 1 \le i \le N$$

$$(6.2.23)$$

For outside values f_0 and f_{N+1}, we use the formula (6.2.17):

$$f_0 = 2 U_0 - f_1, \qquad f_{N+1} = 2 U_1 - f_N$$

Thus, the truncation error of the finite-difference equation (6.2.23) is $O(\Delta x^2)$ for $1 < i < N$ and $O(1)$ for $i = 1$ or $i = N$. The finite-difference problem is the following:

$$(f_{i+1} - 2 f_i + f_{i-1}) - \frac{\delta \Delta x}{2}(f_{i+1} - f_{i-1}) = 0, \qquad 1 < i < N \quad (6.2.24a)$$

$$(f_2 - 3 f_1 + 2 U_0) - \frac{\delta \Delta x}{2}(f_2 + f_1 - 2 U_0) = 0 \qquad\qquad (6.2.24b)$$

$$(2 U_1 - 3 f_N + f_{N-1}) - \frac{\delta \Delta x}{2}(2 U_1 - f_N - f_{N-1}) = 0 \qquad (6.2.24c)$$

The solution is

$$f_i = c_1 q_1^i + c_2 q_2^i$$

where q_1 and q_2 are the roots of the characteristic equation associated with (6.2.24a) and c_1, c_2 are constants determined by the relationships (6.2.24b) and (6.2.24c). The solution f_i is found to be

$$f_i = \frac{U_0 q^N - U_1}{q^N - 1} - \left(1 - \frac{\delta}{2} \Delta x\right)\left(\frac{U_0 - U_1}{q^N - 1}\right)q^i, \qquad 1 \le i \le N$$

with

$$q = \frac{2 + \delta \Delta x}{2 - \delta \Delta x}$$

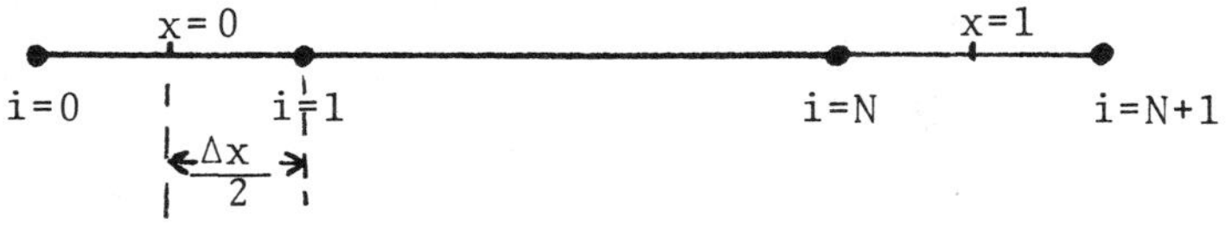

Fig. 6.2.3 Mesh distribution.

After expansion with respect to Δx ($\Delta x \to 0$) by assuming that x_i remains fixed, we get the error E_i:

$$E_i = f_i - f(x_i) = \Delta x^2 (U_0 - U_1)\Psi(x_i) + O(\Delta x^3), \qquad 1 \le i \le N$$

(6.2.25)

with

$$\Psi(x_i) = \frac{\delta^2}{12 \, (e^\delta - 1)^2} [\delta e^{\,\delta}(e^{\,\delta x_i} - 1) - (e^\delta - 1)(\delta x_i - \tfrac{3}{2})e^{\,\delta x_i}]$$

The error E_i is $O(\Delta x^2)$ everywhere, even at point $i = 1$ or $i = N$. In particular, at point $i = 1$ the error is

$$E_1 = -\frac{(U_0 - U_1)}{8} \frac{\delta^2 \, \Delta x^2}{1 - \exp(\delta)} + O(\Delta x^3)$$

(6.2.26)

This result would lead one to think that a simple reflection technique will yield results equivalent to a noncentered consistent approximation. In reality, if the above first- or second-order approximations are used, we obtain an error $O(\Delta x^2)$ which becomes $O(\Delta x^3)$ at points located near the boundary; i.e., to an order higher than (6.2.26). Therefore, it is recommended that the consistent approximations be used. Generally, good results are obtained by using the three-point approximations (6.2.20a) for $\partial v/\partial x$ and for $\partial^2 v/\partial x^2$.

6.2.5 The Poisson equation for pressure

The artificial compressibility method can be considered also as a special iteration technique to solve the steady Navier–Stokes equations using, in particular, a Poisson equation for the pressure. If the velocity at time $n + 1$ in (6.2.5c) is eliminated with the help of Eqs. (6.2.5a) and (6.2.5b), then Eq. (6.2.5c) becomes

$$p_{i,j}^{n+1} - p_{i,j}^n - c^2 \, \Delta t^2 \nabla_h^2 p_{i,j}^n = -c^2 \, \Delta t \, \boldsymbol{\nabla}_h \cdot \left[\mathbf{V} - \Delta t\left(\mathbf{A} - \frac{1}{\mathrm{Re}} \, \nabla_h^2 \mathbf{V}\right)\right]_{i,j}^n$$

(6.2.27)

where

$$\boldsymbol{\nabla}_h = (\Delta_x^1, \Delta_y^1), \qquad \nabla_h^2 = \Delta_{xx} + \Delta_{yy}$$

At convergence $n \to \infty$, when $p_{i,j}^{n+1} = p_{i,j}^n$ then $\boldsymbol{\nabla}_h \cdot \mathbf{V}_{i,j}^n = \boldsymbol{\nabla}_h \cdot (\nabla_h^2 \mathbf{V})_{i,j}^n = 0$, and the resulting equation

$$\nabla_h^2 p_{i,j} = -\boldsymbol{\nabla}_h \cdot \mathbf{A}_{i,j}$$

(6.2.27a)

is an approximation of the Poisson equation for p obtained by taking the divergence of the steady momentum equation (6.2.1a), then using Eq. (6.2.1b) which expresses the fact that $\mathbf{V}$ is a divergence-free vector.

Near a boundary, Eq. (6.2.27) takes a different form because of the use of value u_Γ or v_Γ in Eq. (6.2.5c). For example, near a vertical boundary (Fig. 6.2.4), Eq. (6.2.27) is replaced by

$$p_{1,j}^{n+1} - p_{1,j}^n - c^2\,\Delta t\left[\frac{1}{\Delta x^2}(p_{2,j}^n - p_{1,j}^n) + \frac{1}{\Delta y^2}(p_{1,j+1}^n - 2\,p_{1,j}^n + p_{1,j-1}^n)\right]$$

$$= -c^2\,\Delta t\left[\frac{1}{\Delta x}(u_{3/2,j}^n - \Delta t\,a_{3/2,j}^n + \frac{\Delta t}{Re}\,\nabla_h^2 u_{3/2,j}^n) - \frac{u_\Gamma}{\Delta x}\right. \qquad (6.2.28)$$

$$\left. + \Delta_y^1(v_{1,j}^n - \Delta t b_{1,j}^n + \frac{\Delta t}{Re}\,\nabla_h^2 v_{1,j}^n)\right] \qquad (u_{1/2,j} = u_\Gamma)$$

By comparison of this last equation with Eq. (6.2.27) written at point $(1, j)$, it is easy to see that (6.2.28) can be considered as (6.2.27) written at point $(1, j)$ with the boundary condition

$$\frac{1}{\Delta x}(p_{1,j}^n - p_{0,j}^n) = -a_{1/2,j}^n + \frac{1}{Re}\,\nabla_h^2 u_{1/2,j}^n \qquad (6.2.29)$$

This is nothing more than a discretization of the Neumann condition for pressure obtained by projection of the momentum equation (6.2.1a) on the normal to the boundary Γ.

We must note that, except for the value $v_{0,j\pm1/2}$, the artificial compressibility method does not need consideration of the outside value $u_{-1/2,j}^n$ as would be necessary if Eq. (6.2.29) was used. The advantage of the method is that the

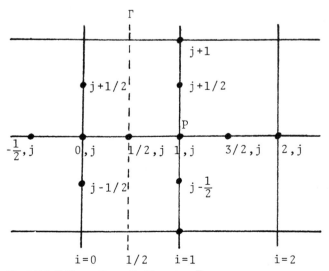

Fig. 6.2.4 Grid near the vertical boundary Γ.

Neumann condition for the pressure is automatically involved in the numerical solution, and this fact is consistent with the mathematical problem.

At this point it is interesting to discuss the effect of the mesh on the numerical solution. In the first mesh (Fig. 6.2.5a) (used by Chorin, 1967), the velocity and the pressure are defined at the nodes of the mesh. The advantages of such a mesh are its simplicity and the fact that the velocity is defined, in particular, on the boundary Γ where this quantity is generally prescribed. On the other hand, one of its disadvantages is the fact that the pressure is also defined on the boundary. Since there is generally no boundary condition for pressure, it is necessary to devise a special technique to compute the pressure on this boundary.

In the second mesh (Fig. 6.2.5b) used by Kuznetsov (1968) or by Fortin et al. (1971), the pressure is defined at nodes, and the velocity at the center point of the cell, but the boundary Γ does not pass through the nodes. Therefore, the pressure is no longer defined on the boundary Γ, and we can use the same formulas to compute the whole pressure field.

Finally, the MAC mesh described above (Fig. 6.2.5c) differs from the previous one by the location of discretization points for velocity. A small

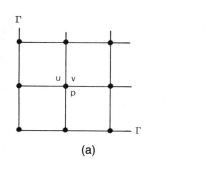

(a)

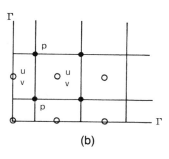

(b)

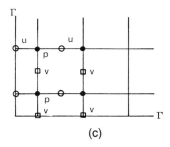

(c)

Fig. 6.2.5 Mesh distributions used by (a) Chorin, (b) Kuznetsov and Fortin et al., and (c) the MAC mesh.

disadvantage of this mesh is that only one of the velocity components is defined on each side of the boundary Γ; so it is necessary to employ noncentered differences near the boundary as explained in Section 6.2.4. But this inconvenience is largely balanced by the advantage of the MAC mesh for the pressure computation.

As a matter of fact, it is very instructive to compare the layout of points involved in the discretization of the Poisson equation and obtained as a result of the use of the artificial compressibility method. In Fig. 6.2.6 the pressure is defined at all grid points and the black dots represent the points involved in the approximation of the Poisson equation at a given point P, when standard centered differences are used. In the first two cases (a) and (b), we can see that there exist two uncoupled networks of pressure points. This fact leads to the existence of two solutions for the pressure. These solutions differ from each other by an arbitrary constant and there is apparently no reason for these two arbitrary constants to be identical. The existence of the two pressure fields could lead to oscillations in the computed pressure; but the pressure gradient itself is not oscillatory. The computed velocity is not oscillatory. However, we must note that it would be possible to couple (weakly) the two fields by way of the special technique needed to define the pressure on Γ in the case of mesh (a) or by a modification of the approximation of $\nabla \cdot \mathbf{V}$ near Γ in the case of mesh (b). It is seen in Fig. 6.2.5c that such a phenomenon of uncoupling does not appear in the case of the MAC mesh—a feature that makes this mesh very convenient.

6.2.6 Other schemes for the artificial compressibility method

In Section 6.2.5 we presented a very simple scheme to solve the equations associated with the artifical compressibility method. In some cases, the conditions of stability can become very restrictive. For example, if Re is very small (Re $<<$ 1), condition (6.2.13) could lead to very small values of Δt and c^2 and, consequently, to a slow convergence toward the steady state. On the other side, if Re is very large, condition (6.2.10) induces a very small value of Δt.

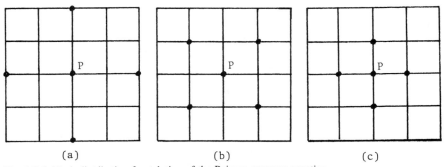

(a)   (b)   (c)

Fig. 6.2.6 Mesh distribution for solution of the Poisson pressure equation.

In fact, numerical experiments have shown that the method is very efficient for Re varying between 1 and 10^3 according to a reasonable choice of the mesh size $\Delta x, \Delta y$.

We now present some schemes that have less restrictive conditions of stability.

6.2.6.1 *The leapfrog DuFort–Frankel scheme*: This scheme was used by Chorin (1967) when he introduced the artificial compressibility method. The scheme can be written

$$\frac{1}{2\,\Delta t}(u^{n+1}_{i+1/2,j} - u^{n-1}_{i+1/2,j}) + a^n_{i+1/2,j} + \Delta^1_x p^n_{i+1/2,j} - \frac{1}{Re}\,\nabla^2_h u^n_{i+1/2,j}$$

$$+ \frac{\Delta t^2}{Re}\left(\frac{1}{\Delta x^2} + \frac{1}{\Delta y^2}\right)\Delta_{tt} u^n_{i+1/2,j} = 0$$

$$\frac{1}{2\,\Delta t}(v^{n+1}_{i,j+1/2} - v^{n-1}_{i,j+1/2}) + b^n_{i,j+1/2} + \Delta^1_y p^n_{i,j+1/2} - \frac{1}{Re}\,\nabla^2_h v^n_{i,j+1/2} \qquad (6.2.30)$$

$$+ \frac{\Delta t^2}{Re}\left(\frac{1}{\Delta x^2} + \frac{1}{\Delta y^2}\right)\Delta_{tt} v^n_{i,j+1/2} = 0$$

$$\frac{1}{2\,\Delta t}(p^{n+1}_{i,j} - p^{n-1}_{i,j}) + c^2(\Delta^1_x u^n_{i,j} + \Delta^1_y v^n_{i,j}) = 0$$

where

$$\Delta_{tt} f^n = \frac{1}{\Delta t^2}(f^{n+1} - 2f^n + f^{n-1})$$

The partially implicit nature of the scheme leads to good stability properties and does not make the computations difficult since $u^{n+1}_{i+1/2,j}$ and $v^{n+1}_{i,j+1/2}$ can be explicitly calculated. Again, the scheme is not consistent during the transient stage: The main part of the truncation error, for the first equation (6.2.30) for example, is

$$\frac{\Delta t^2}{Re}\left(\frac{1}{\Delta x^2} + \frac{1}{\Delta y^2}\right)\frac{\partial^2 u}{\partial t^2}$$

which tends toward zero if a steady state is reached. Due to the presence of three levels in time, the general analysis of stability is complicated. Partial results are obtained by considering the following approximate cases:

(i) For the advection–diffusion equation analogous to (6.2.9), the stability condition is $\sqrt{2}(u_0^2 + v_0^2)^{1/2}\Delta t/\Delta x \leq 1$, where $\Delta y = \Delta x$ (Cushman-Roisin, 1984).

(ii) For the inviscid linearized equations, Chorin has found the conditions $(u_0^2 + v_0^2)^{1/2} < c$ and $\sqrt{2}\,(1 + \sqrt{5})c\,\Delta t/\Delta x < 1$. Note that Chorin's analysis is made on the standard mesh (Fig. 6.2.5a).

One of the disadvantages of the leapfrog scheme is that, if viscous terms are not present, it becomes unstable in the nonlinear case; therefore, numerical instabilities can be experienced in cases of very large Reynolds numbers.

6.2.6.2 *The upwind scheme*: Condition (6.2.10), which can be very restrictive in the case of large Reynolds numbers, can be relaxed if upwind differences are used for the approximation of the convective term $\mathbf{A}(\mathbf{V})$. In this case, however, if two-point noncentered differences are considered, the resulting accuracy is only of first order.

6.2.6.3 *The alternating-direction-implicit (ADI) schemes*: The use of ADI schemes is more expensive because tridiagonal matrices must be inverted, but their stability properties often make them very efficient. Consider equations (6.1.1a) and (6.2.3) in nonconservative form, and let us introduce the notations

$$\mathbf{V}_h = (u_{i+1/2,j},\ v_{i,j+1/2}),\qquad p_h = p_{i,j}$$

Then, the ADI scheme is written

$$\frac{2}{\Delta t}(\tilde{\mathbf{V}}_h - \mathbf{V}_h^n) + u_h^n\,\Delta_x^*\tilde{\mathbf{V}}_h + v_h^n\,\Delta_y^{**}\mathbf{V}_h^n$$

$$-\frac{1}{\mathrm{Re}}(\Delta_{xx}\tilde{\mathbf{V}}_h + \Delta_{yy}\mathbf{V}_h^n) + \nabla_h^+ p_h^n = 0 \tag{6.2.31a}$$

$$\frac{2}{\Delta t}(\mathbf{V}_h^{n+1} - \tilde{\mathbf{V}}_h) + u_h^n\,\Delta_x^{**}\tilde{\mathbf{V}}_h + v_h^n\,\Delta_y^*\mathbf{V}_h^{n+1}$$

$$-\frac{1}{\mathrm{Re}}(\Delta_{xx}\tilde{\mathbf{V}}_h + \Delta_{yy}\mathbf{V}_h^{n+1}) + \nabla_h^+ p_h^n = 0 \tag{6.2.31b}$$

$$\frac{1}{\Delta t}(p_h^{n+1} - p_h^n) + c^2\,\nabla_h^-\cdot\mathbf{V}_h^{n+1} = 0 \tag{6.2.31c}$$

where $\nabla_h^\pm = (\Delta_x^\pm,\ \Delta_y^\pm)$, $\Delta_x^+ f_{l,m} = (f_{l+1,m} - f_{l,m})/\Delta x$, $\Delta_y^+ f_{l,m} = (f_{l,m+1} - f_{l,m})/\Delta y$, $\Delta_x^- f_{l,m} = (f_{l,m} - f_{l-1,m})/\Delta x$, and $\Delta_y^- f_{l,m} = (f_{l,m} - f_{l,m-1})/\Delta y$; u_h and v_h are approximations to u and v as in Eq. (6.2.7a).

Generally, the difference operators corresponding to each direction are identical in both steps, i.e.,

$$\Delta_x^* = \Delta_x^{**},\qquad \Delta_y^* = \Delta_y^{**}$$

and, in particular,

$$\Delta_x^* = \Delta_x^{**} = \Delta_x^0,\qquad \Delta_y^* = \Delta_y^{**} = \Delta_y^0 \tag{6.2.32}$$

where the centered operators Δ_x^0, Δ_y^0 have been defined in Section 6.2.2. In this case, it is possible to make scheme (6.2.31) more implicit by computing at the first step a provisional value $\tilde{p}_h$ with the artificial compressibility equation and then using this value in the second step.

At each time step, the computational effort is to solve linear algebraic sytems with tridiagonal matrices. The inversion is then easily accomplished by the method of factorization. However, a sufficient condition for the factorization

method to work is that the matrix to be inverted satisfies the conditions
(2.2.7a). If $0 \leqslant |u_h^n|\text{Re }\Delta x \leqslant 2$ and $0 \leqslant |v_h^n|\text{Re }\Delta y \leqslant 2$, these conditions
are satisfied. If $|u_h^n|\text{Re }\Delta x > 2$, it is necessary that

$$\Delta t < \frac{2\text{ Re }\Delta x^2}{|u_h^n|\text{Re }\Delta x - 2} \tag{6.2.33}$$

and an analogous condition if $|v_h^n|\text{Re }\Delta y > 2$.

There is a way to partially avoid condition (6.2.33). It consists of considering upwind difference operators for the implicit part and downwind operators
for the explicit part (Peyret, 1971). So at each time step, the matrices are
strictly diagonally dominant, and at convergence—i.e., at steady state—the
accuracy is of second order. More precisely,

$$\Delta_x^* = \frac{1}{2}[(1 - \epsilon_u)\Delta_x^+ + (1 + \epsilon_u)\Delta_x^-],$$

$$\Delta_x^{**} = \frac{1}{2}[(1 + \epsilon_u)\Delta_x^+ + (1 - \epsilon_u)\Delta_x^-] \tag{6.2.34}$$

with $\epsilon_u = \text{sign }(u_h^n)$, $\Delta_x^\pm$ is as defined above, and definitions of Δ_y^* and Δ_y^{**} are
analogous to (6.2.34) with $\Delta_x^\pm$ replaced by $\Delta_y^\pm$ and ϵ_u by $\epsilon_v = \text{sign}(v_h^n)$. During
the transient stage, the scheme is of first-order accuracy, and at convergence
the truncation error is $O(\Delta t\, \Delta x, \Delta t\, \Delta y, \Delta x^2, \ldots)$.

The condition of stability of scheme (6.2.31), whatever the choice of
(6.2.32) or (6.2.34) and with a MAC mesh, is found to be

$$\Delta t < \frac{2\,\Delta x}{c} \quad \text{with } \Delta x = \Delta y \tag{6.2.35}$$

which can be less restrictive than (6.2.33) if c is chosen sufficiently small. In
numerical experiments it has been found necessary to employ the criterion
$\Delta t < 2\text{ Re }\Delta x^2$. Such a limitation is likely related to the effect of boundary
conditions (see Sections 6.5.2 and 6.5.6).

6.3 The Unsteady Navier–Stokes Equations

The most common methods used to solve the unsteady equations deal with a
Poisson equation for the pressure and with the momentum equations for the
computation of velocity. The MAC (marker-and-cell) method (Harlow and
Welsh, 1965) is the prototype of such a method. The MAC method was
initially devised to solve problems with free surfaces, but it can be applied to
any incompressible fluid flow. An analogous method was devised by Williams
(1969) for three-dimensional problems in which the finite-difference Poisson
equation was solved by using trigonometric expansion techniques.

A method has also been presented (Chorin, 1968; Temam 1969b) called the
projection method which, in some cases where explicit schemes are used, is

identical to the MAC method. It appears, however, in a different form than the MAC method.

Another type of method is also successfully used to solve the unsteady Navier–Stokes equations (Peyret, 1976). The method is based upon the use of the artificial compressibility technique applied at each time step. As we have already mentioned, use of the artificial compressibility technique is equivalent to the solution of a Poisson equation for the pressure. This fact, which was shown for the steady case, is also true in the unsteady case where the artificial compressibility method is used as an iterative procedure to solve the algebraic nonlinear system obtained at each time step.

We shall now present the projection method and discuss its relationship with the MAC method for the explicit case. Then, the implicit iterative method will be described. Finally, a few words will be said about a perturbation method that recently has produced interesting results.

6.3.1 The projection and MAC methods

6.3.1.1 *The projection method*: This method was proposed independently by Chorin (1968) and by Temam (1969b), while an explicit version of such a method was presented by Fortin et al. (1971). This explicit method is a fractional step method with first-order accuracy in time. At the first step, we compute explicitly a provisional value $\mathbf{V}^*$ with

$$\frac{\mathbf{V}^* - \mathbf{V}^n}{\Delta t} + \mathbf{A}(\mathbf{V}^n) - \frac{1}{\mathrm{Re}} \nabla^2 \mathbf{V}^n = 0 \qquad (6.3.1)$$

which is the momentum equation without a pressure gradient. Note that only the discretization in time is considered here. Then, at the second step, we correct $\mathbf{V}^*$ by considering the equations

$$\frac{\mathbf{V}^{n+1} - \mathbf{V}^*}{\Delta t} + \nabla p^{n+1} = 0 \qquad (6.3.2a)$$

$$\nabla \cdot \mathbf{V}^{n+1} = 0 \qquad (6.3.2b)$$

By taking the divergence of Eq. (6.3.2a) and by making use of (6.3.2b) which states that $\mathbf{V}^{n+1}$ must be a divergence-free vector, we get the Poisson equation

$$\nabla^2 p^{n+1} = \frac{1}{\Delta t} \nabla \cdot \mathbf{V}^* \qquad (6.3.3a)$$

The boundary condition for p is obtained by projecting the vector equation (6.3.2a) on the outward normal unit $\mathbf{N}$ to the boundary Γ. Thus, we obtain the Neumann condition

$$\left(\frac{\partial p}{\partial N}\right)^{n+1}_{\Gamma} = -\frac{1}{\Delta t} (\mathbf{V}^{n+1}_{\Gamma} - \mathbf{V}^*_{\Gamma}) \cdot \mathbf{N} \qquad (6.3.3b)$$

where $\mathbf{V}^*_{\Gamma}$ is the (not yet defined) value of $\mathbf{V}^*$ on Γ. The condition of compatibility for the Neumann problem is

$$\frac{1}{\Delta t} \int_\Omega \mathbf{\nabla} \cdot \mathbf{V}^* \, ds = -\frac{1}{\Delta t} \int_\Gamma (\mathbf{V}^{n+1} - \mathbf{V}^*) \cdot \mathbf{N} \, ds$$

and it is identically satisfied thanks to condition (6.1.6) which expresses the fact that the velocity on the boundary Γ has a zero total flux. It is important that the discretization with respect to space conserves the above compatibility condition.

To sum up, Eq. (6.3.1) gives $\mathbf{V}^*$, then the solution of the Neumann problem (6.3.3) gives p^{n+1}, and, finally, the velocity $\mathbf{V}^{n+1}$ is computed from (6.3.2a).

Before we discuss the treatment of the boundary condition, and more precisely the condition on $\mathbf{V}^*_\Gamma$, we must say a few words about the discretization in space. Chorin (1968) makes use of the mesh shown in Fig. 6.2.5a; the mesh shown in Fig. 6.2.5b was used by Fortin et al. (1971) and Ladevèze and Peyret (1974) but the method seems to be more efficient if the MAC mesh (Fig. 6.2.5c) is used. In the latter case, the projection method[†] becomes identical to the MAC method as long as the boundary conditions are not considered. If the MAC mesh is used along with centered finite difference as described above, the conditions of stability are given by Eqs. (6.2.10) and (6.2.11).

The essential feature of the projection method is that the numerical solution is independent of the value $\mathbf{V}^*_\Gamma$ and more precisely of $\mathbf{V}^*_\Gamma \cdot \mathbf{N}$, the only component of $\mathbf{V}^*_\Gamma$ which appears when the MAC mesh is used. This assertion is based upon the following two points: (i) $\mathbf{V}^*$ at inner points is independent of $\mathbf{V}^*_\Gamma$ because it is calculated by an explicit scheme; (ii) the value $\mathbf{V}^*_\Gamma \cdot \mathbf{N}$ appears in the Neumann problem (6.3.3) simultaneously in the right-hand side of the Poisson equation (6.3.3a), and in the Neumann condition (6.3.3b) and it cancels identically. In order to prove this assertion, it is sufficient to analyze the discretization of (6.3.3) for points located near the boundary Γ, since $\mathbf{V}^*_\Gamma$ appears in the problem only for these points. Let us consider the point P (Fig. 6.3.1). The approximation of (6.3.3a) at this point ($i = 1, j = m$) can be written

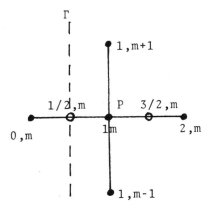

Fig. 6.3.1 Point P near the boundary Γ for the Poisson equation.

[†]The name comes from the fact that the second step can be considered as a projection of the vector field ($\mathbf{V}^*$) onto its subspace with zero divergence and satisfying appropriate boundary conditions.

$$\frac{1}{\Delta x}\left(\frac{p^{n+1}_{2,m} - p^{n+1}_{1,m}}{\Delta x} - \frac{p^{n+1}_{1,m} - p^{n+1}_{0,m}}{\Delta x}\right)$$

$$+ \frac{1}{\Delta y}\left(\frac{p^{n+1}_{1,m+1} - p^{n+1}_{1,m}}{\Delta y} - \frac{p^{n+1}_{1,m} - p^{n+1}_{1,m-1}}{\Delta y}\right) \qquad (6.3.4a)$$

$$= \frac{1}{\Delta t}\left(\frac{u^{*}_{3/2,m} - u^{*}_{\Gamma}}{\Delta x} + \frac{v^{*}_{1,m+1/2} - v^{*}_{1,m-1/2}}{\Delta y}\right)$$

and the Neumann condition (6.3.3b) is approximated by

$$\frac{1}{\Delta x}(p^{n+1}_{1,m} - p^{n+1}_{0,m}) = -\frac{1}{\Delta t}(u^{n+1}_{\Gamma} - u^{*}_{\Gamma}) \qquad (6.3.4b)$$

The quantity u^{*}_{Γ} is the value of u^{*} at point $(1/2, m)$. Now, it is easy to see that, when the value of $p^{n+1}_{1,m} - p^{n+1}_{0,m}$ given by (6.3.4b) is substituted into (6.3.4a), the unknown quantity u^{*}_{Γ} cancels from both sides of Eq. (6.3.4a). From this, one concludes that the solution is independent of the value u^{*}_{Γ}. In particular, we can choose $u^{*}_{\Gamma} = u^{n+1}_{\Gamma}$ and get a zero normal difference for the pressure on Γ. However, it must be clear that this zero-derivative condition is purely numerical and does not imply that the real pressure gradient is zero.

Finally, we must note that if $\mathbf{V}^{*}$ is eliminated from (6.3.1) and (6.3.2a) we obtain the scheme

$$\frac{\mathbf{V}^{n+1} - \mathbf{V}^{n}}{\Delta t} + \mathbf{A}(\mathbf{V}^{n}) + \nabla p^{n+1} - \frac{1}{Re}\nabla^{2}\mathbf{V}^{n} = 0 \qquad (6.3.5)$$

which is a first-order approximation in time of the momentum equation. Therefore, the stability of the projection method is the stability of the difference equation associated with (6.3.5). The stability conditions, discussed above, depend partly on the (centered or upwind) approximation of the convective term $\mathbf{A}(\mathbf{V})$. Equation (6.3.5) is nothing more than the equation to be considered in the MAC method described in Section 6.3.1.2.

As a matter of fact, the projection method is a particular case of the general splitting technique used to solve Eqs. (6.3.5) and (6.3.2b). This point of view permits a large variety of solution methods for the momentum equations. Another possibility is discussed in Section 6.3.1.3.

6.3.1.2 *The MAC method*: In the MAC method proposed in 1965 by Harlow and Welsh, the momentum equation (6.1.1a) is first discretized, then a combination of the two components of the discretized momentum equation yields the Poisson equation which must be satisfied by the pressure. More precisely, Eq. (6.1.1) is discretized as

$$\frac{1}{\Delta t}(u^{n+1}_{i+1/2,j} - u^{n}_{i+1/2,j}) + a^{n}_{i+1/2,j} + \Delta^{1}_{x}p^{n+1}_{i+1/2,j} - \frac{1}{Re}\nabla^{2}_{h}u^{n}_{i+1/2,j} = 0$$

$$(6.3.6a)$$

$$\frac{1}{\Delta t}(v_{i,j+1/2}^{n+1} - v_{i,j+1/2}^n) + b_{i,j+1/2}^n + \Delta_y^1 p_{i,j+1/2}^{n+1} - \frac{1}{\text{Re}} \nabla_h^2 v_{i,j+1/2}^n = 0$$

$$(6.3.6b)$$

$$\Delta_x^1 u_{i,j}^{n+1} + \Delta_y^1 v_{i,j}^{n+1} = 0 \qquad\qquad (6.3.6c)$$

where the operators are defined by Eq. (6.2.6).

The quantities $u_{i+1/2,i}^{n+1}$ as given by (6.3.6a) and $v_{i,j+1/2}^{n+1}$ as given by (6.3.6b) are introduced into the incompressibility condition (6.3.6c) so that (6.3.6c) becomes

$$\nabla_h^2 p_{i,j}^{n+1} = \Delta_x^1 \left(\frac{u_{i,j}^n}{\Delta t} - a_{i,j}^n + \frac{1}{\text{Re}} \nabla_h^2 u_{i,j}^n \right)$$

$$+ \Delta_y^1 \left(\frac{v_{i,j}^n}{\Delta t} - b_{i,j}^n + \frac{1}{\text{Re}} \nabla_h^2 v_{i,j}^n \right) \qquad (6.3.7a)$$

which is nothing more than a discrete approximation of Eq. (6.3.3a). As a result, the only change with respect to the projection method lies in the manner in which the Poisson equation is written, but this change involves a difference in the interpretation of the right-hand sides when points near a boundary are considered. Let us assume that Eq. (6.3.7a) is written at point P (Fig. 6.3.1) with $i = 1$, $j = m$:

$$\frac{1}{\Delta x^2}(p_{2,m}^{n+1} - 2p_{1,m}^{n+1} + p_{0,m}^{n+1}) + \frac{1}{\Delta y^2}(p_{1,m+1}^{n+1} - 2p_{1,m}^{n+1} + p_{1,m-1}^{n+1})$$

$$= \frac{1}{\Delta t}\left[\frac{1}{\Delta x}(u_{3/2,m}^n - u_{1/2,m}^n) + \frac{1}{\Delta y}(v_{1,m+1/2}^n - v_{1,m-1/2}^n) \right] \qquad (6.3.8a)$$

$$- \frac{1}{\Delta x}(\bar{a}_{3/2,m}^n - \bar{a}_{1/2,m}^n) - \frac{1}{\Delta y}(\bar{b}_{1,m+1/2}^n - \bar{b}_{1,m-1/2}^n)$$

where $\bar{a}_{i,j}^n = a_{i,j}^n - \text{Re}^{-1} \nabla_h^2 u_{i,j}^n$ and $\bar{b}_{i,j}^n = b_{i,j}^n - \text{Re}^{-1} \nabla_h^2 v_{i,j}^n$.

At the same time, we consider the boundary condition for the pressure given by the projection of the momentum equation (6.1.1a) on the normal $\mathbf{N}$ to the boundary Γ.

$$\left(\frac{\partial p}{\partial N} \right)_\Gamma^{n+1} = -\left[\frac{\partial \mathbf{V}}{\partial t} + \mathbf{A}(\mathbf{V}) - \frac{1}{\text{Re}} \nabla^2 \mathbf{V} \right]_\Gamma^{n+1} \cdot \mathbf{N} \qquad (6.3.7b)$$

At point P Eq. (6.3.7b) becomes

$$\left(\frac{\partial p}{\partial x} \right)_{1/2,m}^{n+1} = -\left(\frac{\partial u}{\partial t} + a - \frac{1}{\text{Re}} \nabla^2 u \right)_{1/2,m}^{n+1}$$

and discretized it becomes

$$\frac{1}{\Delta x}(p_{1,m}^{n+1} - p_{0,m}^{n+1}) = -\frac{1}{\Delta t}(u_{1/2,m}^{n+1} - u_{1/2,m}^n) - \bar{a}_{1/2,m}^n \qquad (6.3.8b)$$

where $\bar{a}_{1/2,m}$ has been evaluated at level n rather than $n + 1$.

The essential difference between the MAC method and the projection method is that in (6.3.4) the quantity u_Γ^* is considered an unknown to be defined; but the analog of u_Γ^*, which is $u_{1/2,m}^n + \Delta t\, \bar{a}_{1/2,m}^n$, needs a special approximation for $\bar{a}_{1/2,m}^n$ involving the undefined quantities $u_{-1/2,m}^n$ and $v_{0,m\pm1/2}^n$. Generally, the definition of these quantities corresponds to a special noncentered discretization for $\bar{a}_{1/2,m}^n$. Moreover, a simplified form of (6.3.7b) is often used.

However, considering Eq. (6.3.8) we can substitute the expression $p_{1,m}^{n+1} - p_{0,m}^{n+1}$, as given by (6.3.8b), into (6.3.8a) and ascertain that $\bar{a}_{1/2,m}^n$ disappears from both sides of the equation. Therefore, we can conclude that the quantities $u_{-1/2,m}^n$ and $v_{0,m\pm1/2}^n$ could be arbitrarily chosen in the current discretization of $\bar{a}_{1/2,m}$. On the other hand, $v_{0,m\pm1/2}^n$ appears also in $\bar{b}_{1,m\pm1/2}$, but here the problem is that which was discussed in Section 6.2.4 and is associated with the use of the MAC mesh.

In conclusion, we note that the numerical solution given by the MAC method is independent of the evaluation of the normal pressure gradient on Γ, provided the evaluation of $\bar{a}_{1/2,m}^n$ is consistent with the approximation of the right-hand side of the Poisson equation. This condition is essential in order to have the exact analogy with the canceling of u_Γ^* in Eq. (6.3.4). Finally, if $\bar{a}_{1/2,m}^n$ is taken equal to $-(u_\Gamma^{n+1} - u_\Gamma^n)/\Delta t$, which corresponds to a special choice of $u_{-1/2,m}^n$, we again obtain a zero normal pressure gradient on Γ.

The use of a homogeneous boundary condition for the pressure gradient was first considered by Fortin et al. (1971) in the projection method. Then, for the MAC method it was proposed by Easton (1972). And, in a different context, it was considered by Amsden and Harlow (1970) in the SMAC method.

6.3.1.3 *Implicit schemes*:

In the methods presented in the previous sections, the velocity $\mathbf{V}^{n+1}$ is computed by an explicit scheme; therefore, the time step is limited by the constraining parabolic stability criterion. This restriction can be avoided by the use of implicit schemes.

For example, in the case of the MAC method, Pracht (1971) has considered an implicit evaluation of the viscous term and explicit convective terms; Deville (1974, 1975) makes use of an ADI method. In the projection method, the implicit character of the scheme can appear either in the computation of the auxiliary velocity field $\mathbf{V}^*$ or in the approximation in time of the momentum equation analogous to (6.3.5). The first possibility has been considered by Chorin (1968) who used an ADI method, and by Fortin et al. (1971) who used a fractional-step method based upon a splitting of operators according to the directions x or y. Note that when $\mathbf{V}^*$ is computed with an implicit scheme, the question of the boundary condition for the pressure is not clear as in the case of an explicit calculation. Precisely, the previous proof was based on the fact that the provisional value $\mathbf{V}^*$ was computed with an explicit scheme that allowed us to ensure that $\mathbf{V}^*$ at inner points was independent of $\mathbf{V}^*$ on the boundary. However, theoretical results obtained by Temam (1969b, 1977) showed that the numerical solution $\mathbf{V}^{n+1}, p^{n+1}$ tends toward the exact solution when Δt, Δx, and $\Delta y \to 0$, if $\mathbf{V}^* \cdot \mathbf{N}$ on Γ is chosen equal to the exact boundary value $\mathbf{V}_{\Gamma}^{n+1} \cdot \mathbf{N}$. But, we do not know the behavior of the solution if this last condition is not satisfied.

The difficulty associated with the determination of $\mathbf{V}^*$ at the boundary can be avoided if the implicitness is introduced into the scheme before splitting. As an example, consider the implicit second-order accurate scheme (Adams–Bashforth/Crank–Nicolson scheme)

$$\frac{\mathbf{V}^{n+1} - \mathbf{V}^n}{\Delta t} + \frac{3\mathbf{A}^n - \mathbf{A}^{n-1}}{2} + \nabla p^{n+1/2} - \frac{\nabla^2 \mathbf{V}^{n+1} + \nabla^2 \mathbf{V}^n}{2\,\mathrm{Re}} = 0 \quad (6.3.9)$$

$$\nabla \cdot \mathbf{V}^{n+1} = 0 \qquad (6.3.10)$$

Equation (6.3.9) now replaces (6.3.5). The splitting technique is then analogous to the technique already described, i.e.,

$$\frac{\mathbf{V}^* - \mathbf{V}^n}{\Delta t} + \frac{3\mathbf{A}^n - \mathbf{A}^{n-1}}{2} - \frac{\nabla^2 \mathbf{V}^n}{2\,\mathrm{Re}} = 0 \qquad (6.3.11)$$

$$\frac{\mathbf{V}^{n+1} - \mathbf{V}^*}{\Delta t} + \nabla p^{n+1/2} - \frac{\nabla^2 \mathbf{V}^{n+1}}{2\,\mathrm{Re}} = 0 \qquad (6.3.12)$$

$$\nabla \cdot \mathbf{V}^{n+1} = 0 \qquad (6.3.13)$$

The Poisson equation for the pressure is obtained by taking the divergence of Eq. (6.3.12) and taking account of (6.3.13):

$$\nabla^2 p^{n+1/2} = \frac{1}{\Delta t}\, \nabla \cdot \mathbf{V}^* \qquad (6.3.14a)$$

The Neumann condition is again obtained by projecting (6.3.12) onto the unit normal $\mathbf{N}$ to the boundary Γ, i.e.,

$$\left(\frac{\partial p}{\partial \mathbf{N}}\right)_{\Gamma}^{n+1/2} = -\frac{1}{\Delta t}\left[\mathbf{V}^{n+1} - \mathbf{V}^* - \Delta t\,\frac{\nabla^2 \mathbf{V}^{n+1}}{2\,\mathrm{Re}}\right] \cdot \mathbf{N} \qquad (6.3.14b)$$

This boundary condition involves the unknown quantity $(\nabla^2 \mathbf{V}^{n+1}) \cdot \mathbf{N}$ on the boundary Γ. This quantity is expressed by using the continuity equation so that for a boundary $x = \mathrm{const}$, and using the MAC mesh (Fig. 6.3.1):

$$-(\nabla^2 \mathbf{V}^{n+1}) \cdot \mathbf{N}|_{\Gamma} = \left(\frac{\partial^2 u}{\partial x^2}\right)_{1/2,m}^{n+1} + \left(\frac{\partial^2 u}{\partial y^2}\right)_{1/2,m}^{n+1}$$

$$= -\left[\frac{\partial}{\partial x}\left(\frac{\partial v}{\partial y}\right)\right]_{1/2,m}^{n+1} + \left(\frac{\partial^2 u}{\partial y^2}\right)_{1/2,m}^{n+1} \qquad (6.3.14c)$$

The derivative with respect to x is then approximated by a first-order accurate or a second-order accurate noncentered difference involving the known value of $\partial v/\partial y$ at the boundary. Such a procedure has been used by Morchoisne

(1981). It was previously proposed in association with a reflection technique by Pracht (1971).

Concerning the value of $\mathbf{V}^*$ on the boundary Γ, the same reasoning used for the explicit scheme now allows us to claim that the solution $p^{n+1/2}$, $\mathbf{V}^{n+1}$ does not depend on this boundary value $\mathbf{V}_\Gamma^*$.

In the case of the splitting method described by Eqs. (6.3.11)–(6.3.14), the numerical algorithm is the following: (i) Solve Eq. (6.3.11) explicitly for $\mathbf{V}^*$. (ii) Equation (6.3.12) with the associated boundary condition for $\mathbf{V}^{n+1}$ and Eqs. (6.3.14) with arbitrary $\mathbf{V}^*$ on Γ are solved simultaneously and give $\mathbf{V}^{n+1}$ and $p^{n+1/2}$. This can be done iteratively by performing one Gauss-Seidel sweep for each quantity, alternatively.

Concerning the stability of the method, we note that if second-order-accurate centered differences are used for the approximation of the spatial derivatives in the MAC mesh, the resulting scheme for this case has much better properties than fully explicit schemes (see Section 2.7.4).

The implicit method just described has been examined in test cases by the authors. It has been checked that the constraint $\nabla \cdot \mathbf{V} = 0$ is well satisfied. Schemes analogous to Eqs. (6.3.9)–(6.3.10), but associated with a spectral approximation in space, have been considered in various forms of splitting by Orszag and Kells (1980), Taylor and Murdock (1980). The methods used in these works are reported in Chapter 8.

6.3.1.4 *Relationship with the artificial compressibility method*: It is interesting to note the relationship between the projection method (or the MAC method since they are identical when the staggered mesh is used) and the artificial compressibility method described in Section 6.2. Assume that the Poisson equation (6.3.3a) is solved by a simple iterative Jacobi-type procedure:

$$p^{n+1,m+1} - p^{n+1,m} - \lambda \, \nabla_h^2 p^{n+1,m} = -\frac{\lambda}{\Delta t} \, \nabla_h \cdot \mathbf{V}^*, \quad p^{n+1,0} = p^n \quad (6.3.15)$$

where m is the index of iteration and λ is a convergence parameter. The above procedure would be exactly the Jacobi procedure if

$$\lambda = \lambda_0 = \frac{1}{2} \frac{\Delta x^2 \, \Delta y^2}{\Delta x^2 + \Delta y^2}$$

[Note that (6.3.15) converges if $\lambda \le \lambda_0$.] Now, comparing (6.3.15) and (6.2.27), we conclude that if the projection method is used to compute a steady solution of the Navier–Stokes equations and if, at each time step, the Poisson equation for the pressure is not solved exactly but only one iteration $m \equiv 0$ is made, then the method becomes identical to the artificial compressibility method with $\lambda = c^2 \Delta t^2$. Obviously, it is possible to make more than one iteration at each time step, and thus we can obtain an algorithm that could have better convergence properties. In Fortin et al. (1971) and Ladevèze and Peyret (1974), twelve iterations were made at each time step.

6.3.2 An iterative method

An iterative method for solving the unsteady viscous equations can be developed based on (i) implicit discretization of the momentum and continuity equations and (ii) iterative solution of the resulting nonlinear algebraic system (Fortin, 1972; Peyret, 1976). The essential property of such a method is its good stability, so that the time step needs to be chosen only according to the accuracy. Moreover, it is simpler to construct implicit schemes of second-order accuracy in time than explicit schemes.

A very efficient scheme is the Crank–Nicolson scheme. Here, we consider the more general two-level scheme. So that the discretization with respect to time is

$$\frac{\mathbf{V}^{n+1} - \mathbf{V}^n}{\Delta t} + \theta \left[\mathbf{A}(\mathbf{V}^{n+1}) - \frac{\nabla^2 \mathbf{V}^{n+1}}{Re} \right]$$

$$+ (1 - \theta) \left[\mathbf{A}(\mathbf{V}^n) - \frac{\nabla^2 \mathbf{V}^n}{Re} \right] + \nabla p^{n+\tau} = 0 \qquad (6.3.16a)$$

$$\nabla \cdot \mathbf{V}^{n+1} = 0 \qquad (6.3.16b)$$

where θ is a constant, $0 \le \theta \le 1$. If $\theta = 0$ we get the explicit scheme, and if $\theta = 1$ a fully implicit scheme results. The Crank–Nicolson scheme corresponds to $\theta = \frac{1}{2}$. The number τ which defines the time $(n + \tau)\Delta t$ $(\tau > 0)$ is arbitrary and must be chosen in order to ensure the better truncation error associated with (6.3.16a). This is due to the fact that no time derivative of the pressure appears in the Navier–Stokes equations, and the pressure is defined with an arbitrary function of time. The main part of the truncation error (in time) associated with (6.3.16a) is

$$\frac{\Delta t}{2} \frac{\partial}{\partial t} \left\{ \frac{\partial \mathbf{V}}{\partial t} + 2 \theta \left[\mathbf{A}(\mathbf{V}) - \frac{\nabla^2 \mathbf{V}}{Re} \right] + 2\tau \nabla p \right\}$$

The scheme is of second-order accuracy in time if $\theta = \tau = \frac{1}{2}$. If $\theta = 0$ or $\theta = 1$, the scheme is only first-order accurate and τ can be chosen arbitrarily (generally $\tau = 1$).

The discretization in space makes use of the staggered MAC mesh described in Section 6.2.2 and, generally, second-order accurate centered differences as described in Section 6.2.2 are used. If $\theta < \frac{1}{2}$, the conditions of stability for scheme (6.3.16) are

$$\frac{\Delta t}{Re} \frac{\Delta x^2 + \Delta y^2}{\Delta x^2 \Delta y^2} \le \frac{1}{2(1 - 2\theta)}, \qquad (u^2 + v^2) \Delta t \, Re \le \frac{2}{1 - 2\theta}$$

If $\theta \ge \frac{1}{2}$, the scheme is (linearly) unconditionally stable.

After discretization with respect to space as described in Section 6.2.2, Eqs. (6.3.16a) and (6.3.16b) lead, at each time step, to a nonlinear algebraic system for the unknowns which are the values of velocity and pressure at mesh points, say $u_{i+1/2,j}^{n+1}$, $v_{i,j+1/2}^{n+1}$, $p_{i,j}^{n+\tau}$. Let us denote

$$u_h = u_{i+1/2,j}^{n+1}, \qquad v_h = v_{i,j+1/2}^{n+1}, \qquad p_h = p_{i,j}^{n+\tau}$$

then the finite-difference equations resulting from (6.3.16) are written in the symbolic form

$$L_u(u_h, v_h, p_h) = 0 \tag{6.3.17a}$$

$$L_v(u_h, v_h, p_h) = 0 \tag{6.3.17b}$$

$$D(u_h, v_h) = 0 \tag{6.3.17c}$$

where (6.3.17a, b) are the momentum equations and (6.3.17c) is the continuity equation. When these equations are written for all the discretization points of the computation domain, we obtain a nonlinear algebraic system that is solved with an iterative procedure.

The simpler iterative procedure that can be devised is the following one (noting that the index of iterations is m):

$$
\begin{aligned}
u_h^{m+1} - u_h^m + \kappa_1 L_u(u_h^m, v_h^m, p_h^m) = 0, \qquad & u_h^0 = u_{i+1/2,j}^n \\
v_h^{m+1} - v_h^m + \kappa_2 L_v(u_h^m, v_h^m, p_h^m) = 0, \qquad & v_h^0 = v_{i,j+1/2}^n \\
p_h^{m+1} - p_h^m + \lambda D(u_h^{m+1}, v_h^{m+1}) = 0, \qquad & p_h^0 = p_{i,j}^{n+\tau-1}
\end{aligned}
\tag{6.3.18}
$$

where κ_1, κ_2, and λ are parameters that must be chosen in order to ensure the convergence of the procedure. We note that the above iterative procedure is nothing more than the artificial compressibility method applied at each time step. The necessary conditions for convergence of procedure (6.3.18) can be obtained in the same way that stability for the artificial compressibility method was studied. If $\kappa_1 = \kappa_2 = \kappa$, $\Delta x = \Delta y$, and the convective terms are neglected (Stokes approximation), we obtain the following conditions:

$$\frac{\kappa}{\Delta x^2}\left(\frac{4}{\mathrm{Re}}\frac{\theta}{} + \frac{\Delta x^2}{2\,\Delta t} + 2\lambda\right) \le 1, \qquad \kappa > 0, \lambda > 0 \tag{6.3.19}$$

Numerical experience must be utilized to obtain optimal values of κ and λ, which lead to a minimal number of iterations to reach convergence of procedure (6.3.18). No theoretical results are known concerning the choice of optimal parameters. Assuming Δx, Δt, and κ are given, Eq. (6.3.19) shows that there exists a maximal value $\lambda_{\max}$ for λ given by

$$\lambda_{\max} = \frac{\Delta x^2}{2\,\kappa} - \left(\frac{2}{\mathrm{Re}}\frac{\theta}{} + \frac{\Delta x^2}{4\,\Delta t}\right)$$

which necessitates, since $\lambda > 0$, that

$$\frac{\kappa}{\Delta x^2} < \left(\frac{4\,\theta}{\text{Re}} + \frac{\Delta x^2}{2\,\Delta t}\right)^{-1}$$

It is convenient to use values of κ of the order of Δx^2 and experience shows that best convergence is obtained when λ is close (but not equal) to λ_{max}.

Possibly the delicate point of the method lies in the choice of parameters, since the number of iterations necessary to reach convergence is very sensitive to this choice. As a result, it is recommended that one conducts sufficient preliminary tests to determine the optimal values before performing calculations.

In order to improve the convergence (the numerical program is also simpler), the iterative procedure described above can be modified by using in each of (6.3.18) the values of the unknowns at iteration $m + 1$ as soon as they are computed. The procedure then becomes a Gauss–Seidel technique while the previous iterative procedure was of Jacobi type. For this case the study of convergence by a stability analysis becomes difficult analytically because the amplification matrix entries are complex numbers. The study of the amplification matrix eigenvalues can be performed numerically.

Typical values of convergence parameters used in calculations are given in Section 6.4.2.

In conclusion, we note that it is possible to use time discretization other than the two-level scheme (6.3.16a). For example, the three-level second-order accurate scheme

$$\frac{1}{2\,\Delta t}[3\mathbf{V}^{n+1} - 4\mathbf{V}^n + \mathbf{V}^{n-1}] + \mathbf{A}(\mathbf{V}^{n+1}) - \frac{\nabla^2\mathbf{V}^{n+1}}{\text{Re}} + \nabla p^{n+1} = 0$$

$$\mathbf{V}\cdot\mathbf{V}^{n+1} = 0 \tag{6.3.20}$$

with a spatial discretization in the staggered MAC mesh was successfully used in various problems (Childress and Peyret, 1976; Blanschong and Hartmann, 1979; Peyret, 1981).

The fully implicit scheme (6.3.20) is unconditionally stable. If the iterative procedure (6.3.18) is associated with (6.3.20), the conditions of convergence are

$$\frac{\kappa}{\Delta x^2}\left(\frac{4}{\text{Re}} + \frac{3}{4}\frac{\Delta x^2}{\Delta t} + 2\lambda\right) \le 1, \qquad \kappa > 0,\ \lambda > 0 \tag{6.3.21}$$

The advantages of such a scheme compared to the Crank–Nicolson scheme are (i) economy of programming and (ii) better damping of harmonics of short wavelength which can be interesting in cases where the solution is highly variable (Richtmyer and Morton, 1967). On the other hand, as in all the three-level schemes, scheme (6.3.20) does not determine the solution at the first time step, given suitable initial condition. Therefore, the first time step must be treated using another scheme.

6.3.3 Relationship between the various methods

As noted previously for the artificial compressibility method, the iterative procedure just outlined is equivalent to the solution of a Poisson equation for the pressure which can be obtained by combining Eqs. (6.3.18) for u_h^{n+1} and v_h^{n+1}. In order to make clear the relationship between the present method and the methods presented in Section 6.3.2, we outline the construction of this equation. The iterative procedure applied to Eqs. (6.3.16) is written

$$\mathbf{V}^{n+1,m+1} - \mathbf{V}^{n+1,m} + \kappa[\nabla p^{n+\tau,m} + \mathbf{S}_1(\mathbf{V}^{n+1,m}) + \mathbf{S}_0(\mathbf{V}^n)] = 0 \quad (6.3.22a)$$

$$p^{n+\tau,m+1} - p^{n+\tau,m} + \lambda \, \nabla \cdot \mathbf{V}^{n+1,m+1} = 0 \quad (6.3.22b)$$

with

$$\mathbf{S}_1(\mathbf{V}^{n+1}) = \frac{\mathbf{V}^{n+1}}{\Delta t} + \theta\left[\mathbf{A}(\mathbf{V}^{n+1}) - \frac{\nabla^2\mathbf{V}^{n+1}}{\mathrm{Re}}\right]$$

$$\mathbf{S}_0(\mathbf{V}^n) = -\frac{\mathbf{V}^n}{\Delta t} + (1 - \theta)\left[\mathbf{A}(\mathbf{V}^n) - \frac{\nabla^2\mathbf{V}^n}{\mathrm{Re}}\right]$$

Now, after the elimination of $\mathbf{V}^{n+1,m+1}$, using (6.3.22a), Eq. (6.3.22b) becomes

$$p^{n+\tau,m+1} - p^{n+\tau,m} - \kappa\lambda \, \nabla^2 p^{n+\tau,m}$$
$$= -\lambda \nabla \cdot [\mathbf{V}^{n+1,m} - \kappa\mathbf{S}_1(\mathbf{V}^{n+1,m}) - \kappa\mathbf{S}_0(\mathbf{V}^n)] \quad (6.3.23)$$

This last equation represents a special iterative procedure to solve the Poisson equation for pressure which could be obtained by applying the operator $(\nabla \cdot)$ to Eq. (6.3.16a). As a matter of fact at convergence, when $m \to \infty$, Eq. (6.3.23) becomes

$$\nabla^2 p^{n+\tau} = - \nabla \cdot [\mathbf{S}_1(\mathbf{V}^{n+1}) + \mathbf{S}_0(\mathbf{V}^n)] \quad (6.3.24)$$

since $\lim_{m\to\infty} (\nabla \cdot \mathbf{V}^{n+1,m}) = 0$. In particular, in the explicit case $\theta = 0$, we have $\nabla \cdot \mathbf{S}_1(\mathbf{V}^{n+1}) = (\nabla \cdot \mathbf{V}^{n+1})/\Delta t = 0$; therefore we obtain exactly the Poisson equation which is usually considered in the projection method [Eq. (6.3.3a)] or in the MAC method [Eq. (6.3.7a)]. We note that the explicit scheme $\theta = 0$, associated with the artificial compressibility technique, was introduced by Chorin (1968) and used, in particular, by Viecelli (1971) and Liu (1976). Viecelli has shown how it is possible to advance the computations in order to have, instead of a Jacobi-type procedure (6.3.22) a Gauss–Seidel procedure. In order to complete the analogy between the iterative method in the explicit case $\theta = 0$ and the MAC or projection method, we recall that careful treatment of the right-hand side of the Poisson equation and the Neumann condition at a boundary makes the numerical solution independent of the outside value $\mathbf{V}_0$ in the MAC method or the wall value $\mathbf{V}_I^*$ in the projection method. This "similarity" between the treatment of the right-hand side of the Poisson equation and the Neumann condition is automatically satisfied in the present implicit iterative method whatever the value of θ. This result is analogous to those obtained in Section 6.2.5.

6.3.4 A perturbation (penalization) method

The following perturbation method has been proposed by Temam (1968). It makes use of

$$\epsilon p + \nabla \cdot \mathbf{V} = 0, \qquad \epsilon > 0, \ \epsilon \to 0 \qquad (6.3.25)$$

instead of the divergence equation (6.1.1b). It then is possible to eliminate p from the momentum equation, so that (6.1.1a) and (6.1.1b) are replaced by the single equation

$$\frac{\partial \mathbf{V}}{\partial t} + (\mathbf{V} \cdot \nabla)\mathbf{V} - \frac{\nabla(\nabla \cdot \mathbf{V})}{\epsilon} - \frac{\nabla^2 \mathbf{V}}{\mathrm{Re}} = -\frac{(\nabla \cdot \mathbf{V})\mathbf{V}}{2} \qquad (6.3.26)$$

which must be solved with $\epsilon \to 0$. Note that the term on the right-hand side of (6.3.26) has been introduced for stability considerations. It has been proven by Temam (1968) that the solution of (6.3.26) tends toward the solution of (6.1.1) when $\epsilon \to 0$. Although numerical difficulties can appear with the use of very small values of ϵ, the method associated with a finite-element approximation has been successfully used by Hughes et al. (1979) and Bercovier and Engelman (1979) (with $\epsilon \approx 10^{-2} \Delta x^2$) and, for natural convection problems by Marshall et al. (1978) and Upson et al. (1980).

6.4 Example Solutions for Primitive-Variable Formulation

6.4.1 Steady flow over a step

In order to demonstrate a finite-difference solution of the primitive-variable formulation, we consider the case of steady flow in a channel (Taylor and Ndefo, 1971) with a step as shown in Fig. 6.4.1. In studying this flow we employ the splitting method of Yanenko (1971) to solve the primitive-variable equation (6.1.1a) and the pressure equation

$$\nabla^2 p = 2\left(\frac{\partial u}{\partial x} \frac{\partial v}{\partial y} - \frac{\partial v}{\partial x} \frac{\partial u}{\partial y} \right) \qquad (6.4.1)$$

The initial and boundary conditions which are consistent only for the steady state are

$$\text{at } t = 0, \ 0 \le x \le \frac{L}{H}, \ 0 \le y \le 1; \quad u = 1, \ v = 0 \qquad (6.4.2)$$

$$\text{at } x = 0, \ \frac{h}{H} \le y \le 1; \quad u = 1, \ v = 0, \ \frac{\partial p}{\partial x} = \frac{1}{\mathrm{Re}} \frac{\partial^2 u}{\partial x^2} \qquad \text{for } t > 0$$

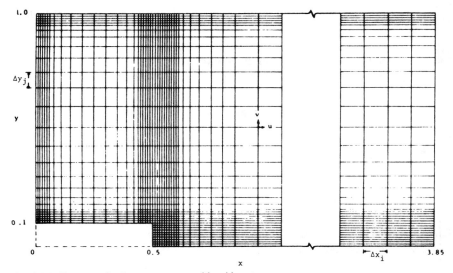

Fig. 6.4.1 Geometry for flow over a step with grid system.

at $x = \dfrac{L}{H}$, $0 \le y \le 1$; $\dfrac{\partial u}{\partial x} = \dfrac{\partial v}{\partial x} = 0$, $\dfrac{\partial p}{\partial x} = -\beta$ (const) for $t > 0$

at $y = \dfrac{h}{H}$ for $0 \le x \le \dfrac{l}{H}$, at $y = 1$ for $0 \le x \le \dfrac{L}{H}$ (6.4.3)

and at $y = 0$ for $\dfrac{l}{H} \le x \le \dfrac{L}{H}$; $u = 0$, $v = 0$, $\dfrac{\partial p}{\partial y} = \dfrac{1}{\mathrm{Re}} \dfrac{\partial^2 v}{\partial y^2}$

at $x = \dfrac{l}{H}$ for $0 \le y \le \dfrac{h}{H}$; $u = 0$, $v = 0$, $\dfrac{\partial p}{\partial x} = \dfrac{1}{\mathrm{Re}} \dfrac{\partial^2 u}{\partial x^2}$

Under these conditions, β is the prescribed pressure gradient for Poiseuille flow, l is the distance from the entrance to the step, L is the total length of the channel, H is the total height and h is the step height.

Applying splitting reduces the momentum equations to the form

Set I: $\dfrac{1}{2}\dfrac{\partial u}{\partial t} + \dfrac{\partial u^2}{\partial x} + \dfrac{\partial p}{\partial x} = \dfrac{1}{\mathrm{Re}}\dfrac{\partial^2 u}{\partial x^2}$

$\dfrac{1}{2}\dfrac{\partial v}{\partial t} + \dfrac{\partial uv}{\partial x} = \dfrac{1}{\mathrm{Re}}\dfrac{\partial^2 v}{\partial x^2}$

$\qquad\qquad\qquad\qquad\qquad\qquad\qquad\qquad$ (6.4.4)

Set II: $\dfrac{1}{2}\dfrac{\partial u}{\partial t} + \dfrac{\partial uv}{\partial y} = \dfrac{1}{\mathrm{Re}}\dfrac{\partial^2 u}{\partial y^2}$

$\dfrac{1}{2}\dfrac{\partial v}{\partial t} + \dfrac{\partial v^2}{\partial y} + \dfrac{\partial p}{\partial y} = \dfrac{1}{\mathrm{Re}}\dfrac{\partial^2 v}{\partial y^2}$

Since the pressure equation has no time derivative, it is not split and must be solved as a two-dimensional equation. Here, this equation is solved at each half-time-step, contrarily to the splitting method of projection (Fortin et al., 1971) in which the pressure is calculated once by time-step.

To each set of one-dimensional equations (6.4.4) is associated a set of boundary conditions deduced from Eq. (6.4.3):

Set I: at $x = 0$; $u = 1, v = 0$,   for all y

at $x = \dfrac{L}{H}$; $\dfrac{\partial u}{\partial x} = \dfrac{\partial v}{\partial x} = 0$ for all y

at $x = \dfrac{l}{H}$; $u = 0, v = 0$ for $0 \le y \le \dfrac{h}{H}$

Set II: at $y = \dfrac{h}{H}$ for $0 \le x \le \dfrac{l}{H}$ and $y = 0$ (6.4.5)

for $\dfrac{l}{H} \le x \le \dfrac{h}{H}$; $u = 0, v = 0$

at $y = 1$ for $0 \le x \le \dfrac{L}{H}$; $u = 0, v = 0$

The solution is obtained according to the following procedure.

1. Using the solution u^n and v^n at time $n \, \Delta t$, a provisional value $\bar{p}$ is computed from Eq. (6.4.1) and the boundary conditions written in (6.4.3). Using $\bar{p}$, Set I is then integrated for a half-time-step and gives intermediate values $\bar{u}$ and $\bar{v}$.

2. The pressure p^{n+1} is computed as in step 1, but using now $\bar{u}$ and $\bar{v}$. Finally, u^{n+1} and v^{n+1} are obtained by integrating Set II for another half-time-step, using $\bar{u}$, $\bar{v}$ and p^{n+1}.

Finite-differences are employed to solve the equations. The velocity and pressure are all defined at the nodes of the mesh. The spatial derivatives are approximated in the nonuniform grid system ($x_{i+1} - x_i = \Delta x_i$, $y_{j+1} - y_j = \Delta y_j$) by central differences in the form

$$\left(\frac{\partial u}{\partial x}\right)_{i,j} = \frac{u_{i+1,j} - u_{i-1,j}}{\Delta x_i + \Delta x_{i-1}} \tag{6.4.6}$$

$$\left(\frac{\partial^2 u}{\partial x^2}\right)_{i,j} = 2 \frac{\Delta x_{i-1} u_{i+1,j} - (\Delta x_i + \Delta x_{i-1}) u_{i,j} + \Delta x_i u_{i-1,j}}{\Delta x_i \Delta x_{i-1}(\Delta x_i + \Delta x_{i-1})} \tag{6.4.7}$$

For simplicity, an explicit time integration scheme is employed so that

$$\frac{\partial u}{\partial t} = 2\,\frac{\bar{u} - u^n}{\Delta t} \quad \text{for Set I,} \qquad \frac{\partial u}{\partial t} = 2\,\frac{u^{n+1} - \bar{u}}{\Delta t} \quad \text{for Set II} \tag{6.4.8}$$

An implicit approximation to the viscous terms could have been employed to increase stability since the split equations are one-dimensional and easily solved by factorization.

The stability conditions of the scheme are of one-dimensional type [Eq. (2.6.19)]. The truncation error (without considering boundary conditions) is $O(\Delta t)$. The incompressibility equation is satisfied to $O(\Delta t^2)$. At steady state, the error remains $O(\Delta t)$: this fact is common to splitting methods. The question of accuracy resulting from the treatment of boundary conditions was considered in Sections 2.7.3 and 2.8.4. In the present example, the use of the straightforward split system for the intermediate values has given good results.

The Neumann condition for pressure was discretized by one-sided differences. The difficulty concerning the pressure at the corner was overcome by extending the normal derivatives up to the corner on each side and averaging the result. The difference equations for pressure were solved both by Jacobi iteration and by over-relaxation. It was found that over-relaxation reduced the computation time by about a factor of 3 from that required by the simple Jacobi procedure.

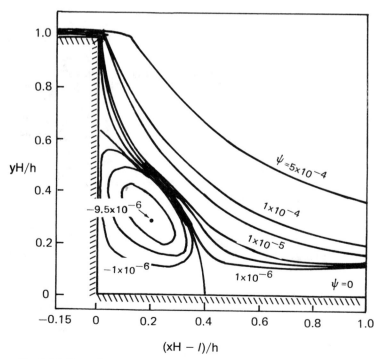

Fig. 6.4.2 Streamline patterns for Re = 100.

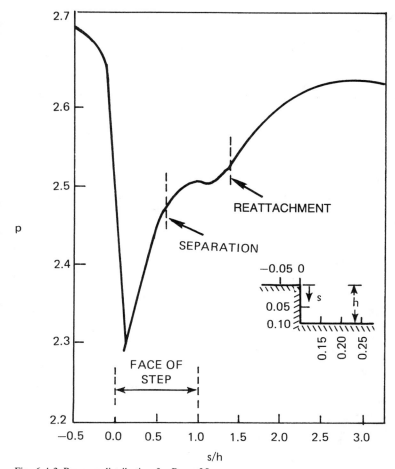

Fig. 6.4.3 Pressure distribution for Re = 25.

The described splitting scheme was utilized to compute the flow for Reynolds numbers of 25 and 100. The downstream pressure gradient β was chosen to have the value

$$\beta = \frac{-12}{Re} \frac{H - h}{H}$$

For the calculations, a step size ranging from $\Delta x = \Delta y = 0.01$ to $\Delta x = 0.1$ and $\Delta y = 0.09$ was used for the step channel. The arrangement of the grid for the step is shown in Fig. 6.4.1. The time step for all the calculations was chosen to be $\Delta t = 0.001$. Selected results obtained for flow at Reynolds number of 25 and 100 are shown in Figs. 6.4.2 and 6.4.3. These results were tested for conservation of mass and they showed a total mass loss of 2% in the complete calculation. In Fig. 6.4.2, the streamline patterns near the base are shown. Note that separation appears to occur at about $2/3$ the step height instead of at the top. Figure 6.4.3 shows the local behavior of the pressure in this region. Note that the pressure on the face of the step first decreases sharply, then begins to increase until the adverse pressure gradient is sufficient to induce separation.

This brief example indicates the basics of the splitting method. More details can be found in Taylor and Ndefo (1971). A special method for treating the singular nature of the flow near the corner can be found in Ladevèze and Peyret (1974).

6.4.2 Unsteady horizontal jet in a stratified fluid

This example (Peyret and Rebourcet, 1981, 1982) solution of the Navier–Stokes equations in primitive variables (using the method of Section 6.3.2) concerns an unsteady laminar jet flowing into a stratified fluid at rest in a semi-infinite channel. The domain in which the flow takes place is shown in Fig. 6.4.4. The fluid initially at rest in the channel of height $2H$ is stably stratified by thermal effects so that

$$T_s(y) = \frac{(T_2 - T_1)y}{2H} + \frac{(T_2 + T_1)}{2} \qquad \text{with } T_2 > T_1$$

The density ρ and the temperature T are connected through the state law:

$$\rho = \rho_0[1 - \beta(T - T_0)] \qquad \text{with } T_0 = \frac{T_1 + T_2}{2}$$

where β is the volume coefficient of thermal expansions. The initial density $\rho_s(y)$ and the pressure $p_s(y)$ in the fluid correspond to the temperature $T_s(y)$. The pressure is hydrostatic so that

$$p_s(y) = -g\rho_0\left[y - \frac{\beta}{4H}(T_2 - T_1)y^2 \right] + \text{const}$$

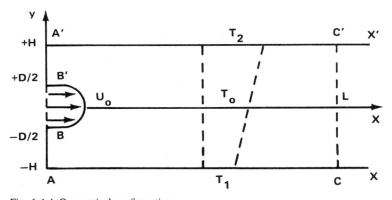

Fig. 6.4.4 Geometrical configuration.

where g is the gravitational acceleration. At $t = 0$, a fluid having the same physical properties as the fluid at rest is injected into the channel with a maximum velocity of U_0 through the slot BB' (of height D). The temperature of the injected fluid is constant and equal to T_0, and its density is ρ_0, i.e., the corresponding values of the stratified fluid at rest at $y = 0$.

As usual in such problems, it is convenient to introduce the perturbation of pressure π and temperature Θ with respect to the corresponding values p_s and T_s in the fluid at rest. Also dimensionless variables are defined by using the following reference quantities: $L_* = 2D$ for the length, L_*/U_0 for the time, U_0 for the velocity, $\rho_0 U_0^2$ for the pressure, and $(T_2 - T_1)/\eta$ for the temperature with $\eta = H/D$. In using the same symbols for the dimensionless variables, the Navier–Stokes within the Boussinesq approximation are

$$\frac{\partial \mathbf{V}}{\partial t} + (\mathbf{V} \cdot \nabla)\mathbf{V} + \nabla\pi - \frac{\nabla^2 \mathbf{V}}{\mathrm{Re}} - \mathrm{Ri}\,\Theta\mathbf{j} = 0 \qquad (6.4.9)$$

$$\nabla \cdot \mathbf{V} = 0 \qquad (6.4.10)$$

$$\frac{\partial \Theta}{\partial t} + \mathbf{V} \cdot \nabla\Theta + \mathbf{V} \cdot \mathbf{j} - \frac{\nabla^2 \Theta}{\mathrm{Re}\,\mathrm{Pr}} = 0 \qquad (6.4.11)$$

where $\mathbf{j}$ is the unit vector in the vertical direction and where the Reynolds number Re, the Richardson number Ri, and the Prandtl number Pr are defined by

$$\mathrm{Re} = \frac{L_* \rho_0 U_0}{\mu}, \qquad \mathrm{Ri} = \beta g \left(\frac{L_*}{U_0}\right)^2 \frac{T_2 - T_1}{2H}, \qquad \mathrm{Pr} = \frac{\mu c_p}{k}$$

The initial condition at $t = 0$ are $u = v = \Theta = 0$, and the boundary conditions are as follows:

On BB': $u = (1 - 16\,y^2)\,\phi(t)$, $v = 0$, $\Theta = -y$

On AB and $B'A'$: $u = v = \dfrac{\partial \Theta}{\partial x} = 0$

On AX and $A'X'$: $u = v = \Theta = 0$

where $\phi(t)$ is a function that allows a progressive intrusion of the fluid: $\phi(0) = 0$, $\phi(t) = 1$ for $t \geq t_0$. Although the particular choice of $\phi(t)$ had no influence on the established flow, its smoothness [i.e., $\phi'(0) = \phi'(t_0) = 0$] has a large effect on the number of iterations needed to solve the problem (by the method described in Section 6.3.2) during the transient state $t \leq t_0$. In the results shown here, the function $\phi(t)$ is chosen as a polynomial of fourth order and $t_0 = 0.25$.

The main numerical difficulty with the present problem is related to the unboundedness of the physical domain in which the flow takes place. The solution used here is to bound the domain with an artificial boundary CC' (Fig.

6.4.4) and determine the flow quantities on this boundary by suitable equations allowing the fluid to freely leave the domain without perturbing the upstream region or creating numerical instabilities. This problem, which is already delicate for nonstratified fluids because of the velocity–pressure formulation of the Navier–Stokes equations, is enhanced here by the stratification of the fluid and the propagation of internal waves which is associated with it.

Near the boundary CC', the flow remains unsteady for the value of time considered here and the presence of eddies propagating downstream makes it very different from a Poiseuille-type flow as in the steady problem studied in Section 6.4.1. Various treatments to determine flow quantities on CC' are reported by Peyret and Rebourcet (1981). The following conditions gave the best results:

$$\frac{\partial^2 \Theta}{\partial x^2} = 0, \qquad \frac{\partial v}{\partial x} = 0$$

$$\frac{\partial u}{\partial x} = -\frac{\partial v}{\partial y}, \qquad \frac{\partial \pi}{\partial y} = -\frac{\partial v}{\partial t} - v\frac{\partial v}{\partial y} + \frac{1}{\mathrm{Re}}\frac{\partial^2 v}{\partial y^2} + \mathrm{Ri}\ \Theta$$

When included in the general solution procedure, the above equations give successive iterative values of Θ, v, u, and π at the downstream boundary. The pressure π is obtained by an integration of $\partial\pi/\partial y$ along CC' performed by imposing $\pi = 1$ at the first point near C.

Another technique that could be used (Kao et al., 1978) consists of mapping the semi-infinite interval $0 \leq x < \infty$ onto the finite one $0 \leq \bar{x} < 1$ by means of a coordinate transformation $\bar{x} = 1 - \exp(-ax)$, $a = \mathrm{const}$. Then, the conditions of Poiseuille-type flow would be imposed at "infinity" $\bar{x} = 1$. A general discussion of the choice of stretched coordinates has been made by Sills (1969).

The equations of motion (6.4.9)–(6.4.11) are solved by using the method described in Section 6.3.2 with the particular choice of the Crank–Nicolson discretization parameters of $\frac{1}{2}$. In the staggered MAC mesh of Fig. 6.2.1, the temperature Θ is defined at the center of the cell and denoted by $\Theta_{i+1/2,j+1/2}$. The forcing term in (6.4.9) appears only in the v equation and is evaluated by the average value between n and $n + 1$ as well as $(i + \frac{1}{2}, j + \frac{1}{2})$ and $(i - \frac{1}{2}, j + \frac{1}{2})$. Equation (6.4.11) is discretized in the same manner as the momentum equations except that the derivatives in the convective term $\mathbf{V} \cdot \nabla\Theta$ are approximated by fourth-order accurate centered five-point differences. The use of such high-order approximation is made in order to minimize the associated truncation error compared to the diffusive term $(\mathrm{Re}\ \mathrm{Pr})^{-1}\ \nabla^2\Theta$.

The unknowns $u_{i+1/2,j}^{n+1}$, $v_{i,j+1/2}^{n+1}$, $p_{i,j}^{n+1/2}$, and $\Theta_{i+1/2,j+1/2}^{n+1}$ denoted, respectively, by u_h, v_h, p_h, and Θ_h are calculated by the Gauss–Seidel version of the iterative procedure described by Eq. (6.3.18) with a supplementary equation for the temperature

$$\Theta_h^{m+1} - \Theta_h^m + \chi E(u_h^{m+1}, v_h^{m+1}, \Theta_h^m) = 0$$

where the parameter χ must satisfy the convergence conditions

$$\frac{2\chi}{\Delta x^2}\left(\frac{1}{\text{Re Pr}} + \frac{\Delta x^2}{4\Delta t}\right) \leq 1, \qquad \chi > 0 \qquad (\Delta y = \Delta x) \qquad (6.4.12)$$

The form of criteria (6.3.19) and (6.4.12) suggests that one use $a = \kappa/\Delta x^2$, $b = \chi/\Delta x^2$, and $c = \lambda/\Delta x$ rather than κ, χ, and λ. Typical values of a, b, and c leading to convergence (determined from numerical tests) of the iterative procedure are given in Table 6.4.1.

Convergence is assumed when $\max\{|L_u|, |L_v|, |D|, |E|\} \leq 0.25 \times 10^{-2}$. The maximum is generally given by the v equation; the divergence equation is then satisfied at less that 10^{-3}. The number of iterations needed to obtain convergence at each time step is about 35 in the established regime $(t > t_0 = 0.25)$ for the case Re $= 250$, Ri $= 64$. This number can reach a few hundred during the start-up phase $(0 < t \leq t_0)$. A calculation with 80×20 points $(\Delta x = \Delta y = 1/16)$ and a final time $t = 2$ $(\Delta t = 1/64)$ requires 45 minutes of CPU time (CDC 6600).

Preliminary numerical tests associated with physical arguments have shown (Peyret and Rebourcet, 1982) that, for the values of the physical parameters considered, the flow remains symmetrical with respect to the axis $y = 0$. Consequently, the hypothesis of symmetry is assumed, so that the calculations are performed only in a half-domain. The results illustrated in Figs. 6.4.5 and 6.4.6 are for the case Re $= 250$, Ri $= 64$, Pr $= 10$, and Ri $= 0$ (neutral stratification). The geometrical parameters are $\eta = H/D = 2.3125$ and $\xi = L/2D = 5$.

Figure 6.4.5 shows the instantaneous streamlines at different times. The effect of stratification is evident if one compares Fig. 6.4.5c to Fig. 6.4.6 which corresponds to flow without gravity effects. In the latter case, the eddies created near the entrance remain attached to the wall and the expansion of the streamlines are prevented only by the walls. On the other hand, when stratification is present, the first eddies are propagated away while others are created near the entrance with opposite rotation and smaller magnitude. As a result, the jet is channeled in the axis region between two rows of eddies. The alternate rotation of the successive eddies makes the streamlines periodically expand or contract in the jet region. Consequently, the corresponding velocity

Table 6.4.1 Table of convergence parameters.

Re	$\Delta x = \Delta y$	Δt	a	b	c
10	1/10	1/50	0.40	0.40	10.50
100	1/16	1/64	0.75	0.75	9.50
250	1/16	1/64	0.82	0.82	8.70

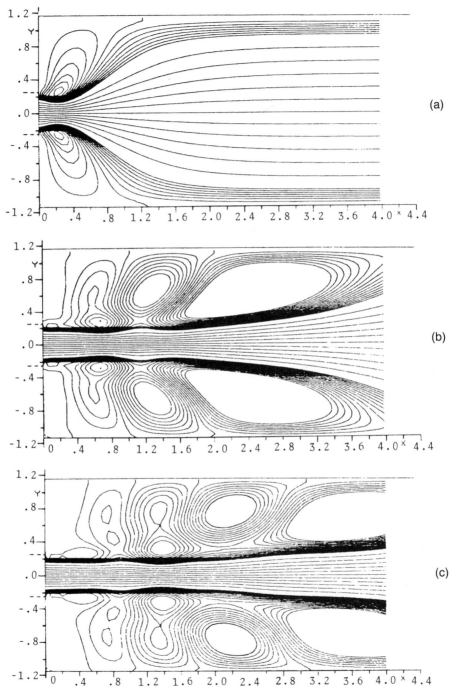

Fig. 6.4.5 Instantaneous streamlines at various time, Re = 250, Ri = 64, Pr = 10; (a) $t = 0.50$, (b) $t = 1.50$, (c) $t = 2.0$.

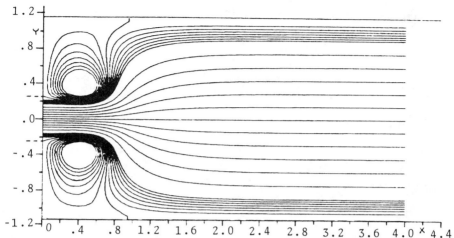

Fig. 6.4.6 Instantaneous streamlines, $t = 2.0$, Re = 250, in the case without gravity effects, Ri = 0.

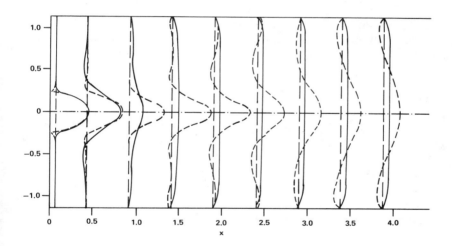

Fig. 6.4.7 Profiles of horizontal velocity u at $t = 2.0$: —— Re = 250, Ri = 0; --- Re = 250, Ri = 64, Pr = 10.

has an oscillatory character with maximal values larger than the entry velocity. Finally, it should also be noted that the magnitude of the velocity outside the jet is rather small (see Fig. 6.4.7) and such a fact is not represented by the above streamlines maps that give only a qualitative picture of the flow.

6.5 The Stream-Function Vorticity Formulation and Solution Approaches

6.5.1 The steady equations

The stream-function vorticity equations can be written in the case of a steady plane flow in the form:

$$\left[L(\Psi) - \frac{1}{\mathrm{Re}} \nabla^2 \right] \omega = 0 \tag{6.5.1a}$$

$$\nabla^2 \Psi + \omega = 0 \tag{6.5.1b}$$

where Ψ is the stream function, ω is the vorticity, and

$$L(\Psi) = u \frac{\partial}{\partial x} + v \frac{\partial}{\partial y}, \qquad \nabla^2 = \frac{\partial^2}{\partial x^2} + \frac{\partial^2}{\partial y^2}$$

where x, y are Cartesian coordinates and u and v are the components of the velocity $\mathbf{V}$ and are expressed in terms of the stream function by $u = \partial\Psi/\partial y$, $v = -\partial\Psi/\partial x$. Suitable boundary conditions as described in Section 6.5.3 must be added to Eq. (6.5.1) to complete the formulation.

6.5.2 The pseudo-unsteady methods

Various methods can be devised to solve Eq. (6.5.1). The first one is to consider the unsteady equation

$$\frac{\partial \omega}{\partial t} + \left[L(\Psi) - \frac{1}{\mathrm{Re}} \nabla^2 \right] \omega = 0 \tag{6.5.2a}$$

$$\nabla^2 \Psi + \omega = 0 \tag{6.5.2b}$$

and to solve this evolution problem in order to obtain the steady solution as $t \to \infty$. Although only a finite computing time is necessary to obtain a sufficiently small time derivative $\partial\omega/\partial t$—depending on the magnitude of the Reynolds number $\mathrm{Re} = UL/\nu$—this time is often still too long to make the method useful.

Another method, similar in fact to the unsteady approach, consists of modifying the unsteady equations in order to have two coupled parabolic equations instead of a parabolic and an elliptic equation. To accomplish this, one employs the equations

$$\frac{\partial \omega}{\partial t} + \left[L(\Psi) - \frac{1}{\mathrm{Re}} \nabla^2 \right] \omega = 0 \tag{6.5.3a}$$

$$\frac{\partial \Psi}{\partial t} - \alpha \left[\nabla^2 \Psi + \omega \right] = 0 \tag{6.5.3b}$$

where $\alpha > 0$ is a constant parameter. The advantage of such a formulation compared to (6.5.2) is that a step-by-step economical scheme can be used, in contrast to formulation (6.5.2) where the elliptic equation (6.5.2b) has to be solved at each time step. Although in this first procedure it is not necessary to solve Eq. (6.5.2b) exactly at each time step since only the steady state solution is of interest.

The use of the pseudo-unsteady system (6.5.3) seems to have been introduced by Burggraf (1966) in association with an explicit-implicit discretization with respect to time. Because of its explicit character, the applied scheme was convergent only for Reynolds number values that were sufficiently small in order to maintain stability.

A very efficient implicit technique for solving the system, which will improve convergence, is the alternating-direction-implicit method (ADI) (Chapter 2). This method is now extensively used in both its original form (Peaceman and Rachford, 1955) and in its generalized form (Douglas and Gunn, 1964) as employed by Mallinson and De Vahl Davis (1973).

Assuming a solution can be computed in a square domain Ω, $0 < x < 1$, $0 < y < 1$, with $\omega_{i,j}^n$, $u_{i,j}^n$, $v_{i,j}^n$, and $\Psi_{i,j}^n$ the respective approximations of ω, u, v, and Ψ at point $x_i = i\Delta x$, $y_j = j\Delta y$, and time $t_n = n\Delta t$, the Peaceman–Rachford ADI scheme for Eqs. (6.5.3) can then be written*

$$\frac{2}{\Delta t}(\tilde{\omega}_{i,j} - \omega_{i,j}^n) + \left(u_{i,j}^n \Delta_x^0 - \frac{\Delta_{xx}}{\text{Re}}\right)\tilde{\omega}_{i,j} + \left(v_{i,j}^n \Delta_y^0 - \frac{\Delta_{yy}}{\text{Re}}\right)\omega_{i,j}^n = 0$$

$$\text{(6.5.4a)}$$

$$\frac{2}{\Delta t}(\omega_{i,j}^{n+1} - \tilde{\omega}_{i,j}) + \left(u_{i,j}^n \Delta_x^0 - \frac{\Delta_{xx}}{\text{Re}}\right)\tilde{\omega}_{i,j} + \left(v_{i,j}^n \Delta_y^0 - \frac{\Delta_{yy}}{\text{Re}}\right)\omega_{i,j}^{n+1} = 0$$

$$\text{(6.5.4b)}$$

$$\frac{2}{\Delta t}(\tilde{\Psi}_{i,j} - \Psi_{i,j}^n) - \alpha(\Delta_{xx}\tilde{\Psi}_{i,j} + \Delta_{yy}\Psi_{i,j}^n + \omega_{i,j}^{n+1}) = 0 \qquad \text{(6.5.5a)}$$

$$\frac{2}{\Delta t}(\Psi_{i,j}^{n+1} - \tilde{\Psi}_{i,j}) - \alpha(\Delta_{xx}\tilde{\Psi}_{i,j} + \Delta_{yy}\Psi_{i,j}^{n+1} + \omega_{i,j}^{n+1}) = 0 \qquad \text{(6.5.5b)}$$

$$u_{i,j}^{n+1} = \Delta_y^0 \Psi_{i,j}^{n+1}, \qquad v_{i,j}^{n+1} = -\Delta_x^0 \Psi_{i,j}^{n+1} \qquad \text{(6.5.5c)}$$

Without imposed boundary conditions this ADI scheme is unconditionally stable in the linearized case. The effect of the boundary conditions on the stability has been studied by Bontoux (1978) and Bontoux et al. (1980), and, more precisely, the effect of coupling between Ψ and ω on the boundary.

Generally, the resulting linear algebraic systems at each step are solved by the method of factorization (Section 2.2.1). This method works well for any Δt if $|u|\,\text{Re}\,\Delta x \leq 2$ and $|v|\,\text{Re}\,\Delta y \leq 2$. But, if these conditions on the mesh

---

*The difference operators are defined in Section 6.2.2.

Reynolds number are not satisfied, the limitation on the time step is given by
Eq. (6.2.33). Although such a condition on Δt is not very restrictive because
Re is large, it can be relaxed by using noncentered approximations in (6.5.4a)
and (6.5.4b) for $\partial\omega/\partial x$ and $\partial\omega/\partial y$ according to the sign of u and v.

In order to preserve the tridiagonal nature of the matrix, such approximations
are two-point differences, therefore, only first-order accurate. In order to
recover the second-order accuracy at steady state, it is possible to use the
alternate difference technique expressed in Eq. (2.8.6) of Chapter 2. Such a
technique has been used for the vorticity equation by Daube and Ta Phuoc Loc
(1979) and Ta Phuoc Loc and Daube (1981).

Finally, note that the ADI scheme (6.5.4)–(6.5.5) can be used also if the
conservative form of Eq. (6.5.1a) is used. In this form, $L(\Psi)\omega =
\partial u\omega/\partial x + \partial v\omega/\partial y$, and the derivatives can be approximated with schemes
described in Section 6.5.6. When these approximations are substituted into
(6.5.4a) and (6.5.4b) the velocity components involved are evaluated at the old
level n.

6.5.3 Boundary conditions

The boundary condition associated with the solution of Eq. (6.5.1) in a
domain Ω with a boundary Γ is deduced from the problem defined in Section
6.1 for the primitive-variable equations. The velocity $\mathbf{V}$ is known on Γ, i.e.,
$\mathbf{V}=\mathbf{V}_\Gamma$ [Eqs. (6.1.4) and (6.1.6)]. In terms of the stream function, the boundary condition becomes

$$\left(\frac{\partial\Psi}{\partial\tau}\right)_\Gamma = \mathbf{V}_\Gamma\cdot\mathbf{N}, \qquad \left(\frac{\partial\Psi}{\partial N}\right)_\Gamma = -\mathbf{V}_\Gamma\cdot\boldsymbol{\tau}$$

where $\mathbf{N}$ and $\boldsymbol{\tau}$ are, respectively, the outward normal and tangent unit vectors
to Γ. By integrating the first equation along Γ, we obtain the general form of
the boundary conditions on Γ

$$\Psi = f(x,y), \qquad \frac{\partial\Psi}{\partial N} = g(x,y) \tag{6.5.6a}$$

where f must satisfy the total flux condition

$$\int_\Gamma (\nabla f\cdot\boldsymbol{\tau})\,ds = 0 \tag{6.5.6b}$$

Because f and g do not depend on time, for the ADI technique we can use
the same boundary values for both steps. Therefore, the knowledge of Ψ on Γ,
i.e., $\tilde{\Psi}_{i,j} = \Psi_{i,j}^{n+1} = f_{i,j}$ if $(x_i, y_j)\in\Gamma_h$, allows us to calculate $\tilde{\Psi}_{i,j}$ and $\Psi_{i,j}^{n+1}$
in the whole domain Ω_h thanks to Eqs. (6.5.5a) and (6.5.5b). The problem is
to derive boundary conditions for the vorticity. This is the main difficulty with
the stream-function formulation.

We now describe three methods to handle this problem. We denote by ϕ^{m+1}
any quantity $\tilde{\phi}$, ϕ^{n+1}, or any $(m + 1)$th iterate. The methods are

(i) Use $\omega_\Gamma^{m+1} = -(\nabla^2\Psi)_\Gamma^m$

(ii) Use $\omega_\Gamma^{m+1} - \omega_\Gamma^m - \gamma[(\partial\Psi/\partial N)_\Gamma^m - g] = 0$, where the parameter of convergence γ is positive.

(iii) Use Green's formula to derive as many supplementary equations for ω as there are boundary points.

These three techniques are now successively examined.

First technique (i). Let us consider the part $y = 0$ of the boundary Γ. The boundary conditions are

$$\Psi(x,0) = f_0(x) \tag{6.5.7a}$$

$$\frac{\partial\Psi(x,0)}{\partial y} = -g_0(x) \tag{6.5.7b}$$

By using $\omega_\Gamma = -(\nabla^2\Psi)_\Gamma$ we have

$$\omega(x,0) = -\frac{\partial^2\Psi(x,0)}{\partial x^2} - \frac{\partial^2\Psi(x,0)}{\partial y^2} \tag{6.5.8}$$

From Eq. (6.5.7a), we have $\partial^2\Psi(x,0)/\partial x^2 = f_0''(x)$, but it is necessary to evaluate the second derivative $\partial^2\Psi(x,0)/\partial y^2$. This is generally done by using Taylor expansions. For example, from the expansion

$$\Psi(x,\Delta y) = \Psi(x,0) + \Delta y\frac{\partial\Psi}{\partial y}(x,0) + \frac{\Delta y^2}{2}\frac{\partial^2\Psi(x,0)}{\partial y^2} + \cdots \tag{6.5.9}$$

we obtain for the point $(x,0)$

$$\frac{\partial^2\Psi}{\partial y^2} = \frac{2}{\Delta y^2}[\Psi(x,\Delta y) - f_0(x)] + \frac{2}{\Delta y}g_0(x)$$

with an error $O(\Delta y)$. By substituting this expression in (6.5.8) we obtain the *first-order-accurate formula*

$$\omega(x,0) = -f_0''(x) - \frac{2}{\Delta y^2}[\Psi(x,\Delta y) - f_0(x)] - \frac{2}{\Delta y}g_0(x) \tag{6.5.10}$$

This formula involves the unknown value $\Psi(x,\Delta y)$ which translates the coupling between ω at the boundary and Ψ into the domain. However, this coupling necessitates evaluation of $\Psi(x,\Delta y)$ at a previous time or at a previous iteration.

The first-order formula (6.5.10) was proposed in 1933 by Thom and has been frequently used since (Bryan, 1963; Greenspan, 1969).

A *second-order-accurate formula* can be derived in the same manner by using the Taylor expansion of $\Psi(x,2\Delta y)$ analogous to (6.5.9). A linear combination between expansions of $\Psi(x,\Delta y)$ and $\Psi(x,2\Delta y)$ allows us to derive an evaluation of $\partial^2\Psi(x,0)/\partial y^2$ accurate to second order. Hence, the following expression of $\omega(x,0)$ is obtained:

$$\omega(x,0) = -f_0''(x) + \frac{1}{2\,\Delta y^2}\left[\Psi(x,2\,\Delta y) - 8\,\Psi(x,\Delta y)\right.$$
$$\left. + 7\,f_0(x)\right] - \frac{3}{\Delta y}\,g_0(x) \qquad (6.5.11)$$

This second-order-accurate formula has been used by Wilkes (1963), Pearson (1965), and others. Other formulas of the same type have been considered by Briley (1971), Roache (1972), Orszag and Israeli (1974), and Gupta and Manohar (1979).

Another *formula with second-order accuracy* which involves the value $\omega(x,\Delta y)$ rather than $\Psi(x,2\,\Delta y)$ has been proposed by Woods (1954). It makes use of expansion (6.5.9) in order to define $\partial^2\Psi(x,0)/\partial y^2$ keeping the third-order term $\partial^3\Psi(x,0)/\partial y^3$. This latter term is then evaluated thanks to the equation

$$\frac{\partial^2\Psi}{\partial y^2} = -\omega - \frac{\partial^2\Psi}{\partial x^2}$$

which is differentiated with respect to y, so that

$$\frac{\partial^3\Psi(x,0)}{\partial y^3} = -\frac{\partial\omega(x,0)}{\partial y} - \frac{\partial^3\Psi(x,0)}{\partial x^2\,\partial y}$$

Noting that

$$\frac{\partial^3\Psi(x,0)}{\partial x^2\,\partial y} = -g_0''(x)$$

and

$$\frac{\partial\omega(x,0)}{\partial y} \cong \frac{1}{\Delta y}[\omega(x,\Delta y) - \omega(x,0)]$$

the following expression is found:

$$\omega(x,0) = -\frac{3f_0''(x)}{2} - \frac{3}{\Delta y^2}[\Psi(x,\Delta y) - f_0(x)] - \frac{\omega(x,\Delta y)}{2}$$
$$- \frac{3}{\Delta y}\,g_0(x) + \frac{\Delta y}{2}\,g_0''(x) \qquad (6.5.12)$$

This formula has been used in particular by Runchal et al. (1969) and Bozeman and Dalton (1973).

All three formulas (6.5.10), (6.5.11), and (6.5.12) necessitate evaluation of $\Psi(x,\Delta y)$, and $\Psi(x,2\,\Delta y)$, or $\omega(x,\Delta y)$ at a previous level m. Moreover, numerical experiments show that it is often necessary to include a relaxation.

Therefore, if formulas (6.5.10)–(6.5.12) are written in the form

$$\omega(x, 0) = F[\Psi(x, \Delta y), \; \Psi(x, 2\Delta y), \; \omega(x, \Delta y)] \qquad (6.5.13)$$

the value ω^{m+1} at the boundary is defined by

$$\hat{\omega}^{m+1} = F[\Psi^m(x, \Delta y), \; \Psi^m(x, 2\Delta y), \; \omega^m(x, \Delta y)]$$
$$\omega^{m+1} = \gamma \hat{\omega}^{m+1} + (1 - \gamma) \omega^m \qquad (6.5.14)$$

where γ is a relaxation parameter ($0 < \gamma \leq 1$). A study of the theoretical error associated with these various formulas has been conducted by Orszag and Israeli (1974) for a one-dimensional linear problem. In particular, it is found that the first-order formula (6.5.10) gives a numerical solution with an error of second order with respect to the exact solution. This is a new example of the fact (Kreiss, 1972) that a low-order treatment of the boundary conditions does not lead necessarily to a loss of the theoretical accuracy. However, this fact is true only if the order is considered, but as shown in Section 6.6, the results given by the various formulas can differ. An extensive comparison between results given by these formulas in the case of the driven cavity flow has been made by Gupta and Manohar (1979). Analogous comparisons are given in Section 6.5.8.

Second technique (*ii*). This seems to have been introduced by Dorodnytsin and Meller (1968) and can be written

$$\omega^{m+1}(x, 0) = \omega^m(x, 0) - \gamma \left[\frac{\Psi^m(x, \Delta y) - f_0(x)}{\Delta y} + g_0(x) \right] \qquad (6.5.15)$$

if a first-order-accurate difference is used to approximate the normal derivative $\partial \Psi(x, 0)/\partial y$. Obviously, a three-point second-order-accurate difference can also be used. At convergence when $\omega^{m+1} = \omega^m$, the condition $(\partial \Psi/\partial N)_\Gamma = g$ is satisfied. The parameter γ must be chosen in order to ensure convergence. An analysis of such a choice has been made by Israeli (1970, 1972).

Note the relationship between the two techniques just described. As a matter of fact, if formula (6.5.10) is combined with the relaxation procedure (6.5.14), we obtain

$$\omega^{m+1}(x, 0) = \omega^m(x, 0) - \frac{2\gamma}{\Delta y} \left\{ \left[\frac{\Psi^m(x, \Delta y) - f_0(x)}{\Delta y} + g_0(x) \right] \right.$$
$$\left. + \frac{\Delta y}{2} [f_0''(x) + \omega^m(x, 0)] \right\}$$

This last expression is identical with expression (6.5.15) (with γ replaced by $2\gamma/\Delta y$) except for the Δy-order term; but we recall that both formulas (6.5.10) or (6.5.15) are accurate to first order only.

Third technique (iii). This makes use of Green's formula applied to Eq. (6.5.1b) in the domain Ω' of Fig. 6.5.1. Green's formula is written

$$p\Psi(\lambda,\mu) = \int_{\Omega'} G(x,y;\lambda,\mu)\,\nabla^2\Psi(x,y)\,d\sigma$$
$$+ \int_{\Gamma\cup\Gamma'} \left[\frac{\partial G(x,y;\lambda,\mu)}{\partial N}\,\Psi(x,y) \right.$$
$$\left. - G(x,y;\lambda,\mu)\frac{\partial\Psi(x,y)}{\partial N} \right] ds \qquad (6.5.16)$$

where G is the fundamental solution of the Laplace operator, i.e., in two dimensions

$$G(x,y;\lambda,\mu) = \tfrac{1}{2}\ln\left[(x-\lambda)^2 + (y-\mu)^2\right]$$

The point (λ,μ) is any point of the plane (x,y) and

$$p = \begin{cases} 2\pi, & \text{if } (\lambda,\mu)\in\Omega' \\ \pi, & \text{if } (\lambda,\mu)\in\Gamma\cup\Gamma' \\ 0, & \text{if } (\lambda,\mu)\notin\overline{\Omega'} \end{cases}$$

The normal derivatives in the curvilinear integral refer to the (x,y) variables. If we consider the point (λ,μ) belonging to Γ and if we use Eq. (6.5.1b) and the boundary conditions (6.5.6), Eq. (6.5.16) becomes

$$\pi f = -\int_{\Omega'} G\omega\,d\sigma + \int_{\Gamma}\left(\frac{\partial G}{\partial N}f - Gg\right)ds + \int_{\Gamma'}\left(\frac{\partial G}{\partial N}\Psi - G\frac{\partial\Psi}{\partial N}\right)ds$$
$$(6.5.17)$$

Now, if this last equation is written for all boundary points (e.g., N_Γ points),

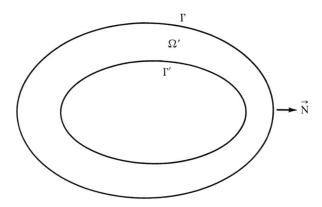

Fig. 6.5.1 Boundaries of domain Ω'.

we obtain N_Γ supplementary equations for ω which can be considered as boundary conditions. Here again, the values of Ψ and $\partial\Psi/\partial N$ in the curvilinear integral along Γ' are not known, but they can be evaluated at the previous time or previous iteration. A numerical approximation of the surface integral in (6.5.17) leads to an expression of the form

$$\int_{\Omega'} G(x,y;\lambda,\mu)\,\omega(x,y)\,d\sigma \cong \sum A_{i,j}(\lambda,\mu)\,\omega_{i,j}$$

where the summation has to be taken on all (inner Ω' and boundary $\Gamma\cup\Gamma'$) points. Then Eq. (6.5.17) yields

$$\sum A_{i,j}(\lambda,\mu)\,\omega_{i,j} = F^m(\lambda,\mu) \qquad \text{for } (\lambda,\mu)\in\Gamma \tag{6.5.18}$$

where F^m includes, in particular, the contribution of $\Psi^m_{\Gamma'}$ and $(\partial\Psi/\partial N)^m_{\Gamma'}$ at the previous level m.

A simple technique is to calculate (6.5.18) at each boundary point in order to obtain the boundary value ω^{m+1}_Γ through the iterative procedure.

$$\omega^{m+1}_\Gamma = \omega^m_\Gamma + \gamma\left(\sum A_{i,j}\,\omega^m_{i,j} - F^m\right) \tag{6.5.19}$$

Here again, γ is a parameter of convergence which must be positive.

Another technique is to again evaluate (6.5.18) for all boundary points in order to get an algebraic system which can be written

$$\mathbf{A}_\Gamma \mathbf{X}_\Gamma + \mathbf{A}_I \mathbf{X}_I = \mathbf{F}^m \tag{6.5.20}$$

where $\mathbf{X}_\Gamma$ is the vector whose elements are the N_Γ boundary values ω_Γ; $\mathbf{X}_I$ is the vector corresponding to the N_I inner values $\omega_{i,j}$; $\mathbf{A}_\Gamma$ and $\mathbf{A}_I$ are, respectively, $N_\Gamma \times N_\Gamma$ and $N_I \times N_I$ matrices; and $\mathbf{F}^m$ is the vector corresponding to $F^m(\lambda,\mu)$ in Eq. (6.5.18). Then, the following iterative procedure (with possible relaxation)

$$\mathbf{A}_\Gamma \mathbf{X}^{m+1}_\Gamma = -\mathbf{A}_I \mathbf{X}^m_I + \mathbf{F}^m$$

allows one to calculate $\mathbf{X}^{m+1}_\Gamma$ with only one inversion of matrix, i.e., $\mathbf{A}^{-1}_\Gamma$ which can be made only once.

The use of Green's formula to derive supplementary boundary conditions was first considered by Wu and Wahbah (1976). The inner boundary Γ' was not considered, so that the surface integral was calculated in the whole domain except in regions where the vorticity ω was zero. The method based on the use of the inner boundary Γ', that saves computer time, has been introduced and applied to the Navier–Stokes equations by Céa et al. (1981).

6.5.4 The iterative method

Instead of introducing a fictitious time and solving a parabolic system with a time-dependent method, the elliptic system (6.5.1) can be solved directly by means of an iterative procedure. First, Eqs. (6.5.1) are approximated by the difference equations

$$\left[L_h(\Psi_{i,j}) - \frac{\nabla_h^2}{\mathrm{Re}} \right] \omega_{i,j} = 0 \tag{6.5.21a}$$

$$\nabla_h^2 \Psi_{i,j} + \omega_{i,j} = 0 \tag{6.5.21b}$$

where the finite-difference operator $L_h(\Psi_{i,j})$ defined by

$$\begin{aligned}
L_h(\Psi_{i,j}) &= (\Delta_y^0 \Psi_{i,j}) \, \Delta_x^0 - (\Delta_x^0 \Psi_{i,j}) \, \Delta_y^0 \\
&= u_{i,j} \, \Delta_x^0 + v_{i,j} \, \Delta_y^0
\end{aligned} \tag{6.5.22}$$

is a centered approximation of the convective operation $\mathbf{V} \cdot \nabla$, and $\nabla_h^2 = \Delta_{xx} + \Delta_{yy}$. The accuracy of the finite-difference equations (6.5.21) is of second order.

Equations (6.5.21) are solved by the following iterative procedure (Greenspan, 1969; Runchal et al., 1969; Roache, 1975) which is characterized by index m. Let us assume that $\Psi_{i,j}^m$ and $\omega_{i,j}^m$ are known. The quantities $\Psi_{i,j}^{m+1}$ and $\omega_{i,j}^{m+1}$ are determined in the following way:

1. Compute a provisional value $\tilde{\Psi}_{i,j}^{m+1}$ by

$$\nabla_h^2 \tilde{\Psi}_{i,j}^{m+1} = -\omega_{i,j}^m \qquad \text{for } (x_i, y_j) \in \Omega_h \tag{6.5.23a}$$

and

$$\tilde{\Psi}_{i,j}^{m+1} = f_{i,j} \qquad \text{for } (x_i, y_j) \in \Gamma_h \tag{6.5.23b}$$

2. Define the final value $\Psi_{i,j}^{m+1}$ by the relaxation formula

$$\Psi_{i,j}^{m+1} = \alpha \tilde{\Psi}_{i,j}^{m+1} + (1 - \alpha) \Psi_{i,j}^m, \qquad 0 < \alpha \le 1, \text{ for } (x_i, y_j) \in \Omega_h \tag{6.5.24}$$

3. Compute a provisional value $\tilde{\omega}_{i,j}^{m+1}$ by

$$\left[L_h(\Psi_{i,j}^{m+1}) - \frac{\nabla_h^2}{\mathrm{Re}} \right] \tilde{\omega}_{i,j}^{m+1} = 0, \qquad (x_i, y_j) \in \Omega_h \tag{6.5.25a}$$

$$\tilde{\omega}_{i,j}^{m+1} = F_{i,j}(\Psi^{m+1}, \omega^m), \qquad (x_i, y_j) \in \Gamma_h \tag{6.5.25b}$$

where the term $F_{i,j}$ corresponds to the boundary evaluation of the vorticity ω.

4. Define the final value $\omega_{i,j}^{n+1}$ by the relaxation formulas

$$\begin{aligned}
\omega_{i,j}^{m+1} &= \beta \, \tilde{\omega}_{i,j}^{m+1} + (1 - \beta) \, \omega_{i,j}^m, \qquad (x_i, y_j) \in \Omega_h \\
\omega_{i,j}^{m+1} &= \gamma \, \tilde{\omega}_{i,j}^{m+1} + (1 - \gamma) \, \omega_{i,j}^m, \qquad (x_i, y_j) \in \Gamma_h
\end{aligned} \tag{6.5.26}$$

where the relaxation parameters β and γ corresponding, respectively, to the inner points and the boundary points are not necessarily equal ($0 < \beta, \gamma \leq 1$).

The iterative process is stopped when convergence is obtained. The question of criteria of convergence is delicate. Various criteria can be considered. Simple criteria of convergence are

$$R_{\Psi}^{(1)} = \max_{i,j} \left(\left| \Psi_{i,j}^{m+1} - \Psi_{i,j}^{m} \right| \right) \leq \epsilon_{\Psi}$$

$$R_{\omega}^{(1)} = \max_{i,j} \left(\left| \omega_{i,j}^{m+1} - \omega_{i,j}^{m} \right| \right) \leq \epsilon_{\omega}$$

where ϵ_{Ψ} and ϵ_{ω} must be sufficiently small with respect to the mean value of the corresponding quantity. It can be useful to consider relative criteria such as

$$R_{\phi}^{(2)} = \frac{\max \left(\left| \phi_{i,j}^{m+1} - \phi_{i,j}^{m} \right| \right)}{\max \left(\left| \phi_{i,j}^{m+1} \right| \right)} \leq \epsilon_{\phi} \qquad (\phi = \Psi \text{ or } \omega)$$

which takes account of the magnitude of the solution.

It can happen, in some complex flows, that the iterative solution converges well everywhere except at isolated points, for example, near singularities or more generally in regions of very large gradients in which an accurate solution cannot be expected. In this case, it is best to define averaged criteria of the type:

$$R_{\phi}^{(3)} = \frac{1}{N} \sum_{i,j} \left| \phi_{i,j}^{m+1} - \phi_{i,j}^{m} \right| \leq \epsilon_{\phi}$$

or

$$R_{\phi}^{(4)} = \left(\frac{1}{N} \sum_{i,j} \left| \phi_{i,j}^{m+1} - \phi_{i,j}^{m} \right|^2 \right)^{\frac{1}{2}} \leq \epsilon_{\phi}$$

where N is the total number of points on which the sum is taken. If such criteria are used, possibly normalized with the maximal value of $\left| \phi_{i,j}^{m+1} \right|$, it is recommended that the maximum criteria defined above be tested.

Finally, great care must be taken in evaluating the effect of the various relaxation parameters and of the form of the finite-difference formulas used in the effective computation (e.g., multiplication by Re, by a power of the mesh size, etc.). Thus, it is recommended that when convergence is assumed to be obtained, the residues of the equations be tested; for example, in the maximum norm

$$R_{\omega}^{(5)} = \max_{i,j} \left\{ \left| \left[L_h(\Psi_{i,j}) - \frac{\nabla_h^2}{Re} \right] \omega_{i,j} \right| \right\}$$

$$R_{\Psi}^{(5)} = \max_{i,j} \left(\left| \nabla_h^2 \Psi_{i,j} + \omega_{i,j} \right| \right)$$

The magnitude of these residues, which must be evaluated in comparison with the magnitude of the solution, gives a measure of the degree of accuracy with which the finite-difference equations are solved.

At each step of the iterative procedure (6.5.23)–(6.5.26), two linear algebraic systems must be solved. A priori, these systems can be solved by any type of method—direct or iterative. However, it is recommended that iterative methods be used here because the solution of these systems is only an intermediate solution included in a global iterative procedure. As this global iterative procedure converges, the variation between two successive iterated values decreases and, consequently, the number of iterations needed to solve the linear algebraic system (6.5.23) or (6.5.25) is continuously decreasing. Moreover, this number can be generally very small (of the order of 1 during the last stage of the global iterative process) if the demanded degree of convergence for the iterative solution of systems (6.5.23) and (6.5.25) is not too high (taking into account the fact that it is only an intermediate solution).

The solution (6.5.23) usually does not lead to difficulty because it is simply a Laplacian problem. On the other hand, problems of convergence arise for the solution of (6.5.25) when $|u_{i,j}| \, \mathrm{Re} \, \Delta x > 2$ or $|v_{i,j}| \, \mathrm{Re} \, \Delta y > 2$, because the matrix associated with system (6.5.25) is no longer diagonally dominant. Such a phenomenon, which occurs when the Reynolds number is too high compared to the inverse of the mesh size, has long been an obstacle to the numerical solution of the Navier–Stokes equations. The usual technique to surmount this obstacle is described in the next section.

6.5.5 The problem of high Reynolds numbers

The approach to reduce the difficulty associated with the loss of diagonal dominance is to use an upwind noncentered approximation for the convective term $\mathbf{V} \cdot \nabla \omega$ [or $\nabla \cdot (\mathbf{V}\omega)$ in its conservative form]. Schemes using such an approximation are due to Barakat and Clark (1966), Runchal et al. (1967, 1969), Gosman et al. (1969), Torrance (1968), and Greenspan (1969). These include both an iterative solution of the steady equations or, with a discretization in time, of the unsteady equation.

The centered operator $L_h(\Psi_{i,j})$ in Eq. (6.5.21a) is replaced by

$$L_h'(\Psi_{i,j}) = (\Delta_y^0 \Psi_{i,j}) \Delta_x^* - (\Delta_x^0 \Psi_{i,j}) \Delta_y^* \qquad (6.5.27)$$

with

$$\Delta_x^* = \frac{(1 - \epsilon_{i,j}^u)\Delta_x^+}{2} + \frac{(1 + \epsilon_{i,j}^u)\Delta_x^-}{2} \equiv \Delta_x^0 - \frac{\Delta x}{2} \epsilon_{i,j}^u \Delta_{xx} \qquad (6.5.28)$$

where $\epsilon_{i,j}^u = \mathrm{sign}\,(u_{i,j}) = \mathrm{sign}\,(\Delta_y^0 \Psi_{i,j})$, and with an analogous definition for Δ_y^*:

$$\Delta_y^* = \frac{(1 - \epsilon_{i,j}^v)\Delta_y^+}{2} + \frac{(1 + \epsilon_{i,j}^v)\Delta_y^-}{2} \equiv \Delta_y^0 - \frac{\Delta y}{2} \epsilon_{i,j}^v \Delta_{yy} \qquad (6.5.29)$$

where $\epsilon_{i,j}^v = \mathrm{sign}\,(v_{i,j}) = \mathrm{sign}\,(-\Delta_x^0 \Psi_{i,j})$.

As discussed in Chapter 2, such a discretization gives a diagonally dominant matrix whatever the Reynolds number; however, the accuracy is only first order. Second-order accuracy can be recovered by using the correction technique described in Section 2.4. Equation (6.5.25a) is then replaced by

$$\left[L_h'\,(\Psi_{i,j}^{\,m+1}) - \frac{\nabla_h^2}{\mathrm{Re}} \right] \tilde\omega_{i,j}^{\,m+1} = -\Lambda_h\,(\Psi_{i,j}^{\,m+1})\,\omega_{i,j}^{\,m} \qquad (6.5.30)$$

with the operator $\Lambda_h\,(\Psi_{i,j}^{\,m+1})$ defined by

$$\Lambda_h\,(\Psi_{i,j})\,\omega_{i,j} = \tfrac{1}{2}\Delta x\,\epsilon_{i,j}^{u}\,u_{i,j}\,\Delta_{xx}\,\omega_{i,j} + \tfrac{1}{2}\Delta y\,\epsilon_{i,j}^{v}\,v_{i,j}\,\Delta_{yy}\,\omega_{i,j}$$

$$+ C_1\,\Delta_{xx}\,\omega_{i,j} + C_2\,\Delta_{yy}\,\omega_{i,j} \qquad (6.5.31)$$

$$C_1 = \tfrac{1}{4}\Delta x^2\,\phi_u\,\Delta_x^0\,u_{i,j} + \tfrac{1}{8}\Delta x^3\,\chi_u\,\Delta_{xx}\,u_{i,j}$$

$$C_2 = \tfrac{1}{4}\Delta x^2\,\phi_v\,\Delta_y^0\,v_{i,j} + \tfrac{1}{8}\Delta x^3\,\chi_v\,\Delta_{yy}\,v_{i,j} \qquad (6.5.32)$$

If the centered approximation (6.5.22) is to be recovered at convergence, we must choose $C_1 = C_2 = 0$ or

$$\phi_u = \chi_u = \phi_v = \chi_v = 0 \qquad (6.5.33)$$

This choice corresponds to the correction introduced by Dennis and Chang (1969) and used, in particular, by Veldman (1973) and Ta Phuoc Loc (1975).

If an artificial viscosity is needed, the following choice can be made:

$$\phi_u = -2\,\alpha_u\,\bar\epsilon_{i,j}^{\,u} \qquad \text{with } \alpha_u = \text{const} > 0 \text{ and } \bar\epsilon_{i,j}^{\,u} = \text{sign}\,(\Delta_x^0\,u_{i,j})$$

$$\qquad (6.5.34)$$

$$\chi_u = -2\,\beta_u\,\bar{\bar\epsilon}_{i,j}^{\,u} \qquad \text{with } \beta_u = \text{const} > 0 \text{ and } \bar{\bar\epsilon}_{i,j}^{\,u} = \text{sign}\,(\Delta_{xx}\,u_{i,j})$$

Analogous definitions hold for the quantity corresponding to the y direction. Let

$$\phi_v = -2\,\alpha_v\,\bar\epsilon_{i,j}^{\,v} \qquad \text{and} \qquad \chi_v = -2\,\beta_v\,\bar{\bar\epsilon}_{i,j}^{\,v} \qquad (6.5.35)$$

One choice of the ϕ and χ values is particularly interesting when recirculating flow in a cavity (see Fig. 6.5.2) is computed. With this choice, characterized by

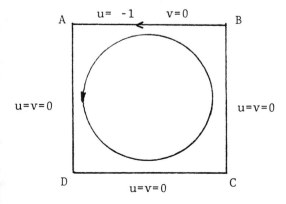

Fig. 6.5.2 Flow in a cavity.

$$\phi_u = -2 \, \bar{\epsilon}^u_{i,j}, \qquad \chi_u = 2 \, \bar{\bar{\epsilon}}^u_{i,j}$$
$$\phi_v = -2 \, \bar{\epsilon}^v_{i,j}, \qquad \chi_v = 2 \, \bar{\bar{\epsilon}}^v_{i,j}$$

$$(6.5.36)$$

the values of vorticity on the wall do not appear in the approximation of $\mathbf{V} \cdot \nabla\omega$ at first points near the boundary. This fact avoids the divergence of the iterative procedure or the appearance of oscillations in the solution due to the very large values of the vorticity on the side AB and near the corners A and B.

6.5.6 Approximation of the vorticity equation in conservative form
The conservative form of the vorticity equation is

$$\mathbf{V} \cdot (\mathbf{V}\omega) - \frac{\nabla^2 \omega}{\mathrm{Re}} = 0 \tag{6.5.37}$$

and its discretization is written

$$\tilde{\nabla}_h \cdot (\mathbf{V}\omega)_{i,j} - \frac{\nabla^2_h \, \omega_{i,j}}{\mathrm{Re}} = 0 \tag{6.5.38}$$

Two types of approximation for the nonlinear term $\mathbf{V} \cdot (\mathbf{V}\omega)$ are generally considered. The first type of centered conservative differencing is defined by

$$\tilde{\nabla}_h (\mathbf{V}\omega)_{i,j} = \Delta^0_x (u\omega)_{i,j} + \Delta^0_y (v\omega)_{i,j} \tag{6.5.39}$$

The associated upwind differencing is

$$\tilde{\nabla}_h (\mathbf{V}\omega)_{i,j} = \Delta^*_x (u\omega)_{i,j} + \Delta^*_y (v\omega)_{i,j} \tag{6.5.40}$$

with

$$\Delta^*_x (u\omega)_{i,j} = \tfrac{1}{2}[(1 - \epsilon^u_{i,j}) \Delta^+_x (u\omega)_{i,j} + (1 + \epsilon^u_{i,j}) \Delta^-_x (u\omega)_{i,j}]$$
$$= \Delta^0_x (u\omega)_{i,j} - \tfrac{1}{2}\Delta x \, \epsilon^u_{i,j} \, \Delta_{xx} (u\omega)_{i,j} \tag{6.5.41}$$

Δ^*_y is analogous. Concerning the diagonal dominance of the matrix $\mathscr{A} = [\alpha_{l,m}]$ associated with Eq. (6.5.38) it should be noted that the following inequality holds

$$\frac{1}{|\alpha_{l,l}|} \sum_{\substack{m \\ m \neq l}} |\alpha_{l,m}| \leq 1 + O(\mathrm{Re} \, \Delta x^3), \qquad \Delta x = \Delta y$$

The usual condition of diagonal dominance excludes the term $O(\mathrm{Re} \, \Delta x^3)$. From a practical point of view, no divergence problems have been determined in applications of the method (Section 6.5.8).

Note that the remark made at the end of the Section 6.5.5 concerning the flow in the driven cavity applies to the above conservative discretization, centered or not (Roux et al., 1980).

The general corrector term corresponding to Λ_h of the right-hand side of Eq. (6.5.30) is defined by

$$\Lambda_h(\Psi_{i,j})\,\omega_{i,j} = \tfrac{1}{2}\Delta x\; \epsilon^u_{i,j}\, \Delta_{xx}(u\omega)_{i,j} + \tfrac{1}{2}\Delta y\; \epsilon^v_{i,j}\, \Delta_{yy}(v\omega)_{i,j}$$
$$+ C_1\, \Delta_{xx}\, \omega_{i,j} + C_2\, \Delta_{yy}\, \omega_{i,j} \tag{6.5.42}$$

with C_1 and C_2 given by Eq. (6.5.32) and with same choice for the coefficient ϕ_u, ϕ_v, χ_u, χ_v.

The second centered discretization of the nonlinear term $\nabla \cdot (\mathbf{V}\omega)$ is

$$\tilde{\nabla}_h\cdot(\mathbf{V}\omega)_{i,j} = \Delta^1_x(u\omega)_{i,j} + \Delta^1_y(v\omega)_{i,j} \tag{6.5.43}$$

where Δ^1_x and Δ^1_y are defined by

$$\Delta^1_x f_{i,j} = \frac{f_{i+1/2,j} - f_{i-1/2,j}}{\Delta x}$$

$$\Delta^1_y f_{i,j} = \frac{f_{i,j+1/2} - f_{i,j-1/2}}{\Delta y}$$

with

$$f_{i+1/2,j} = \tfrac{1}{2}(f_{i+1,j} + f_{i,j}), \qquad f_{i,j+1/2} = \tfrac{1}{2}(f_{i,j+1} + f_{i,j})$$

An upwind differencing scheme for the conservation equation has been introduced by Runchal et al. (1967, 1969) and by Torrance (1968). It is more complicated than the upwind differencing of the first type but it has the advantage of giving a strongly diagonally dominant matrix. The construction is explained by considering the approximation of the derivative $\partial(u\omega)/\partial x$:

$$\left(\frac{\partial(u\omega)}{\partial x}\right)_{i,j} = \begin{cases} \dfrac{u_{i+1/2,j}\,\omega_{i,j} - u_{i-1/2,j}\,\omega_{i-1,j}}{\Delta x}, & \text{if } u_{i+1/2,j} \geq 0 \text{ and } u_{i-1/2,j} \geq 0 \\[2mm] \dfrac{u_{i+1/2,j}\,\omega_{i+1,j} - u_{i-1/2,j}\,\omega_{i,j}}{\Delta x}, & \text{if } u_{i+1/2,j} < 0 \text{ and } u_{i-1/2,j} < 0 \\[2mm] \dfrac{(u_{i+1,j} - u_{i-1,j})\,\omega_{i,j}}{2\,\Delta x}, & \text{if } u_{i+1/2,j} \geq 0 \text{ and } u_{i-1/2,j} < 0 \\[2mm] \dfrac{u_{i+1/2,j}\,\omega_{i+1,j} - u_{i-1/2,j}\,\omega_{i-1,j}}{\Delta x}, & \text{if } u_{i+1/2,j} < 0 \text{ and } u_{i-1/2,j} \geq 0 \end{cases}$$

The last two differencing formulas are consistent only if $|u|$ is sufficiently near zero to be neglected (that is possible since u changes sign).

With an analogous definition for the approximation of the derivative $\partial(v\omega)/\partial y$, the discretization of $\nabla \cdot (\mathbf{V}\omega)$ can be written in condensed form:

$$\tilde{\nabla}_h\cdot(\mathbf{V}\omega)_{i,j} = \Delta^1_x(u\omega)_{i,j} - \tfrac{1}{2}\Delta x\, \Delta^1_x(\epsilon^u_{i,j} u_{i,j}\, \Delta^1_x \omega_{i,j})$$
$$+ \Delta^1_y(v\omega)_{i,j} - \tfrac{1}{2}\Delta y\, \Delta^1_y(\epsilon^v_{i,j} v_{i,j}\, \Delta^1_y \omega_{i,j}) \tag{6.5.44}$$

The corrector operator Λ_h is defined by the following expression:

$$\Lambda_h(\Psi_{i,j})\omega_{i,j} = \tfrac{1}{2}\Delta x \, \Delta_x^1(\epsilon_{i,j}^u u_{i,j} \, \Delta_x^1 \omega_{i,j}) + \tfrac{1}{2}\Delta y \, \Delta_y^1(\epsilon_{i,j}^v v_{i,j} \, \Delta_y^1 \omega_{i,j})$$
$$+ \, C_1 \, \Delta_{xx} \, \omega_{i,j} + C_2 \, \Delta_{yy} \, \omega_{i,j} \qquad\qquad (6.5.45)$$

with C_1 and C_2 given by Eq. (6.5.32).

Finally, note that the addition of the viscosity term $C_1 \, \Delta_{xx} \, \omega_{i,j} + C_2 \, \Delta_{yy} \, \omega_{i,j}$ destroys the conservative character of the schemes.

In Section 6.5.8, results obtained by the various schemes listed above are discussed. Critical evaluation of the above centered and noncentered schemes, as well as other upwind schemes ("upstream-weighted schemes," Raithby and Torrance, 1974; "skew-upstream schemes," Raithby, 1976a) can be found in papers by Torrance (1968), Runchal (1972), De Vahl Davis and Mallinson (1976), and Raithby (1976b).

6.5.7 The unsteady equations

The unsteady equations

$$\frac{\partial \omega}{\partial t} + \mathbf{V} \cdot \nabla \omega - \frac{\nabla^2 \omega}{\text{Re}} = 0 \qquad\qquad (6.5.46)$$

$$\nabla^2 \Psi + \omega = 0 \qquad\qquad (6.5.47)$$

can be solved by a variety of numerical methods adapted to parabolic equations. However, whatever the method, a solution of a Poisson equation (for the stream function) has to be performed at each time step. Explicit schemes deduced from those presented in Chapter 2 can be used (Fromm, 1964, 1969; Thoman and Szewszyk, 1969). However, as was already mentioned, the use of implicit schemes are generally preferable and among them the alternating-direction-implicit method (ADI) presents the following advantages:

(i) Second-order accuracy in time and space.

(ii) Good stability properties.

(iii) Easy solution by inversion of tridiagonal matrices.

The ADI method and its properties have been described in Chapter 2 as well as in the present chapter. Here, we present the iterative procedure needed to solve the problem.

Let us assume that Eq. (6.5.46) is approximated with the scheme defined by Eqs. (2.8.4) of Chapter 2 where f, A, and B are, respectively, replaced by ω, u, and v. For the various parameters appearing in Eqs. (2.8.4), the following choices have been made:

$$c_0 = \tfrac{1}{4}, \; c_1 = c_2 = 0, \quad \gamma_0 = \tfrac{1}{4}, \; \gamma_1 = \gamma_2 = 0 \qquad \text{(Pearson, 1966)}$$

$$c_0 = 0, \; c_1 = c_2 = -\tfrac{1}{2}, \quad \gamma_0 = 0, \; \gamma_1 = \gamma_2 = -\tfrac{1}{2} \quad \text{(Briley, 1971)}$$

$$c_0 = \tfrac{1}{2}, \; c_1 = c_2 = 0, \quad \gamma_0 = \tfrac{1}{2}, \; \gamma_1 = \gamma_2 = 0 \qquad \text{(Aziz and Hellums, 1967)}$$

The ADI scheme associated with Eqs. (6.5.46) and (6.5.47) can be written in symbolic form

$$\Lambda_1(u_{i,j}^{n+1})\,\tilde{\omega}_{i,j} + M_1(v_{i,j}^{n+1})\,\omega_{i,j}^n = 0 \tag{6.5.48a}$$

$$\Lambda_2(u_{i,j}^{n+1})\,\omega_{i,j}^{n+1} + M_2(v_{i,j}^{n+1})\,\tilde{\omega}_{i,j} = 0 \tag{6.5.48b}$$

$$\nabla_h^2\Psi_{i,j}^{n+1} + \omega_{i,j}^{n+1} = 0 \tag{6.5.49a}$$

$$u_{i,j}^{n+1} = \Delta_y^0\Psi_{i,j}^{n+1}, \qquad v_{i,j}^{n+1} = -\Delta_x^0\Psi_{i,j}^{n+1} \tag{6.5.49b}$$

Note the apparent dependence of the finite-difference operator $\Lambda_1(u_{i,j}^{n+1})$, $M_1(v_{i,j}^{n+1})$, etc., on the solution at time $n+1$, depending on the choice of the approximation for the velocity components as mentioned above.

Let us assume that the solution is determined in a square domain Ω, $0 < x < 1$, $0 < y < 1$, discretized in Ω_h with $x_i = i\,\Delta x$, $y_j = j\,\Delta y$, $i,j = 1,\ldots,N$, with the initial condition

$$\mathbf{V}(x,y,0) = \mathbf{V}^0(x,y) \qquad (x,y)\in\Omega \tag{6.5.50}$$

and the boundary conditions

$$\Psi(x,y,t) = f(x,y,t) \tag{6.5.51a}$$

$$\frac{\partial\Psi}{\partial N}(x,y,t) = g(x,y,t) \qquad (x,y)\in\Gamma \tag{6.5.51b}$$

From (6.5.50) we can derive the initial condition for ω

$$\omega(x,y,0) = \omega^0(x,y) \tag{6.5.51c}$$

With Eq. (6.5.49) are associated the Dirichlet conditions (6.5.51a). On the other hand, the solution of (6.5.48a) requires knowledge of $\tilde{\omega}_{i,j}$ on the boundaries $x = 0$ and $x = 1$ and the solution of (6.5.48b) for $\omega_{i,j}^{n+1}$ on the boundaries $y = 0$ and $y = 1$. For these we may use one of the techniques developed in Section 6.5.3. The resulting formula is symbolically written

$$\omega_{i,0}^{n+1} = F(\omega_{i,1}^{n+1},\Psi_{i,1}^{n+1}) \equiv F_0^{n+1}$$
$$\omega_{i,N+1}^{n+1} = F(\omega_{i,N}^{n+1},\Psi_{i,N}^{n+1}) \equiv F_{N+1}^{n+1} \tag{6.5.52}$$

where $j = 0$ denotes $y = 0$ and $j = N + 1$ denotes $y = 1$.

For obtaining the intermediate values of $\tilde{\omega}_{i,j}$ on the boundaries, the more accurate way is to use formula (2.8.5), when possible; i.e., when the approximations of u in both steps are identical. Assuming this, we obtain

$$\tilde{\omega}_{i,j} = \frac{\omega_{i,j}^{n+1} + \omega_{i,j}^n}{2} + \frac{\Delta t}{4}\left[(v_2\Delta_y^0 - \frac{1}{\mathrm{Re}}\Delta_{yy})\,\omega_{i,j}^{n+1}\right.$$

$$\left. - (v_1\Delta_y^0 - \frac{1}{\mathrm{Re}}\Delta_{yy})\,\omega_{i,j}^n\right] \tag{6.5.53}$$

for $i = 0$ ($x = 0$) and $i = N + 1$ ($x = 1$). v_1 and v_2 are approximations to v as explained in Section 2.8.2. In (6.5.53) noncentered differences can be used near the corners to avoid the need of ω at these points.

We note that when Eq. (6.5.53) does not hold, the simple way to define the vorticity on the boundary, which gives sufficiently accurate results (Bontoux, 1978), is

$$\tilde{\omega}_{i,j} = \tfrac{1}{2}(\omega_{i,j}^{n+1} + \omega_{i,j}^{n}) \tag{6.5.54}$$

In Eqs. (6.5.53) or (6.5.54) $\omega_{i,j}^{n}$ and $\omega_{i,j}^{n+1}$ on the boundary are given by formulas analogous to (6.5.52). Therefore, we can write in symbolic form

$$\tilde{\omega}_{0,j} = G(\omega_{1,j}^{n+1}, \omega_{1,j}^{n}) \equiv G_0^{n+1}$$
$$\tilde{\omega}_{N+1,j} = G(\omega_{N,j}^{n+1}, \omega_{N,j}^{n}) \equiv G_{N+1}^{n+1} \tag{6.5.55}$$

Taking into account all the formulas, the iterative solution of the problem is the following one. Knowing $u_{i,j}^{n}$, $v_{i,j}^{n}$, $\omega_{i,j}^{n}$, the solution $u_{i,j}^{n+1}$, $v_{i,j}^{n+1}$, $\omega_{i,j}^{n+1}$ is determined by way of the iteration procedure characterized with the index m:

1. Equations (6.5.48a) and (6.5.55) give $\tilde{\omega}_{i,j}^{m+1}$:

$$\Lambda_1(u_{i,j}^{n+1,m})\,\tilde{\omega}_{i,j}^{m+1} + M_1(v_{i,j}^{n+1,m})\,\omega_{i,j}^{n} = 0$$
$$\tilde{\omega}_{0,j}^{m+1} = \gamma G_0^{n+1,m} + (1 - \gamma)\,\tilde{\omega}_{0,j}^{m} \tag{6.5.56a}$$
$$\tilde{\omega}_{N+1,j}^{m+1} = \gamma G_{N+1}^{n+1,m} + (1 - \gamma)\,\tilde{\omega}_{N+1,j}^{m}, \qquad j = 1, \ldots, N$$

2. Equations (6.5.48b) and (6.5.52) give $\omega_{i,j}^{n+1,m+1}$:

$$\Lambda_2(u_{i,j}^{n+1,m})\,\omega_{i,j}^{n+1,m+1} + M_2(v_{i,j}^{n+1,m})\,\tilde{\omega}_{i,j}^{m+1} = 0$$
$$\omega_{i,0}^{n+1,m+1} = \gamma F_0^{n+1,m} + (1 - \gamma)\,\omega_{i,0}^{n+1,m} \tag{6.5.56b}$$
$$\omega_{i,N+1}^{n+1,m+1} = \gamma F_{N+1}^{n+1,m} + (1 - \gamma)\,\omega_{i,N+1}^{n+1,m} \qquad i = 1, \ldots, N$$

3. Equations (6.5.49a) and (6.5.51a) give $\psi_{i,j}^{n+1,m+1}$:

$$\nabla_h^2 \Psi_{i,j}^{n+1,m+1} = -\omega_{i,j}^{n+1,m+1}$$
$$\Psi_{i,j}^{n+1,m+1} = f_{i,j}^{n+1}, \qquad (x_i, y_j) \in \Gamma_h \tag{6.5.57}$$

4. Equations (6.5.49b) give $u_{i,j}^{n+1,m+1}$:

$$u_{i,j}^{n+1,m+1} = \alpha \Delta_y^0 \Psi_{i,j}^{n+1,m+1} + (1 - \alpha)u_{i,j}^{n+1,m}$$
$$v_{i,j}^{n+1,m+1} = -\alpha \Delta_x^0 \Psi_{i,j}^{n+1,m+1} + (1 - \alpha)v_{i,j}^{n+1,m} \tag{6.5.58}$$

In (6.5.56) and (6.5.58), γ and α are parameters of relaxation ($0 < \gamma$, $\alpha \leq 1$). Note that Step 4 does not exist if u and v are evaluated in (6.5.56) by using only values at previous times as Briley (1971) did. Also it is not really necessary to solve exactly system (6.5.57) at each global iteration. In the case where an iterative technique is used to solve this system, only a few iterations can be performed at each global iteration.

Generally, the iterative procedure is initialized by using the values at level n. Another approach could be to use an extrapolation from the values at n and $n-1$. Such an initialization is efficient if the solution is really unsteady, but it has been found to become less and less efficient as the steady solution is reached.

We refer to Bontoux et al. (1980) for the questions of stability associated with the treatment of the boundary conditions in ADI methods. They found a restriction on the time step of the type $\Delta t / \Delta x^2 < a$ where a depends on the Reynolds number. A similar limitation was found (Section 6.2.6.3) for the ADI method applied to the velocity–pressure formulation.

The use of an ADI technique for the conservative form of the vorticity equation can be considered. Schemes with first-order accuracy in time (u and v at each time step evaluated at level n) have been used, for example, by Torrance (1968), Ta Phuoc Loc (1980). Results concerning second-order-accurate schemes are given in Section 6.6.2. Finally, upwind noncentered schemes, with or without a corrector term (Sections 2.4 and 6.5.5), can also be used in association with ADI techniques (Bonnet and Alziary de Roquefort, 1976; Alziary de Roquefort and Grillaud, 1978).

6.6 Example Solutions for Stream-Function Vorticity Formulation

6.6.1 Steady flow in a square cavity

For several years, steady flow in a square cavity has become a popular example for testing and comparing numerical methods. In most of the works (e.g., Burggraf, 1966; Greenspan, 1969; Fortin et al., 1971; Roache, 1975; De Vahl Davis and Mallinson, 1976; Rubin and Khosla, 1977; Gupta and Manohar, 1979), the fluid velocity is zero on three sides of the square and is tangent to the fourth side with a constant value equal to 1. Because of the discontinuity of the velocity at the corner, the solution of the Navier–Stokes equations is singular at these points (the vorticity becomes infinite). As a matter of fact, it is difficult to measure with precision the effect of a singularity on the accuracy of the solution. This is particularly true when the mesh is refined so that the computation points are drawn nearer and nearer to the corners even if the values of the vorticity at these points are not involved in the numerical scheme as is usual for finite-difference approximations.

Here we consider a regular solution of the Navier–Stokes equations which is the cavity flow (Fig. 6.6.1) proposed by Bourcier and François (1969), in which the velocity u on the side AB is no longer a constant but is defined by

$$u(x) = -16\,x^2(1-x)^2$$

so that $u(0) = u(1) = u'(0) = u'(1) = 0$. In terms of the stream function Ψ, the boundary conditions are

$$\Psi = 0, \quad \frac{\partial \Psi}{\partial y} = -16\,x^2(1-x)^2 \qquad \text{on } AB$$

$$\Psi = 0, \quad \frac{\partial \Psi}{\partial N} = 0 \qquad \text{on } BC,\ CD,\ DA$$

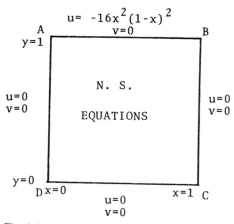

Fig. 6.6.1 Geometry and boundary conditions for example.

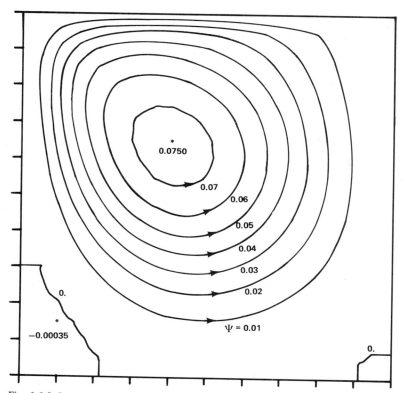

Fig. 6.6.2 Streamlines for flow in cavity, Re = 400.

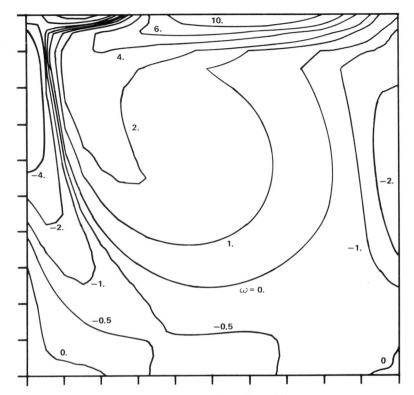

Fig. 6.6.3 Constant-vorticity lines for cavity flow, Re = 400.

where **N** is the unit normal to the boundary.

Computations have been performed with the iterative method described in Sections 6.5.4 and 6.5.5, and two types of results are compared here. The first results were obtained by varying the form (conservative or not) of the vorticity equation and the discretization of the nonlinear term. The first-order formula (6.5.10) was used for the boundary vorticity. The second set of results were obtained by varying the treatment of the boundary conditions for the vorticity. In this part, the computations were performed with the corrected, second-order accurate, scheme deduced from (6.5.40) and (6.5.42) with $C_1 = C_2 = 0$.

In order to illustrate the flow patterns, the streamlines and isovorticity lines for a case Re = 400 are displayed in Figs. 6.6.2 and 6.6.3. These contour maps correspond to results obtained for the mesh $\Delta x = \Delta y = 1/20$, with the conservative form of the vorticity equation discretized by the second-order accurate scheme (6.5.40), (6.5.42), and $C_1 = C_2 = 0$; the vorticity at the boundary was computed from the first-order formula (6.5.10).

Table 6.6.1 Comparison of various methods for cavity flow predictions Re = 400, $\Delta x = \Delta y = 1/20$ (Boundary Vorticity (6.5.10)).

Run	Form of ω equation	Scheme	Results		
			$\max\limits_{x,y} \Psi(x,y)$[b]	$\int_0^1 \omega(x,1)\, dx$	$\max\limits_{y} u(0.5, y)$
1	NC[a]	noncentered (6.5.27)	0.0632	7.62	0.132
2	NC	corrected (6.5.30), (6.5.33)	0.0584	9.38	0.140
3	NC	corrected (6.5.30), (6.5.36)	0.0661	8.00	0.155
4	NC	corrected (6.5.30), (6.5.34), (6.5.35) $\alpha_u = \alpha_v = 1, \beta_u = \beta_v = 0$	0.0652	8.27	0.153
5	C	noncentered, (6.5.40)	0.0593	7.99	0.116
6	C	corrected (6.5.40), (6.5.42) $C_1 = C_2 = 0$	0.0750	6.94	0.188
7	C	corrected (6.5.40), (6.5.42), (6.5.34), (6.5.35) $\alpha_u = \alpha_v = 1, \beta_u = \beta_v = 0$	0.0718	7.55	0.173
8	C	noncentered (6.5.44)	0.0677	7.35	0.141
9	C	corrected (6.5.44), (6.5.45) $C_1 = C_2 = 0$	0.0669	8.15	0.164
10	NC	fourth-order Hermitian[c]	0.0844	8.10	0.229

[a]NC, nonconservative form; C, conservative form.

[b]Obtained for x = 0.35, y = 0.75 (Run 1); x = 0.35, y = 0.70 (Run 2, 4, 8); x = 0.40, y = 0.65 (Runs 3, 5, 6, 7, 9, 10).

[c]Bontoux et al. (1978).

Results for Re $= 400$, $\Delta x = \Delta y = 1/20$ are presented in Table 6.6.1 giving the maximal value of the stream function Ψ; the averaged value of the vorticity ω on the side AB (total shear force), the integral being evaluated by means of the trapezoidal rule with $\omega(0, 1) = \omega(1, 1) = 0$; and the maximal value of the velocity component $u(0.5, y)$. Results of Runs 1–9 are for various schemes with formula (6.5.10) for the boundary vorticity. The accuracy of these results can be measured by comparing them to those given by a fourth-order accurate Hermitian method (Bontoux et al., 1978) taken here as a reference result (Run 10). Figure 6.6.4 illustrates the difference between results given by various schemes for the velocity $u(0.5, y)$.

As can be observed for the present problem, and more generally in several comparison tests (e.g., Rubin and Khosla, 1977), the use of the conservative form of the equations leads to more accurate results than the nonconservative form. Moreover, the centered discretization of the type (6.5.39) (Run 6) (the scheme of run 6 becomes (6.5.39) at convergence) seems preferable to the discretization of the type (6.5.43) (Run 9) (the scheme of run 9 becomes (6.5.43) at convergence) as long as the average values of ω on AB are not considered. It should be noted that such values are of weak significance because ω largely varies between negative and positive values (-5.0 to 25.0).

The results shown in Table 6.6.1 (Runs 1–9) correspond to the criterion of convergence $\text{Max}(R_\Psi^{(1)}, R_\omega^{(1)}) < 10^{-5}$ (see Section 6.5.4), then $\text{Max}(R_\Psi^{(5)}, R_\omega^{(5)}) < 6 \times 10^{-5}$. They were obtained by starting the iterative procedure [with $\alpha = 1$, $\beta = \gamma = 0.5$ as relaxation parameters defined in Eqs. (6.5.24) and (6.5.26)] using the results for Re $= 100$ as an initial condition. For that reason, no precise information about the speed of convergence is given, except that the use of the conservative form of the vorticity equation leads to the best convergence; this superiority is generally observed in comparison tests for other problems as well.

An estimation of the cost of such an iterative method is given by the case with Re $= 100$, $\Delta x = \Delta y = 1/20$ (Run 11 in Table 6.6.2). The use of the corrected scheme, defined by (6.5.40) and (6.5.42) with $C_1 = C_2 = 0$ and $\alpha = 1$, $\beta = 0.5$ as relaxation parameters [Eqs. (6.5.24) and (6.5.26)] associated with the boundary vorticity formula (6.5.10) and $\gamma = 0.5$ [Eq. (6.5.26)], requires 138 global iterations to obtain a degree of convergence characterized by the following quantities (see Section 6.5.4):

$$R_\Psi^{(1)} = 8 \times 10^{-6}, \quad R_\omega^{(1)} = 6 \times 10^{-4}, \quad R_\Psi^{(2)} = 10^{-4}, \quad R_\omega^{(2)} = 5 \times 10^{-5},$$

$$R_\psi^{(3)} = 2 + 10^{-6}, \quad R_\omega^{(3)} = 1.6 \times 10^{-4}, \quad R_\Psi^{(4)} = 3 \times 10^{-6}$$

$$R_\omega^{(4)} = 2 \times 10^{-4}, \quad R_\Psi^{(5)} = 2 \times 10^{-3}, \quad R_\omega^{(5)} = 5 \times 10^{-3}$$

At each global iteration, the linear equations are solved by the successive relaxation method with the relaxation parameters equal to 1.5 for the Ψ equation and to 1.0 for the ω equation. Because these equations are solved approximately (the relaxation solution is stopped when the difference between two

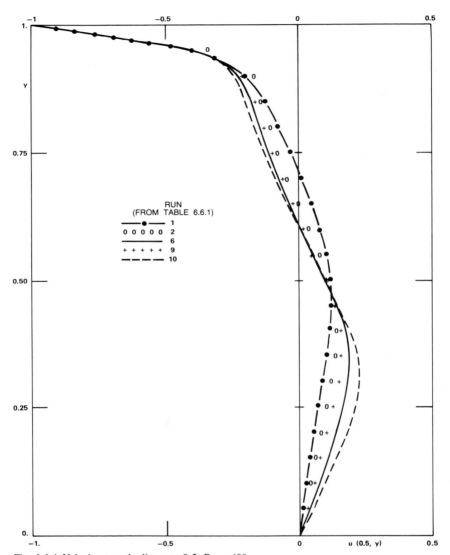

Fig. 6.6.4 Velocity u on the line $x = 0.5$, Re $= 400$.

iterated values is less than 10^{-2}) at each global iteration, the complete computation effort to reach the final convergence is better measured by the total number of Gauss–Seidel type sweeps (144 sweeps for the Ψ equation and 737 for the ω equation). The CPU time was 18 s on a IBM 3033.

Using the scheme just mentioned with the same relaxation parameters, we next make a comparison of some of the techniques described in Section 6.5.3 for handling the vorticity at a boundary. Recall that such a scheme yields

Table 6.6.2 Comparison of boundary vorticity approximations for cavity flow Re = 100 (scheme (6.5.40), (6.5.42), $C_1 = C_2 = 0$).

Run	Mesh size $\Delta x = \Delta y$	Boundary approximation for ω (Runs 11–21) Method (Runs 22 and 23)	Number of global iterations	Total number of Gauss–Seidel sweeps[b] Ψ	ω	Results Max $\Psi(x,y)$[c]	Max$_x \omega(x,1)$[d]	$\int_0^1 \omega(x,1)\,dx$	Max$_y u(0.5,y)$
11	1/10	(6.5.10)	62[a]	71	388	0.0735	9.84	4.44	0.133
12	1/20	(6.5.10)	138	144	737	0.0806	12.61	4.80	0.153
13	1/10	(6.5.11)	50	80	301	0.0792	11.63	4.85	0.145
14	1/20	(6.5.11)	130	148	719	0.0829	13.07	4.94	0.158
15	1/10	(6.5.12)	48	58	292	0.0799	11.95	4.85	0.146
16	1/20	(6.5.12)	127	144	865	0.0829	13.14	4.93	0.158
17	1/10	(6.5.18), $\Omega' = \Omega$	77	86	285	0.0796	11.99	5.17	0.145
18	1/20	(6.5.18), $\Omega' = \Omega$	117	119	635	0.0826	13.01	5.01	0.157
19	1/10	(6.5.18), $\Omega' = \{(0.2, 0.9), (0.1, 0.8)\}$	78	84	316	0.0792	11.89	5.11	0.143
20	1/20	(6.5.18), $\Omega' = \{(0.2, 0.9), (0.1, 0.8)\}$	142	146	697	0.0825	12.99	5.00	0.157
21	1/20	(6.5.18), $\Omega' = \{(0.05, 0.95), (0.05, 0.8)\}$	143	147	726	0.0825	12.95	4.98	0.159
22	1/20	fourth-order Hermitian[e]	—	—	—	0.0835	13.31	5.14	0.163
23	1/20	fourth-order Hermitian[f]	—	—	—	0.0829	13.16	5.20	0.162

[a]Test of convergence Max $(R_\Psi^{(4)}, R_\omega^{(4)}) < \epsilon$: $\epsilon = 5 \times 10^{-5}$ for 1/10 mesh; $\epsilon = 10^{-5}$ for 1/20 mesh.
[b]Test of convergence = 10^{-2}.
[c]Obtained for x = 0.4, y = 0.8.
[d]Obtained for x = 0.6.
[e]Stream-function vorticity (Bontoux et al., 1978).
[f]Velocity–pressure (Elsaesser and Peyret, 1979).

centered differencing (6.5.39) when the convergence is reached. The comparison is conducted for the case $Re = 100$ and for two mesh sizes: $\Delta x = \Delta y = 1/10$ and $\Delta x = \Delta y = 1/20$. Runs 11–16 (Table 6.6.2) were computed using first- and second-order formulas for the boundary vorticity. Results for Runs 17–20 were obtained (Céa et al., 1981) by using Green's formula (6.5.18) with $\gamma = 8$ as a convergence parameter [Eq. (6.5.19)]. In this case, the domain Ω' is either Ω itself or a ring (Fig. 6.6.5) defined by $\Omega' = \{(x_1, x_2), (y_1, y_2)\}$. The integrals involved are evaluated by means of the trapezoidal rule associated with a centered differencing of $\partial\Psi/\partial N$ on Γ'. The results are compared to those given by fourth-order-accurate Hermitian methods (Runs 22 and 23) applied to the stream-function vorticity equations (Bontoux et al., 1978) or to the velocity–pressure equations (Elsaesser and Peyret, 1979).

In the present case of a rather small Reynolds number ($Re = 100$), the variation of the vorticity ω along the side AB (Fig. 6.6.1) lies between -1.50 and 13.31, so that its averaged value as well as its maximal value on AB are more significant than they would be for $Re = 400$.

From various results shown in Table 6.6.2, it is observed that the second-order formulas (6.5.11), (6.5.12), and Green's formula (6.5.18) give comparable results. However, it is important to note that results obtained from Green's formula depend on the manner in which the integrals are evaluated. The same conclusion concerning the accuracy of formula (6.5.11) was drawn by Gupta and Manohar (1979) for the case of the driven cavity flow with $u(x, 1) = -1$. However, these authors found that Wood's formula (6.5.12) was not very accurate and furnished overestimated values of the maximum stream function and vorticity on AB. That is not the case here and in other works (Roache, 1975; Dennis et al., 1979).

Also observe that the speed of convergence resulting from the various

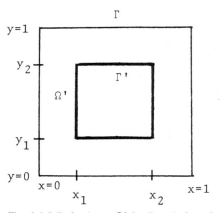

Fig. 6.6.5 Defined area Ω' for Green's formula.

treatments of the boundary vorticity is similar (the use of the Green's formula is the most consuming in computer time). Such a conclusion cannot be general, however, since it depends on several factors, in particular, on the values of the Reynolds number, the relaxation parameter γ, or the manner in which the linear equations are solved at each global iteration. Gupta and Manohar (1979) have found that the second-order formula (6.5.11) was the most expensive in terms of the number of iterations needed to reach convergence.

Finally, we point out that no comparison has been made between the efficiency of the iterative method employed for the present example and the ADI technique [Eqs. (6.5.4) and (6.5.5)]. It is possible that such an ADI method leads to faster convergence toward the final state if the parameters Δt and α are changed according to the rule of optimal parameters (Wachspress, 1966).

6.6.2 Unsteady flow around a circular cylinder

As an example of the computation of unsteady flow using the stream-function vorticity formulation, we now present results obtained by Ta Phuoc Loc (1980) which display the generation of secondary vortices in starting a flow around a circular cylinder. The numerical method employed makes use of the ADI technique described in Sections 6.5.2 and 6.5.7 associated with a fourth-order Hermitian approximation of the stream-function equation.

The configuration in the physical plane is sketched in Fig. 6.6.6. At time $t < 0$, a circular cylinder of radius a is located in a fluid at rest. At time $t \geq 0$, a uniform velocity U_∞ is applied to the fluid at infinity. The unsteady Navier–Stokes equations are solved in the infinite domain exterior to the boundary Γ, with no slip on Γ and uniform-flow conditions at infinity. Moreover, the flow is assumed to remain symmetrical with respect to the direction of the oncoming flow.

The difficulty in the problem is the unboundedness of the physical domain. Such a difficulty is common to many types of flows, for example, the case of channel flows which was considered in Section 6.4 using the velocity–pressure formulation. We refer also to Chapters 10 and 11 for a general discussion of the problem with application to compressible flows. In the present case, the problem is resolved by introducing new coordinates ξ, η by the transformation

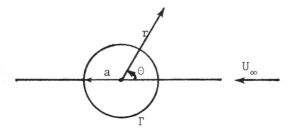

Fig. 6.6.6 Physical geometry.

$$r = ae^{\pi\xi}, \qquad \theta = \pi\eta \tag{6.6.1}$$

so that the resulting equations have to be solved in the half-strip $0 \le \xi < \infty$, $0 \le \eta < 1$ (Fig. 6.6.7). In fact, the domain is limited to a finite distance ξ_∞ and suitable boundary conditions are imposed at $\xi = \xi_\infty$.

Ta Phuoc Loc (1980) assumes the flow to be inviscid potential at $\xi = \xi_\infty$. This assumption limits the calculation to the first stage of the flow, i.e., before the vorticity created by the body has reached the external boundary. The symmetry hypothesis can also impose a time limitation on the possible growth of the perturbations inducing oscillations of the wake. Obviously, this last limitation could be avoided by relaxing the symmetry condition. The first limitation could be avoided by applying a condition representing convection of vorticity ω through a part of the downstream limit, i.e., say

$$\frac{\partial \omega}{\partial t} + \mathbf{V} \cdot \boldsymbol{\nabla} \, \omega = 0 \qquad \text{at } \xi = \xi_\infty, \; \theta_0 \le \theta \le 2\pi - \theta_0 \tag{6.6.2}$$

An equation similar to (6.6.2), but with the vorticity ω replaced by the normal derivative of the stream function $\partial\Psi/\partial\xi$, could be used as a time-dependent boundary condition for Ψ. Such conditions (with $\mathbf{V}$ replaced by $\mathbf{V}_\infty$) were successfully applied by Lugt and Haussling (1974) for unsteady computations around an elliptic cylinder at incidence.

Note that in steady-flow computations the conditions $\partial^2 \omega/\partial\xi^2 = \partial^2\Psi/\partial\xi^2 = 0$ are frequently used with success. However, it must be pointed out that the most satisfactory method, but not the easiest to implement, is the use of an asymptotic solution valid at large distance as Takami and Keller (1969) did for

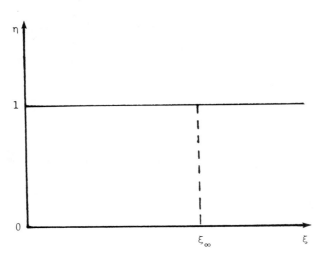

Fig. 6.6.7 Computation domain.

low-Reynolds-number steady flow around a circular cylinder. Unknowns involved in the asymptotic behavior are related to the drag and determined as a part of the solution by means of an iterative procedure.

In addition to the transformation of coordinates (6.6.1), the variables are made dimensionless by means of the characteristic quantities: U_∞/a for time, $U_\infty a$ for the stream function, and a/U_∞ for the vorticity. The equations of motion are then

$$g\frac{\partial\omega}{\partial t} + \frac{\partial(U\omega)}{\partial\eta} + \frac{\partial(V\omega)}{\partial\xi} - \frac{2}{Re}\left(\frac{\partial^2\omega}{\partial\eta^2} + \frac{\partial^2\omega}{\partial\xi^2}\right) = 0 \qquad (6.6.3)$$

$$\frac{\partial^2\Psi}{\partial\eta^2} + \frac{\partial^2\Psi}{\partial\xi^2} + g\omega = 0 \qquad (6.6.4)$$

where $Re = 2\,aU_\infty/\nu$, $g = \pi^2 e^{2\pi\xi}$, and

$$U = \frac{\partial\Psi}{\partial\xi}, \qquad V = -\frac{\partial\Psi}{\partial\eta} \qquad (6.6.5)$$

These equations are solved in the domain $0 < \xi < \xi_\infty$, $0 < \eta < 1$ with the initial condition $\mathbf{V}(\xi, \eta, 0) = 0$; therefore $\omega(\xi, \eta, 0) = 0$ and the boundary conditions

$$\Psi = \frac{\partial\Psi}{\partial\xi} = 0 \qquad \text{on } \xi = 0, \ 0 \le \eta \le 1 \qquad (6.6.6)$$

$$\left.\begin{array}{l} \Psi = -2\sinh\pi\xi\sin\pi\eta \\[2mm] \omega = 0 \end{array}\right\} \quad \text{on } \xi = \xi_\infty, \ 0 \le \eta \le 1 \qquad (6.6.7)$$

and with the symmetry condition

$$\Psi = \omega = 0 \qquad \text{on } \eta = 0 \text{ and } \eta = 1, \ 0 \le \xi \le \xi_\infty \qquad (6.6.8)$$

The computational domain is discretized according to $\xi_i = i\,\Delta\xi$, $\eta_j = j\,\Delta y$ $(i, j = 0, \ldots, N+1)$. The numerical solution at point ξ_i, η_j and time $t_n = n\,\Delta t$ is denoted by $\Psi_{i,j}^n$, $\omega_{i,j}^n$. The equation of vorticity (6.6.3) is discretized by means of an ADI scheme similar to (6.5.4a) and (6.5.4b), except that the nonlinear terms are considered in conservative form.

At the first step, the η derivatives are evaluated implicitly and the quantity U involved in the approximation of $\partial(U\omega)/\partial\eta$ was defined in three different ways

Scheme 1: $U \cong U^n$

Scheme 2: $U \cong \frac{1}{2}(3U^n - U^{n-1})$

Scheme 3: $U \cong \frac{1}{2}(U^{n+1} + U^n)$

At the second-step, the ξ derivatives are evaluated implicitly. The quantity V in $\partial(V\omega)/\partial\xi$ is defined according to the above three schemes. Scheme 1 is first-order accurate in time and Schemes 2 and 3 are second-order accurate. The discretized boundary conditions are deduced from (6.6.6)–(6.6.8) with the use of the second-order formula (6.5.12) to determine the vorticity on the body. The value of the stream function appearing in this formula is considered at level n for schemes 1 and 2 and $n+1$ for scheme 3.

The calculation of the stream function $\Psi_{i,j}^{n+1}$ is carried out by means of a fourth-order Hermitian approximation of Eq. (6.6.4) associated with an ADI iterative procedure. The unknowns are then the values of $\Psi_{i,j}^{n+1}$ and derivatives $(\partial^2\Psi/\partial\eta^2)_{i,j}^{n+1}$, $(\partial^2\Psi/\partial\xi^2)_{i,j}^{n+1}$ which will be denoted below without the super-script $n+1$ for the sake of simplicity. Let m be the index of iteration in the ADI procedure. The first step is written

$$\lambda_m \tilde{\Psi}_{i,j} - \left(\frac{\partial^2\tilde{\Psi}}{\partial\eta^2}\right)_{i,j} = \lambda_m \Psi_{i,j}^m + \left(\frac{\partial^2\Psi}{\partial\xi^2}\right)_{i,j} + g_i\,\omega_{i,j}^{n+1} \qquad (6.6.9a)$$

$$\frac{12}{\Delta\eta^2}\left(\tilde{\Psi}_{i,j+1} - 2\tilde{\Psi}_{i,j} + \tilde{\Psi}_{i,j-1}\right) - \left[\left(\frac{\partial^2\tilde{\Psi}}{\partial\eta^2}\right)_{i,j+1} + 10\left(\frac{\partial^2\tilde{\Psi}}{\partial\eta^2}\right)_{i,j}\right.$$
$$\left. + \left(\frac{\partial^2\tilde{\Psi}}{\partial\eta^2}\right)_{i,j-1}\right] = 0 \qquad (6.6.9b)$$

The second step is

$$\mu_m \Psi_{i,j}^{m+1} - \left(\frac{\partial^2\Psi}{\partial\xi^2}\right)_{i,j}^{m+1} = \mu_m \tilde{\Psi}_{i,j} + \left(\frac{\partial^2\tilde{\Psi}}{\partial\eta^2}\right)_{i,j} + g_i\,\omega_{i,j}^{n+1} \qquad (6.6.10a)$$

$$\frac{12}{\Delta\xi^2}\left(\Psi_{i+1,j}^{m+1} - 2\Psi_{i,j}^{m+1} + \Psi_{i-1,j}^{m+1}\right) - \left[\left(\frac{\partial^2\Psi}{\partial\xi^2}\right)_{i+1,j}^{m+1} + 10\left(\frac{\partial^2\Psi}{\partial\xi^2}\right)_{i,j}^{m+1}\right.$$
$$\left. + \left(\frac{\partial^2\Psi}{\partial\xi^2}\right)_{i-1,j}^{m+1}\right] = 0 \qquad (6.6.10b)$$

At each step, the above equations lead to the solution of a 2×2 block tridiagonal system that can be reduced to a simple tridiagonal system for the values of the stream function by eliminating the second-order derivatives as mentioned in Section 2.5.1. Boundary conditions associated with the first step are $\tilde{\Psi}_{i,j} = (\partial^2\tilde{\Psi}/\partial\eta^2)_{i,j} = 0$ deduced from the symmetry conditions (6.6.8) and from Eq. (6.6.6). At the second step, the boundary values of $\Psi_{i,j}^{m+1}$ and $(\partial^2\Psi/\partial\xi^2)_{i,j}^{m+1}$ are deduced from (6.6.6) and (6.6.7). The parameters λ_m and μ_m in Eqs. (6.6.9) and (6.6.10) are chosen according to the formulas given by Wachspress (1966).

Finally, $U_{i,j}^{n+1} = (\partial\Psi/\partial\xi)_{i,j}^{n+1}$ and $V_{i,j}^{n+1} = -(\partial\Psi/\partial\eta)_{i,j}^{n+1}$ are determined from the values of $\Psi_{i,j}^{n+1}$ by means of Hermitian formulas of the type (2.5.9).

In the case of Schemes 1 and 2, the vorticity and the stream function are computed one after the other successively. Once $\omega_{i,j}^{n+1}$ has been determined, eight iterations of the ADI procedure (6.6.9) and (6.6.10) are necessary to obtain convergence defined by a difference less than 10^{-5}. For Scheme 3, the solution of ω and Ψ is coupled and is obtained by using an iterative procedure. In fact, this procedure is included in the ADI solution of $\Psi_{i,j}^{n+1}$; i.e., the values of U^{n+1} and V^{n+1} are re-evaluated each four iterations of the ADI procedure (6.6.9) and (6.6.10).

Figure 6.6.8 shows the distribution of vorticity on the cylinder for Schemes 1, 2, and 3 in the case where Re = 300, $\xi_\infty = 0.954$ (therefore $r_\infty = 20\ a$), $\Delta\xi = 0.954/40$, $\Delta\eta = 1/40$, and $\Delta t = 0.05$. Note that in the front of the cylinder the differences are indistinguishable in the scale of the figure; these differences are not large on the remainder of the body and tend to diminish as time is increased.

However, a conclusion concerning the similarity between results given by the first-order and second-order-accurate schemes cannot be extended to any flow and is mainly related to the size of the truncation error with respect to time.

Figure 6.6.9 is relative to the case Re = 550, with $\xi_\infty = 0.954$,

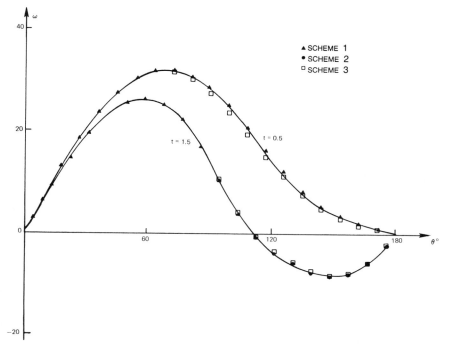

Fig. 6.6.8 Distribution of vorticity on cylinder surface, Re = 300; $\Delta\xi = 0.954/40$, $\Delta\eta = 1/40$, $\Delta t = 0.05$; ▲, scheme 1; ●, scheme 2; ☐, scheme 3. (Courtesy of Ta Phuoc Loc.)

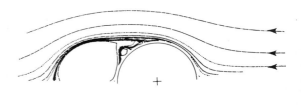

Fig. 6.6.9 Comparison of experimental (Coutanceau and Bouard, 1979) and computed streamline results at $t = 5.0$, Re $= 550$. (Courtesy of M. Coutanceau and R. Bouard, and Ta Phuoc Loc.)

$\Delta \xi = 0.954/50$, $\Delta \eta = 1/50$, and $\Delta t = 0.04$. The CPU time on UNIVAC 1110 is 1.50 s for each time step. The figure shows a comparison of computed instantaneous streamlines at $t = 5$ (from Scheme 1) and experimental visualization obtained by Coutanceau and Bouard (1979). The presence of a secondary vortex is evident and can be considered as an illustration of the ability of numerical methods to describe the details of complex unsteady flows.

References

Alziary de Roquefort, T., and Grillaud, G. *Comput. Fluids* **6**, 259–269 (1978).

Amsden, A. A., and Harlow, F. H. The SMAC Method: A Numerical Technique for Calculating Incompressible Fluid Flows. Los Alamos Research Laboratory Report No. LA-4370, May (1970).

Aziz, K., and Hellums, J. D. *Phys. Fluids* **10**, 314–324 (1967).

Barakat, H. Z., and Clark, J. A. Analytical and Experimental Study of the Transient Laminar Natural Convection Flows in Partially Filled Liquid Containers. In: *Proceedings of the Third International Heat Transfer Conference*, Vol. 2, p. 152, AICE, New York, (1966).

Bercovier, M., and Engelman, M. *J. Comput. Phys.* **30**, 181–201 (1979).

Blanschong, G. P., and Hartmann, C. *J. Méc.* **18**, 713–744 (1979).

Bonnet, J. P., and Alziary de Roquefort, T., *J. Méc.* **15**, 373–397 (1976).

Bontoux, P. Contribution à l'Etude des Écoulements Visqueux en Milieu Confiné. Thèse Doctorat d'Etat, IMFM, Université d'Aix–Marseille (1978).

Bontoux, P., Gilly, B., and Roux, B. *J. Comput. Phys.* **15**, 417–427 (1980).

Bourcier, M., and François, C. *Rech. Aérosp.* No. 131, 23–33 (1969).

Bozeman, J. D., and Dalton, C. *J. Comput. Phys.* **12**, 348–363 (1973).

Bramble, J. H., and Hubbard, B. E. *Numer. Math.* **4**, 313–327 (1962).

Briley, R. N. *J. Fluid Mech.* **47**, 713–736 (1971).

Bryan, K. *J. Atmos. Sci.*, **20**, 594–606 (1963).

Burggraf, O. R. *J. Fluid Mech.* **24**, 113–151 (1966).

Céa, J., Lhomme, B., and Peyret, R. The use of Green's formula for vorticity boundary values. Rapport IMAN, P-28, Université de Nice, June (1981).

Childress, S., and Peyret, R. *J. Méc.* **15**, 753–779 (1976). [Erratum: **16**, 803 (1976)].

Chorin, A. J. *J. Comput. Phys.* **2**, 12–26 (1967).

Chorin, A. J. *Math. Comput.* **22**, 745–762 (1968).

Coutanceau, M., and Bouard, R. *C. R. Acad. Sci, B* **288**, 485–488 (1979).

Cushman-Roisin, B. *J. Comput. Phys.* **53**, 227–239 (1984).

Daube, O., and Ta Phuoc Loc, A Mixed Compact Hermitian Method for the Numerical Study of Unsteady Viscous Flow Around and Oscillating Airfoil. In: *Proceedings of the 3rd GAMM Conference on Numerical Methods in Fluid Mechanics*, p. 56–66, Vieweg & Sons, Cologne, (1979).

Dennis, S. C. R., and Chang, G. Z. *Phys. Fluids* **12**, Suppl. II, II88–II93 (1969).

Dennis, S. C. R., Ingham, D. B., and Cook, R. N. *J. Comput. Phys.* **33**, 325–339 (1979).

De Vahl Davis, G., and Mallinson, G. D. *Comput. Fluids* **4**, 29–43 (1976).

Deville, M. O. *J. Comput. Phys.* **15**, 362–374 (1974).

Deville, M. O. *J. Méc.* **14**, 161–187 (1975).

Dorodnytsin, A. A., and Meller, N. A. *USSR Comput. Math. math. Phys.* **8**, 205–217 (1968).

Douglas, J., and Gunn, J. E. *Numer. Math.* **6**, 428–453 (1964).

Easton, C. R. *J. Comput. Phys.* **9**, 375–379 (1972).

Elsaesser, E., and Peyret, R. Méthodes Hermitiennes pour la Résolution Numérique des Equations de Navier–Stokes. In: *Méthodes Numériques dans les Sciences de l'Ingénieur*, E. Absi and R. Glowinski, Eds. p. 249–258, Dunod, Paris (1979).

Fortin, M. Calcul Numérique des Ecoulements des Fluides de Bingham et des Fluides Newtoniens par la Méthode des Eléments Finis. Thèse d'Etat, Université Paris VI (1972).

Fortin, M., Peyret, R., and Temam, R. *J. Méc.* **10**, 357–390 (1971). [*See also Lecture Notes in Physics*, Vol. 8, pp. 337–342, Springer-Verlag, New York (1971)].

Fromm, J. E. The time dependent flow of an incompressible viscous fluid. In: *Methods in Computational Physics*, Vol. 3, pp. 345–382, Academic Press, New York (1964).

Fromm, J. E. *Phys. Fluids* **12**, Suppl. II., II113–II119 (1969).

Ganoulis, J., and Thirriot, C. *Lecture Notes in Physics*, Vol. 59, pp. 191–196, Springer-Verlag, New York (1976).

Gosman, A. D., Pun, W. M., Runchal, A. K., Spalding, D. B., and Wolfshtein, M. *Heat and Mass Transfer in Recirculating Flows*, Academic Press, London and New York (1969).

Greenspan, D. *Comput. J.* **12**, 89–96 (1969).

Gupta, M. M., and Manohar, R. P. *J. Comput. Phys.* **31**, 265–288 (1979).

Harlow, F. H., and Welsh, J. E. *Phys. Fluids* **8**, 2182–2189 (1965).

Hughes, T. J. R., Liu, W. T., and Brooks, A. *J. Comput. Phys.* **30**, 1–60 (1979).

Israeli, M. *Stud. Appl. Math.* **49**, 327–349 (1970).

Israeli, M. *Stud. Appl. Math.* **51**, 67–71 (1972).

Kao, T. W., Park, C., and Pao, H. P. *Phys. Fluids* **21**, 1912–1922 (1978).

Kreiss, H. O. *Math. Comput.* **26** 605–624 (1972).

Kuznetsov, B. G., *Fluid Dynam. Trans.* **4**, 85–89 (1968).

Ladevèze, J., and Peyret, R. *J. Méc.* **13**, 367–396 (1974).

Liu, N. *Lecture Notes in Physics,* Vol. 59, pp. 300–305, Springer-Verlag, New York (1976).

Lugt, H. J., and Haussling, H. J. *J. Fluid Mech.* **65**, 711–734 (1974).

Mallinson, G. D., and De Vahl Davis, G. I. *J. Comput. Phys.* **12**, 435–461 (1973).

Marshall, R. S., Heinrich, J. C., and Zienkiewicz, O. C. *Numer. Heat Trans.* **1**, 315–330 (1978).

Morchoisne, Y. AIAA Paper No. 81-0109 (1981).

Orszag, S. A., and Israeli, M. in *Annual Review of Fluid Mechanics,* Vol. 6, pp. 281–318, Annual Reviews Inc., Palo Alto, CA (1974).

Orszag, S. A., and Kells, L. C. *J. Fluid Mech.* **96**, 159–205 (1980).

Peaceman, D. W., and Rachford, H. H. *J. Soc. Indus. Appl. Math.* **3**, 28–41 (1955).

Pearson, C. E. *J. Fluid Mech.* **21**, 611–622 (1965).

Pearson, C. E. *J. Fluid Mech.* **28**, 323–336 (1966).

Peyret, R. *C. R. Acad. Sci. A,* **272**, 1274–1277 (1971).

Peyret, R. *J. Fluid Mech.* **28**, 49–63 (1976).

Peyret, R. C. Numerical Studies of Nonhomogeneous Fluid Flows. In: *Advances in Fluid Mechanics*, Lecture Notes in Physics, Vol. 148, pp. 330–361, Springer–Verlag, New York (1981).

Peyret, R., and Rebourcet, B. In: *Proceedings of the 2nd International Conference on Numerical Methods in Laminar and Turbulent Flow,* pp. 1229–1239, Pineridge, Press, Swansea, U. K. (1981).

Peyret, R. and Rebourcet, B. *J. Méc. Théor. Appl.* **1**, No. 3 (1982).

Pracht, W. E. *J. Comput. Phys.,* **7**, 46–60 (1971).

Raithby, G. D. *Comput. Meth. Appl. Mech. Eng.* **9**, 75–102 (1976a).

Raithby, G. D. *Comput. Meth. Appl. Mech. Eng.* **9**, 153–164 (1976b).

Raithby, G. D., and Torrance, K. E. *Comput. Fluids* **2**, 191–206 (1974).

Richtmyer, R. D., and Morton, K. W. *Difference Methods for Initial Value Problems,* Interscience, New York (1967).

Roache, P. J. *Computational Fluid Dynamics,* Hermosa Publishing, Albuquerque, NM (1972).

Roache, P. *Comput. Fluids,* **3**, 179–195 (1975).

Roux, B., Bontoux, P., Ta Phuoc Loc, and Daube, O. In: *Approximation Methods for Navier–Stokes Problems*. Lecture Notes in Mathematics, Vol. 771, pp. 450–468. Springer-Verlag, New York (1980).

Rubin, S. G., and Khosla, P. K. *J. Comput. Phys.* **24**, 217–246 (1977).

Runchal, A. K., Spalding, D. B., and Wolfshtein, M. The Numerical Solution of the Elliptic Equations for Transport of Vorticity, Heat, and Matter in Two-Dimensional Flows. Imperial College, London, Dept. Mechanical Engineering Rept. No. SF/TN/2 (1967).

Runchal, A. K., Spalding, D. B., and Wolfshtein, M. *Phys. Fluids,* **12**, Suppl. II, II21–II28 (1969).

Runchal, A. K. *Int. J. Numer. Meth. Eng.* **4**, 541–550 (1972).

Sills, J. *AIAA J.* **7**, 117–123 (1969).

Takami, H., and Keller, H. B. *Phys. Fluids* **12**, Suppl. II, II51–II56 (1969).

Ta Phuoc Loc, *J. Méc.* **14**, 109–134 (1975).

Ta Phuoc Loc, *J. Fluid Mech.* **100**, 111–128 (1980).

Ta Phuoc Loc, and Daube, O. In *Vortex Flows,* N. L. Swift and P. S. Barna, Eds., pp. 155–171, ASME, New York (1981).

Taylor, T. D., and Murdock, J. W. In: *Approximation Methods for Navier–Stokes Problems.* Lecture Notes in Mathematics, Vol. 771, pp. 519–537, Springer-Verlag, New York (1980).

Taylor, T. D., and Ndefo, E. In: *Lecture Notes in Physics,* No. 8, pp. 356–364, Springer-Verlag, New York (1971).

Temam, R. *Bull. Soc. Math. Fr.* **96**, 115–152 (1968).

Temam, R. *Archiv. Ration. Mech. Anal.* **32**, 135–153 (1969a).

Temam, R. *Archiv. Ration. Mech. Anal.* **32**, 377–385 (1969b).

Temam, R., *Navier–Stokes Equations,* North-Holland, Amsterdam (1977).

Thom, A. *Proc. R. Soc. Lond. A* **141**, 651–666 (1933).

Thoman, D. C., and Szewczyk, A. A. *Phys. Fluids* **12**, Suppl. II, II76–II86 (1969).

Torrance, K. E. *J. Res. Natl. Bur. Stand. B* **72B**, 281–301 (1968).

Upson, C. D., Gresho, P. M., and Lee, R. L. Finite-Element Simulation of Thermally Induced Convection in an Enclosed Cavity. Lawrence Livermore Laboratory Rept. UCID-18602, March (1980).

Veldman, A. E. P. *Comput. Fluids* **1**, 251–271 (1973).

Viecelli, J. A. *J. Comput. Phys.* **8**, 119–143 (1971).

Wachspress, E. L. *Iterative Solution of Elliptic Systems,* Prentice-Hall, Englewood Cliffs, NJ (1966).

Wilkes, J. O. The Finite-Difference Computation of Natural Convection in an Enclosed Rectangular Cavity. Ph.D. Thesis, University of Michigan, Ann Arbor (1963).

Williams, G. L. *J. Fluid Mech.* **37**, 727–750 (1969).

Woods, L. C., *Aeronaut. Q.* **5**, 176–184 (1954).

Wu, J. C., and Wahbah, M. M. In: *Lecture Notes in Physics,* Vol. 59, pp. 448–453, Springer-Verlag, New York (1976).

Yanenko, N. N. *The Method of Fractional Steps,* Springer-Verlag, New York (1971) [in French, *Méthodes à Pas Fractionnaires,* Armand Colin, Paris (1968)].

Zabusky, N. J., and Deem, G. S. *J. Fluid Mech.* **47**, 353–379 (1971).

CHAPTER 7

Finite-Element Methods Applied to Incompressible Flows

The solution of two-dimensional flow problems by finite-element methods can employ either the stream-function vorticity or the primitive-variable formulations along with least-squares or Galerkin techniques. The number of possible combinations including element geometry, function approximations, and formulation become extremely large. As a result, it is difficult to assimilate and review all these approaches as they appear in the literature. Baker and Soliman (1979), Baker (1979),Giraut and Raviart (1979), and Thomasset (1981), as well as Chung (1978), provide recent compilations of progress in the finite-element field for the reader interested in full details. In the following discussion we have attempted to reduce the detail and outline the essential points significant in applying the finite-element method. In the research on finite elements a number of facts are beginning to become apparent. The first is that the primitive-variable formulation is preferable to the stream-function vorticity approach in terms of efficiency and ease of application. For inviscid flows the reason is clear because of the lower order of differentiation required in the primitive-variable formulation compared to the stream-function approach. For viscous flows the reason appears to be difficulty in satisfying vorticity boundary conditions. Hutton (1975), for example, pointed out that the time required to iterate a solution to convergence was disappointing compared to the time required to solve a problem by the pressure–velocity formulation. In fact, Hutton recommended giving up the stream-function vorticity approach in favor of the pressure–velocity solution method.

Other aspects of finite elements can be anticipated by analogy with finite differences. For example, one can expect that standard finite-element approximations will behave similar to centered approximations in finite differences. As a result one can expect the finite-element approach to encounter difficulty with inviscid or strongly inertia-dominated flows. Most solutions in the literature for the Navier–Stokes equations are demonstrated for Reynolds numbers less than 1000. To push the solutions beyond this limit introduces very-small-element requirements or upwind finite-element approximation. Other aspects evolving in the field are (1) the least-squares approach which seems to be more stable than the Galerkin technique for the nonlinear problems (Fletcher, 1976); (2) rectangular elements which yield more efficient solutions than triangular elements for both linear and quadratic shape functions; and (3) the quadratic shape function with the rectangular element of the Serendipity family which yields results with half the error of linear triangular elements. For the reader unfamiliar with the Serendipity quadratic nomenclature it refers simply to the rectangular element with eight nodes as shown in Fig. 7.1*. The shape functions for the corner nodes have the form

$$N_i = \tfrac{1}{4}(1 + \xi\xi_i)(1 + \eta\eta_i)(\xi\xi_i + \eta\eta_i - 1) \tag{7.1}$$

and for the midside nodes

$$N_i = \tfrac{1}{2}(1 - \xi^2)(1 + \eta\eta_i) \qquad \text{at } \xi_i = 0 \tag{7.2}$$

$$N_i = \tfrac{1}{2}(1 + \xi\xi_i)(1 - \eta^2), \qquad \text{at } \eta_i = 0 \tag{7.3}$$

Note that ξ_i and η_i take on the values $(-1, 0, 1)$ depending on the nodal point. In addition to the above findings, Fletcher found that when numerically evaluating the integrals involving shape functions, such as $\iint N_i N_j \, dx \, dy$, equally accurate results could be obtained by reducing the accuracy of the quadrature formulas which produces as much as a factor of 10 reduction in computer time.

One further point to be noted in the finite-element approach is that some type of artificial viscosity may have to be used to stabilize the solution of inviscid flow problems. One-sided approximations may be introduced to accomplish this. Gresho and Lee (1980), however, indicate that this approach requires great caution due to accuracy questions. Also, using an unsteady formulation and relaxing the solution to a steady state can be useful instead of trying to develop iterative or relaxation techniques to solve the algebraic equations

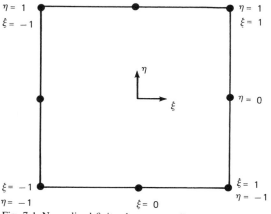

Fig. 7.1 Normalized finite-element coordinates.

---

*Quadratic isoparametric elements like this one can be distorted into curved quadrilateral elements and allow to handle domain with arbitrary shape.

obtained from the steady formulation. Fletcher (1977) found it necessary to employ a least-squares formulation of finite elements in place of the Galerkin formulation for an inviscid flow in order to avoid an unstable iterative solution of the algebraic equations.

By examing the available literature on finite elements, it appears that the most practical way to attack a finite-element solution is

(a) Use a primitive-variable formulation.
(b) Use unsteady relaxation to solve the equations.
(c) Employ a quadratic shape function.
(d) Use rectangular or quadrilateral elements when possible.
(e) Use simplified numerical integration when possible.

These rules should not be considered absolute, but they are offered as a guide to the reader. The field is rapidly changing and new knowledge is becoming available which may modify these recommendations.

In the solutions of incompressible flows, both the least-squares approach and Galerkin scheme are being employed. The approach for including the pressure in each of the calculations varies. Some try to avoid the problem by utilizing the stream-function–vorticity formulation while others introduce the penalty function approach discussed earlier (Section 6.3.4). The most recent finding, by Gresho et al. (1979), Bercovier and Pironneau (1979), Glowinski and Pironneau (1979), indicate that if the pressure is included in a Galerkin formulation the most success is obtained if use is made of a pressure shape function one order lower than the velocity shape function, the velocity shape functions to weight the momentum equations, and the pressure shape functions to weight the continuity equation.

7.1 The Galerkin Approach

If we integrate most of the findings noted above into a Galerkin finite-element scheme, we would begin with expansions of the form

$$u = \sum_{i=1}^{N} u_i(t)F_i(x, y) \qquad\qquad (7.1.1a)$$

$$v = \sum_{i=1}^{N} v_i(t)F_i(x, y) \qquad\qquad (7.1.1b)$$

$$p = \sum_{i=1}^{N} p_i(t)G_i(x, y) \qquad\qquad (7.1.1c)$$

Next, these expansions would be used with the primitive-variable Galerkin formulation in the form

$$\int_\Omega \left(\frac{\partial u}{\partial t} + u \frac{\partial u}{\partial x} + v \frac{\partial u}{\partial y} \right) F_j \, dx \, dy =$$

$$-\int_\Omega \frac{\partial p}{\partial x} F_j \, dx \, dy + \frac{1}{\mathrm{Re}} \int_\Omega \nabla^2 u \cdot F_j \, dx dy \quad (7.1.2)$$

$$\int_\Omega \left(\frac{\partial v}{\partial t} + u \frac{\partial v}{\partial x} + v \frac{\partial v}{\partial y} \right) F_j \, dx \, dy =$$

$$-\int_\Omega \frac{\partial p}{\partial y} F_j \, dx \, dy + \frac{1}{\mathrm{Re}} \int_\Omega \nabla^2 v \cdot F_j \, dx \, dy \quad (7.1.3)$$

$$\int_\Omega \left(\frac{\partial u}{\partial x} + \frac{\partial v}{\partial y} \right) G_j \, dx \, dy = 0 \quad (7.1.4)$$

Applying the divergence theorem to the right-hand side of (7.1.2), (7.1.3) and introducing the expansions for u, v and p, one obtains the finite-element equations:

$$A_{i,j} \frac{du_i}{dt} + (B_{i,j} + C_{i,j})u_i + D_{i,j}v_i + K_{i,j}p_i = \int_\Gamma F_j f_x \, ds \quad (7.1.5)$$

$$A_{i,j} \frac{dv_i}{dt} + D_{i,j}u_i + (B_{i,j} + C_{i,j})v_i + L_{i,j}p_i = \int_\Gamma F_j f_y \, ds \quad (7.1.6)$$

$$K_{i,j}u_i + L_{i,j}v_i = 0 \quad (7.1.7)$$

where

$$A_{i,j} = \int_\Omega F_i F_j \, dx \, dy$$

$$B_{i,j} = \int_\Omega \frac{1}{\mathrm{Re}} \left(2 \frac{\partial F_j}{\partial x} \frac{\partial F_i}{\partial x} + \frac{\partial F_j}{\partial y} \frac{\partial F_i}{\partial y} \right) dx \, dy$$

$$C_{i,j} = \sum_k \left(u_k \int_\Omega F_k F_j \frac{\partial F_i}{\partial x} + v_k \int_\Omega F_k F_j \frac{\partial F_i}{\partial y} \right) dx \, dy$$

$$D_{i,j} = \int_\Omega \frac{1}{\mathrm{Re}} \frac{\partial F_j}{\partial y} \frac{\partial F_i}{\partial x} \, dx \, dy,$$

$$K_{i,j} = \int_\Omega \frac{\partial F_j}{\partial x} G_i \, dx \, dy, \quad f_x = n_x \left(-p + 2 \frac{1}{\mathrm{Re}} \frac{\partial u}{\partial x} \right) + n_y \frac{1}{\mathrm{Re}} \left(\frac{\partial u}{\partial y} + \frac{\partial v}{\partial x} \right)$$

$$L_{i,j} = \int_\Omega \frac{\partial F_j}{\partial y} G_i \, dx \, dy, \quad f_y = n_y \left(-p + 2 \frac{1}{\mathrm{Re}} \frac{\partial v}{\partial y} \right) + n_x \frac{1}{\mathrm{Re}} \left(\frac{\partial u}{\partial y} + \frac{\partial v}{\partial x} \right)$$

These equations represent the standard Galerkin finite-element equations. Their solution can be developed by predictor-corrector techniques. Because of the continuity equation, the predictor can be explicit in time but the corrector must be implicit. The primary difficulty is in solving the implicit equations.

Gresho et al. (1979) solved this set of equations for flow past a cylinder at a Reynolds number of 110 by employing an Adams–Bashforth predictor.

$$y^{n+1} = y^n + \frac{\Delta t_n}{2}\left[\left(2 + \frac{\Delta t_n}{\Delta t_{n-1}}\right)h^n - \frac{\Delta t_n}{\Delta t_{n-1}}h^{n-1}\right] \qquad (7.1.8)$$

and a modified Euler corrector

$$y^{n+1} = y^n + \Delta t_n \frac{h^n + h^{n+1}}{2} \qquad (7.1.9)$$

where y is either u_i or v_i and h the nonderivative terms. These are applied to Eqs. (7.1.5)–(7.1.6). The resulting corrector equations were solved by a one-step Newton iteration along with the continuity equation to obtain u_i^{n+1}, v_i^{n+1}, and p_i^{n+1}. Gresho et al. generously supplied us with the streamline results from their calculations which is shown in Fig. 7.1.1. For the calculation, a nine-node biquadratic velocity and a four-node bilinear pressure element was employed with the mesh shown in Fig. 7.1.2. The mesh contained 196 elements and 850 nodes. The boundary conditions are also shown in Fig. 7.1.2a. Note that f_n and f_τ are defined as*

$$f_n = -p + 2\frac{1}{\text{Re}}\frac{\partial u_n}{\partial n}$$

$$f_\tau = \frac{1}{\text{Re}}\left(\frac{\partial u_n}{\partial \tau} + \frac{\partial u_\tau}{\partial n}\right)$$

where n and τ denote normal and tangential quantities, respectively. These results demonstrate the general Galerkin approach. There are other in-vestigations where results for flow past cavities, steps, and cylinders are presented for Reynolds numbers of about 100. We will not attempt to review them since no significant advance or improvement in calculational ability can be demonstrated at this time. However, there is one area of finite-element work where significant advances have been seen. This is in the area of least-squares finite-element application. The approach that has been developed to an ad-vanced stage can only be outlined here since the mathematics employed are quite complex.

7.2 The Least-Squares Approach

The least-squares finite-element approach has been developed by a team of scientists in France: Glowinski and Pironneau (1979), Bristeau et al, (1978, 1979, 1980). Here we attempt to outline the method without details. Consider the unsteady Navier–Stokes equations

---

*n is used to denote the normal to remain consistent with Gresho's work.

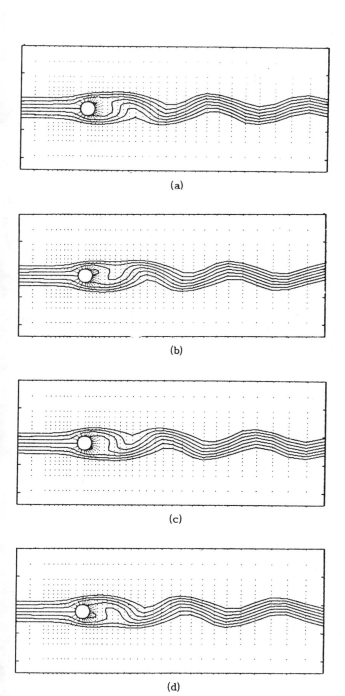

Fig. 7.1.1 Streamlines for vortex shedding flow past a cylinder at Re = 110. (Courtesy of P. Gresho.)
(a) $t = 532.00$, (b) $t = 533.55$, (c) $t = 535.10$, and (d) $t = 536.65$.

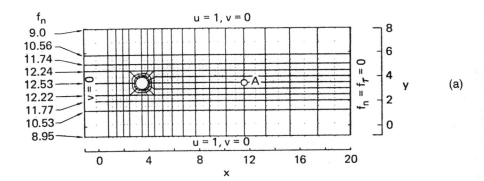

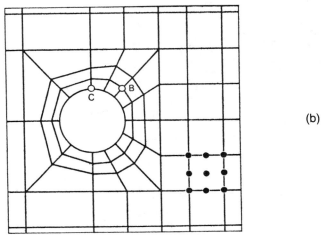

Fig. 7.1.2 (a) Mesh system and boundary conditions employed by Gresho et al. to compute flow past a cylinder. (b) Details of mesh system near cylinder and distribution of nodal points in an element. (Courtesy of P. Gresho.)

$$\frac{\partial \mathbf{V}}{\partial t} - \frac{\nabla^2 \mathbf{V}}{\mathrm{Re}} + (\mathbf{V} \cdot \nabla)\mathbf{V} + \nabla p = \mathbf{f}_e$$

$$\nabla \cdot \mathbf{V} = 0$$

$$\mathbf{V} = \mathbf{V}^0 \quad \text{at } t = 0$$

$$\mathbf{V} = \mathbf{g} \qquad \text{on } \Gamma$$

where Γ is the boundary of domain Ω. The solution to these equations is developed using the following alternating direction scheme starting from an initial field $\mathbf{V}^0$. For $n \geq 0$, from $\mathbf{V}^n$ seek $(\mathbf{V}^{n+1/2}, p^{n+1/2})$ and $\mathbf{V}^{n+1}$ by the following steps:

$$2\frac{\mathbf{V}^{n+1/2} - \mathbf{V}^n}{\Delta t} - \frac{\theta}{\mathrm{Re}} \nabla^2 \mathbf{V}^{n+1/2} + \nabla p^{n+1/2}$$

$$= \mathbf{f}_e^{n+1/2} - (\mathbf{V}^n \cdot \nabla) \mathbf{V}^n + \frac{(1-\theta)}{\mathrm{Re}} \nabla^2 \mathbf{V}^n$$

$$\nabla \cdot \mathbf{V}^{n+1/2} = 0 \qquad \text{in } \Omega$$

$$\mathbf{V}^{n+1/2} = \mathbf{g}^{n+1/2} \qquad \text{on } \Gamma$$

Then use

$$2\frac{\mathbf{V}^{n+1} - \mathbf{V}^{n+1/2}}{\Delta t} - \frac{(1-\theta)}{\mathrm{Re}} \nabla^2 \mathbf{V}^{n+1} + (\mathbf{V}^{n+1} \cdot \nabla) \mathbf{V}^{n+1}$$

$$= \mathbf{f}_e^{n+1} + \frac{\theta}{\mathrm{Re}} \nabla^2 \mathbf{V}^{n+1/2} - \nabla p^{n+1/2} \qquad \text{in } \Omega$$

$$\mathbf{V}^{n+1} = \mathbf{g}^{n+1} \qquad \text{on } \Gamma$$

The optimal value of θ is $\frac{1}{2}$. At the first step, the Stokes-type problem is solved by using an artificial compressibility method. At the second step, the solution of the nonlinear equations is developed by forming a nonlinear least-squares profit function formulation of the general equation

$$\alpha \mathbf{V} - \beta \nabla^2 \mathbf{V} + (\mathbf{V} \cdot \nabla) \mathbf{V} = \mathbf{f} \qquad \text{in } \Omega$$

$$\mathbf{V} = \mathbf{g} \qquad \text{on } \Gamma$$

with $\alpha, \beta > 0$ and then employing the conjugate gradient method with scaling to minimize the profit function. The finite-element implementation of this approach is described in Glowinski (1984) and Glowinski et al. (1981). The procedure requires the solution of a number of Dirichlet problems to arrive at the Navier–Stokes solution. The reader is referred to the references if the details are desired. Before reading these, however, it will be necessary to develop a knowledge of the conjugate-gradient method, Choleski matrix factorization, and Sobolev and Hilbert space definitions.

The results obtained by employing the conjugate-gradient least-squares approach are impressive for the examples demonstrated thus far. Glowinski and Periaux, of the French group, provided results of some of the group's calculations which are displayed here. The first result is for flow over the step shown in Fig. 7.2.1. There were 619 nodes with 1109 elements in the computational mesh. Note that the mesh shown has been truncated on the right, and in the actual calculation the mesh was extended in the same pattern to include about 100 more rectangles. The Reynolds number was Re = 100 and 191. Poiseuille flow was prescribed upstream with $\mathbf{V} = 0$ at the walls. Linear shape functions were employed. An unsteady formulation with $\Delta t = 0.4$ was applied to obtain

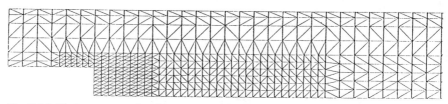

Fig. 7.2.1 Mesh system employed to compute flow over a step. (Courtesy of R. Glowinski and J. Periaux.)

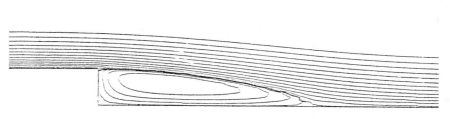

Re = 100

Fig. 7.2.2 Streamlines computed for flow over a step at Re = 100. (Courtesy of R. Glowinski and J. Periaux.)

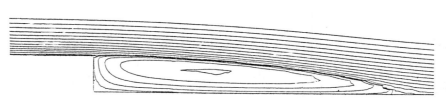

Re - 191

Fig. 7.2.3 Streamlines computed for flow over a step at Re = 191. (Courtesy of R. Glowinski and J. Periaux.)

the final steady result.* Typical streamline results are shown in Figs. 7.2.2 and 7.2.3. These were checked against an independent stream-function vorticity calculation of Hutton (1975) and the agreement was found to be good. A more interesting result is flow past the nozzle at angle attack shown in Fig. 7.2.4. The calculation was conducted with 1458 elements, 795 nodes, an angle of attack of 40°, and a Reynolds number of 250. The initial conditions were a Stokes flow solution. The calculation required several hours of IBM 370/168

---

*Using the above methodology but associated with a one-step scheme (Bristeau et al., 1980).

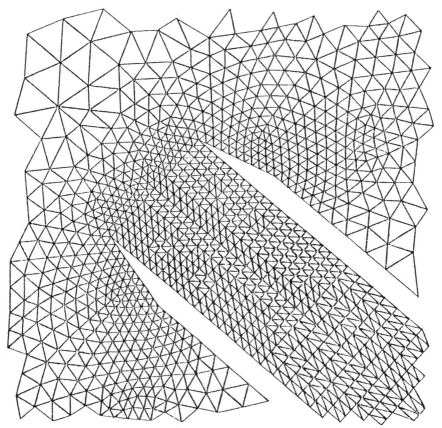

Fig. 7.2.4 Mesh system employed to compute flow past an inlet at Re = 250. (Courtesy of R. Glowinski and J. Periaux.)

time to run 100 time steps with $\Delta t = 0.1$. The calculated streamlines of the flow are shown in Fig. 7.2.5 from $t = 0$ to $t = 100$. The results demonstrate the ability of the method to predict the separated flow at modest Reynolds numbers. The authors point out, however, that the scheme gives difficulty with increasing Reynolds number, and they have investigated the use of approximations which yield equivalent noncentered differences. This is equivalent to putting in damping as the finite-difference investigators have shown in the past.

The examples displayed in this section give a reasonable picture of the achievements that finite elements yield at this time. The principal problems, as with finite differences, are still the cost and stability of calculations at Reynolds numbers above 1000. Since most practical problems are in this large-Reynolds-number range, the finite-element approach needs further improvement to find a large number of applications in incompressible flow.

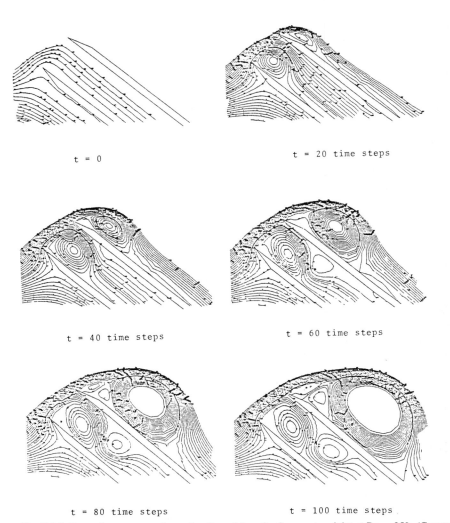

t = 0

t = 20 time steps

t = 40 time steps

t = 60 time steps

t = 80 time steps

t = 100 time steps

Fig. 7.2.5 Streamlines computed as a function of time for flow past an inlet at Re = 250. (Courtesy of R. Glowinski and J. Periaux.)

References

Baker, A. J., and Soliman, M. O. *J. Comput. Phys.* **32,** 289–324 (1979).

Baker, A. J. *Finite Element Computational Fluid Mechanics.* Short Course Notes, University of Texas, Austin, (1979).

Bercovier, M., and Pironneau, O. *Numer. Math.* **33,** 211–224 (1979).

Bristeau, M. O., et al. IRIA Report No. 294, LeChesnay, France, April (1978).

Bristeau, M. O., et al. *Comp. Meth. Appl. Mech. Eng.* **17/18,** 619–657 (1979).

Bristeau, M. O., et al. In: *Approximation Methods for Navier–Stokes Problems*. Lecture Notes in Mathematics, Vol. 771, pp. 78–128, Springer–Verlag, New York (1980).

Chung, T. J. *Finite Element Methods in Fluid Dynamics,* McGraw-Hill, New York (1978).

Fletcher, C. A. J. Dept. of Defense Weapons Research Establishment Rept. WRE-TN-1606 (WR&D), Salisbury, S. Australia, May (1976).

Fletcher, C. A. J. Dept. of Defense Weapons Research Establishment Rept. WRE-TR-1858(W), Salisbury, S. Australia, August (1977).

Giraut, V., and Raviart, P. A. *Finite Element Approximation of the Navier–Stokes Equations,* Lecture Notes in Mathematics, Vol. 749, Springer-Verlag, New York (1979).

Glowinski, R., and Pironneau, O. *Numer. Math.* **33,** 397–424 (1979).

Glowinski, R. *Numerical Methods for Nonlinear Variational Problems,* 2nd ed. Springer–Verlag, New York (1984).

Glowinski, R., Mantel, B., and Periaux, J. *Proceedings of Conference on Numerical Methods in Aeronautical Fluid Dynamics,* University of Reading, March (1981).

Gresho, P. M., Lee, R. L., and Sani, R. L. Lawrence Livermore Laboratory Rept. UCRL-83282, September (1979).

Gresho, P. M., and Lee, R. L. Don't Suppress the Wiggles They are Telling You Something, UCRL Preprint 82979, April (1980). *Comput. Fluids,* **9,** 223–253 (1981).

Hutton, A. G. A General Finite Element Model for Vorticity and Stream Function Applied to a Laminar Separated Flow. Central Electric Generating Board Rept. RD/B/N3050, August (1975).

Thomasset, F. *Implementation of Finite Element Methods for Navier–Stokes Equations,* Springer-Verlag, New York (1981)

CHAPTER 8

Spectral-Method Solutions for Incompressible Flows

The field of spectral methods in fluid mechanics is still very much in the developmental phase. As a result, application of the method is still somewhat of an art. At this time, the spectral method has been applied to solve primitive-variable, stream-function vorticity and stream-function only formulations. In each of these formulations, the unsteady equations were solved and the primary difficulty occurs in satisfying the proper boundary conditions. In some basic studies, one can neglect the boundary-condition questions by simply imposing periodic boundary conditions that are naturally satisfied by employing a Fourier spectral expansion. However, if examining flows with prescribed boundary conditions, then the straightforward Fourier expansion approach can become unsatisfactory.

In this chapter, we outline approaches that have been successful in solving both two-dimensional viscous and inviscid flow problems by the spectral technique.

8.1 Inviscid Flows

The first area to be discussed is the solution of an inviscid incompressible flow by the stream-function vorticity formulation with the equations

$$\frac{D\omega}{Dt} = 0 \tag{8.1.1}$$

$$\nabla^2 \Psi = -\omega \tag{8.1.2}$$

These equations have been solved by Myers et al. (1981) using a Chebyshev expansion method for the movement of a vortex pair in a confined region. Due to the symmetry, it is necessary only to solve the problem for one-half of the region of interest. The problem is displayed in Fig. 8.1.1. The vortices are given an initial vorticity distribution according to the relationship (8.1.13a). Elsewhere, the vorticity is zero. The stream function is defined to be zero on the boundary. This eliminates a need for a boundary condition on ω. The problem solution can be attempted by two approaches. In each, it can be assumed that the solution has the form

$$\omega = \sum_{n=0}^{N} \sum_{m=0}^{M} a_{n,m}(t) \, T_n^*(x) \, T_m^*(y) \tag{8.1.3}$$

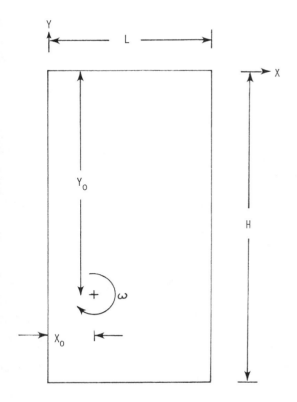

Fig. 8.1.1 Vortex moving in a box.

$$\Psi = \sum_{n=0}^{N} \sum_{m=0}^{M} b_{n,m}(t) \, T_n^*(x) \, T_m^*(y) \tag{8.1.4}$$

where T_n^* denotes the Chebyshev polynomial for $0 \le x \le 1$. The a and b terms are then determined by two separate approaches. In the first approach, the expansions are substituted directly into Eqs. (8.1.1) and (8.1.2) so that one obtains

$$\frac{da_{n,m}}{dt} = c_{n,m} \tag{8.1.5}$$

$$-a_{n.m} = (b^{xx} + b^{yy})_{n,m} \tag{8.1.6}$$

where $b_{n,m}^{xx}$ and $b_{n,m}^{yy}$ satisfy

$$\frac{\partial^2 \Psi}{\partial x^2} = \frac{\partial^2}{\partial x^2} \sum_{n=0}^{N} \sum_{m=0}^{M} b_{n,m} \, T_n^*(x) \, T_m^*(y)$$

$$= \sum_{n=0}^{N} \sum_{m=0}^{M} b_{n,m}^{xx} \, T_n^*(x) \, T_m^*(y)$$

$$\frac{\partial^2 \Psi}{\partial y^2} = \sum_{n=0}^{N} \sum_{m=0}^{M} b_{n,m}^{yy} \, T_n^*(x) \, T_m^*(y)$$

$c_{n,m}$ is derived from the expansion of

$$u \frac{\partial \omega}{\partial x} + v \frac{\partial \omega}{\partial y} = \frac{\partial \Psi}{\partial y} \frac{\partial \omega}{\partial x} - \frac{\partial \Psi}{\partial x} \frac{\partial \omega}{\partial y} = F(x,y,t) \tag{8.1.7}$$

in a double Chebyshev expansion

$$F(x,y,t) = \sum_{n=0}^{N} \sum_{m=0}^{M} c_{n,m} T_n^*(x) T_m^*(y) \tag{8.1.8}$$

Note that the evaluation of $c_{n,m}$ is only reasonable if the values of F are known at time t. As a result, one must resort to a predictor-corrector time-integration procedure to integrate Eq. (8.1.5). There is a large variety of these schemes, the most widely used being the Runge–Kutta type methods. The simplest has the form

$$a_{n,m}(\bar{t}) = a_{n,m}(t) + \tfrac{1}{2}\Delta t \, c_{n,m}(t)$$
$$a_{n,m}(t + \Delta t) = a_{n,m}(t) + \Delta t \, c_{n,m}(\bar{t}) \tag{8.1.9}$$

These approaches, as well as the Adams–Bashforth method

$$a_{n,m}(t + \Delta t) = a_{n,m}(t) - \tfrac{1}{2}\Delta t \, c_{n,m}(t - \Delta t) + \tfrac{3}{2}\Delta t \, c_{n,m}(t) \tag{8.1.10}$$

have been successfully applied.

Once $a_{n,m}(t)$ is known, then one must solve Eq. (8.1.6) for $b_{n,m}^{xx}$ and $b_{n,m}^{yy}$ and satisfy the boundary conditions on the stream function. The boundary conditions require that

$$\sum_{n=0}^{N} \sum_{m=0}^{M} b_{n,m} T_m^*(y) T_n^*(0) = 0, \qquad 0 < y < 1 \tag{8.1.11a}$$

$$\sum_{n=0}^{N} \sum_{m=0}^{M} b_{n,m} T_m^*(y) T_n^*(1) = 0, \qquad 0 < y < 1 \tag{8.1.11b}$$

$$\sum_{n=0}^{N} \sum_{m=0}^{M} b_{n,m} T_m^*(0) T_n^*(x) = 0, \qquad 0 < x < 1 \tag{8.1.11c}$$

$$\sum_{n=0}^{N} \sum_{m=0}^{M} b_{n,m} T_m^*(1) T_n^*(x) = 0, \qquad 0 < x < 1 \tag{8.1.11d}$$

These conditions serve to define the unknown constants $b_{n,m}$ for $n = N-1, N$ and $m = M-1, M$ since Eq. (8.1.6) serves only to define the values of $b_{n,m}$ for $n < N-1$ and $m < M-1$. The reason for this was explained in Chapter 3. When Eq. (8.1.6) is combined with the boundary conditions (8.1.11) a complete set of equations for solving for the $b_{n,m}$ values is obtained. To complete the solution, one must develop a computer routine to relate the $b_{n,m}^{xx}$ and $b_{n,m}^{yy}$ to $b_{n,m}$. This is accomplished by the relationships given by Gottlieb and Orszag (1977) which by transforming to the interval $0 \le x \le 1$ gives

$$f(x) = \sum_{n=0}^{\infty} b_n T_n^*(x)$$

then

$$f'' = \sum_{n=0}^{\infty} b_n^{xx} T_n^*(x)$$

where

$$\alpha_n b_n^{xx} = 4 \sum_{\substack{p=n+2 \\ p+n \text{ even}}}^{\infty} p(p^2 - n^2) b_p, \qquad \alpha_0 = 2, \ \alpha_n = 1 \text{ for } n > 0.$$

With these equations, one then must solve the matrix of the $b_{n,m}$ terms. Numerous techniques exist for this and the reader is referred to Gottlieb and Orszag (1977) or to the recent work of Haidvogel and Zang (1979) for the details of efficient techniques. To discuss these details at this point would overcomplicate the discussion.[†]

Once one has the solutions for the $b_{n,m}$ terms, it is possible to construct the function $F(x, y, t)$ of Eq. (8.1.7) using both the expansions for $\omega(x, y, t)$ and $\Psi(x, y, t)$. This is accomplished by computing $F(x, y, t)$ from these expansions at the extrema x_i, y_j of $T_N^*(x)$ and $T_M^*(y)$, respectively. The values of $F(x_i, y_j, t)$ can then be employed to compute the $c_{n,m}$ terms of the expansion

$$F(x_i, y_j, t) = \sum_{n=0}^{N} \sum_{m=0}^{M} c_{n,m}(t) \, T_n^*(x_i) \, T_m^*(y_j) \qquad (8.1.12)$$

by utilizing a two-dimensional fast Fourier transform routine[†] as (explained in Cooley and Tukey, 1965) a matrix inversion method. Given the $c_{n,m}$ terms, it is then possible to calculate $u_{n,m}$ at a new time $t + \Delta t$ and proceed step-by-step to construct the numerical solution of both Ψ and ω.

The second approach that can be utilized to solve for Ψ and ω is the real-space solution for ω instead of a spectral solution. For this solution technique, one still utilizes the same forms of Ψ and ω. Also, the solution for Ψ can be the same. For ω, however, one employs the equation

$$\omega(x, y, t + \Delta t) - \omega(x, y, t) = -\int_t^{t+\Delta t} \left(u \frac{\partial \omega}{\partial x} + v \frac{\partial \omega}{\partial y} \right) dt \qquad (8.1.13)$$

and only utilizes the expansions for Ψ and ω for the right-hand side. The procedure follows these steps:

1. Given an initial distribution of Ψ and ω.

[†]It has been recently established that the conjugate-gradient method is an effective way to solve such equations. In addition, it was established that preconditioning the matrix to make it triangular, on the first time step, speeded the direct solution for the other time steps (see Taylor et al. (1981) and Hirsh et al. (1982)).

2. Compute u, $\partial\omega/\partial x$, v, $\partial\omega/\partial y$ from Chebyshev expansions of the initial distribution of Ψ and ω.

3. Integrate Eq. (8.1.13) using an explicit time-integration method as described previously.

4. Given $\omega(x_i, y_j, t + \Delta t)$, develop a Chebyshev expansion for $\omega(x, y, t + \Delta t)$.

5. Solve for Ψ as described previously, given $\omega(x, y, t + \Delta t)$.*

Both of the approaches described have been applied to solve the vortex problem. The initial vorticity was taken to be

$$\omega = \frac{\Gamma}{2\pi r_0^2} \exp\left[-\frac{(x-x_0)^2 + (y-y_0)^2}{2r_0^2} \right] \tag{8.1.13a}$$

where $\Gamma = 13.76 \text{ m}^2/\text{s}$, $x_0 = 15 \text{ m}$, $y_0 = -70 \text{ m}$, and $r_0 = 5 \text{ m}$. The overall width of the computation box was 45 m, the height was 90 m, and the time step was taken as $\Delta t = 0.2 \min(\Delta x_i/|u_{i,j}|, \Delta y_j/|v_{i,j}|)$. The expansions employed 17 modes in each direction. In order to obtain results without significant Gibbs-type oscillation phenomena, it is usually necessary to filter the results either during or after the calculations. During the calculations, different filters are possible. Haidvogel et al. (1980) employed a filter of the form

$$a_{n,m} = a_{n,m} f_n f_m \tag{8.1.14}$$

where $f_n = 1.0 \exp[-B(N^2 - n^2)]$ with B being an adjustable parameter. Orszag and Gottlieb (1980) suggest the form

$$f_n = \begin{cases} 1 & \text{for } n < n_0 \\ \exp[-B(n - n_0)^4] & \text{for } n > n_0 \end{cases} \tag{8.1.15}$$

with n_0 being one-half to two-thirds the maximum n and B selected to damp from e^{-1} to e^{-10} in the high modes. Myers et al. (1981), after trying various approaches, found the best calculation procedure in an inviscid incompressible flow was to calculate ω in real space, Eq. (8.1.13) and filter with the procedure described in Section 3.6 for incompressible flows.

Vorticity results obtained for the vortex in a box without filtering are shown at four separate times in Figs. 8.1.2 and 8.1.3. For the calculations, the use of Eq. (8.1.13) to compute the vorticity yielded the least difficulty. Comparisons of results for one time with and without filtering are shown in Fig. 8.1.4.

The results of the spectral methods for the problem were also compared with finite-difference calculations. The results demonstrated that the spectral method was at least a factor of 10, and sometimes as much as a factor of 30, faster than a finite-difference approach for the same accuracy. Haidvogel et al.

---

*Note that one can also employ a finite-difference predictor and a spectral corrector to solve for Ψ in a manner similar to Orszag (1980) and Morchoisne (1981).

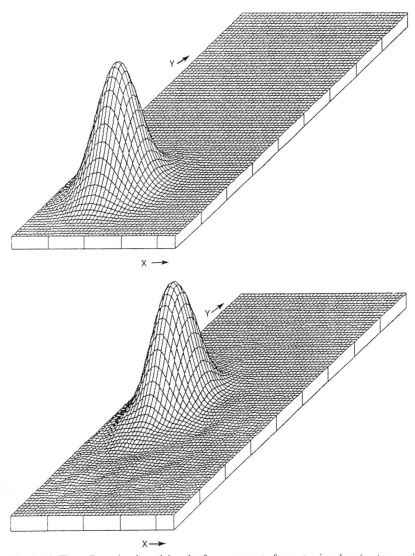

Fig. 8.1.2 Three-dimensional vorticity plot for movement of a vortex in a box (vortex moving up).

(1980) performed a similar comparison for finite-difference, finite-element, and Chebyshev these approaches by solving the stream-function vorticity equations for inviscid open-ocean problems. These authors found the same advantage of spectral methods over both finite-element and finite-difference approaches. They also found it necessary to filter the results periodically in time to maintain stable calculations.

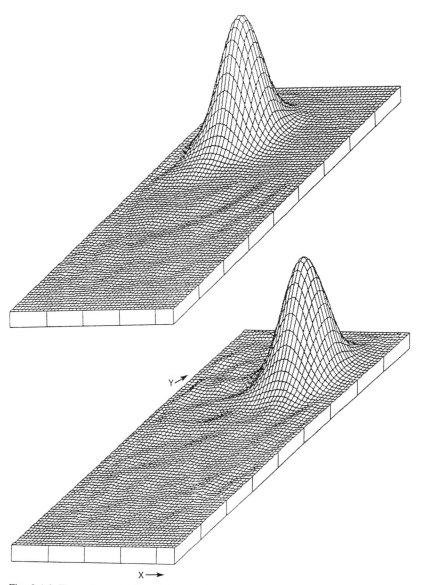

Fig. 8.1.3 Three-dimensional vorticity plot for movement of a vortex in a box (vortex at top and beginning to move downward).

8.2 Viscous Flows—Laminar and Transition

The calculation of two-dimensional incompressible viscous flows by spectral methods is still in a growth phase and all of the details for satisfying boundary conditions are not clearly resolved. However, there has been considerable

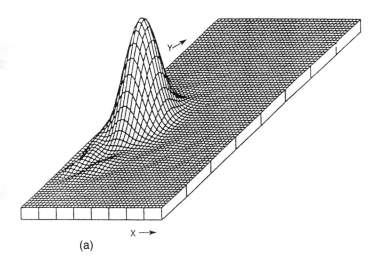

(a)

X →

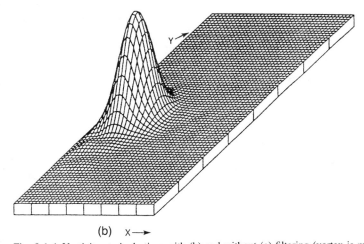

(b)   X →

Fig. 8.1.4 Vorticity at single time with (b) and without (a) filtering (vortex is moving upward).

progress in applying these methods. Results now exist for flow in a box (Morchoisne, 1981) flow in a channel (Deville and Orszag, 1980; Orszag and Kells, 1980), flat-plate flows (Murdock and Taylor, 1977, 1979, 1980), and shear layers flows (Riley and Metcalfe, 1980; Metcalfe and Riley, 1981). The solutions for these flows have been developed for both the primitive-variable and stream-function vorticity formulations. The three-dimensional cases frequently employed periodic boundary conditions in one or more directions. These solutions give insight into the method and fundamental aspects of flows.

The most interesting aspect of the spectral approach is its apparent speed and resolution advantage over the finite-difference and classical finite-element approaches. These advantages have led investigators to attempt to predict the onset of transition from laminar to turbulent flow and also to calculate a turbulent fluctuating flow. For spectral method applications to viscous flow, the optimum approach is to utilize primitive variables to avoid the boundary-condition problems associated with the stream function vorticity formulation. In addition, the primitive-variable formulation has lower-order derivatives than the stream-function only equation. Even with primitive variables, one can encounter problems with the boundary conditions.

The two most troublesome boundary conditions to prescribe and satisfy are:

1. Downstream flow conditions.
2. Pressure conditions at a solid surface.

At this time, there is no optimum way to treat downstream or outflow boundary conditions for viscous incompressible flows unless you know the solution ahead of time. Typical approaches employed to deal with the downstream condition are:

1. Ignore it by neglecting the viscous terms which require that a downstream condition be imposed (boundary-layer-type approximations).
2. Introduce a boundary condition that will permit the flow to exit the boundary with minimal interference.
3. Assume the flow is periodic.

Assumptions 1 and 3 of these conditions are the simplest to implement with spectral methods. Assumption 2, which tends to be the least restrictive on the physics of the flow, exhibits the most problems in implementation. This condition usually requires that a condition which does not induce a boundary layer at the outflow boundary be imposed. This is difficult to accomplish, however. No clear-cut way has been set forth for accomplishing this and the subject remains a research topic in the area of spectral-method applications (see Haidvogel, 1979).

The determination of pressure in an incompressible flow has been a troublesome area due to the lack of understanding of conditions on pressure at boundaries. In the past, many researchers employed the Poisson equation to determine the pressure, but this requires that conditions on the pressure be specified at each boundary. Recently, however, Moin and Kim (1980) pointed out that, in two dimensions, if one uses a primitive-variable formulation and spectral expansions for all dependent variables, then boundary conditions on pressure are not required. The only requirement is to specify a pressure value for normalization. The formulation of Moin and Kim is in spectral space and this makes the analysis somewhat complicated if nonperiodic boundary conditions on velocity occur. In order to properly satisfy such nonperiodic conditions, one will be faced with either an iterative explicit time-relaxation method or solution of an implicit equation set. As a result, it is possibly better to employ ideas similar to those of Moin and Kim and employ a pseudospectral approach using both real and spectral space. Such an approach is outlined in the following discussion.

The attractiveness of avoiding pressure boundary-condition problems becomes clear if one attempts to solve for the pressure field in a problem by employing Poisson's equation. For such a solution, one must prescribe pressure conditions on all boundaries and these are strongly related to the velocity field. As a consequence, an explicit time integration to obtain velocities which will specify the pressure boundary conditions may not yield a stable solution without iteration at each time of integration. For those interested in the Poisson-equation approach, a tested pseudospectral procedure is outlined below. It is important to note that the continuity equation replaces a momentum equation for calculation of one velocity component.

1. Integrate the u momentum equation with a pseudospectral scheme using old time values of u, v, and p to obtain u.
2. Compute v from the u spectral expansion and continuity.
3. Using the new time u and v, compute the right-hand side of the pressure equation and the gradients of p at the boundaries.
4. Solve Poisson's equation for p.
5. Replace the old time value of p by the new p and proceed to Step 1 if the new and old p are not of sufficient accuracy.

This approach may appear complicated but experience has shown that the first time step may require a number of iterations, but for the next time steps the iteration requirement is typically two for four-figure accuracy.

An alternate to the procedure just outlined is to attempt to compute the pressure directly from the primitive equations without employing Poisson's equation in a manner different from Moin and Kim (1980). A suggested pseudospectral approach to accomplish this is as follows:

1. Compute u from the x momentum equation.
2. Compute v from the continuity equation.
3. Compute $\partial p / \partial y$ from the y momentum equation using an implicit time-difference equation such as

$$v(t + \Delta t) - v(t) = \tfrac{1}{2}[F(t + \Delta t) + F(t)]\Delta t \tag{8.2.1}$$

$$F = -\left(\frac{\partial p}{\partial y} + u\frac{\partial v}{\partial x} + v\frac{\partial v}{\partial y}\right) + \frac{1}{Re}\left(\frac{\partial^2 v}{\partial x^2} + \frac{\partial^2 v}{\partial y^2}\right)$$

4. Compute p by analytic integration of the spectral form of $\partial p / \partial y$. This may require integration of $\partial p / \partial x$ for one value of $y = y_0$ to obtain p at one integration limit.*

Orszag and Kells (1980) as well as Deville and Orszag (1980) have applied another approach in the form

---

*This requires the pressure to be specified at one point on $y = y_0$.

Step I: (pseudospectral)

Compute a velocity field from the inviscid convective terms neglecting pressure gradients.

$$u^{\mathrm{I}}(t + \Delta t) - u(t) = -\int_{t}^{t+\Delta t} \left(u\frac{\partial u}{\partial x} + v\frac{\partial u}{\partial y} \right) dt \qquad (8.2.2a)$$

$$v^{\mathrm{I}}(t + \Delta t) - v(t) = -\int_{t}^{t+\Delta t} \left(u\frac{\partial v}{\partial x} + v\frac{\partial v}{\partial y} \right) dt \qquad (8.2.2b)$$

This is usually explicit in time so the terms in brackets are evaluated at t using spectral expansions and the integrals are evaluated by the Adams–Bashforth formula.

Step II: (spectral)

Correct the velocities for pressure so that continuity is satisfied

$$u^{\mathrm{II}}(t + \Delta t) - u^{\mathrm{I}}(t + \Delta t) = -\int_{t}^{t+\Delta t} \frac{\partial p}{\partial x} \, dt \qquad (8.2.3a)$$

$$v^{\mathrm{II}}(t + \Delta t) - v^{\mathrm{I}}(t + \Delta t) = -\int_{t}^{t+\Delta t} \frac{\partial p}{\partial y} \, dt \qquad (8.2.3b)$$

$$\frac{\partial u^{\mathrm{II}}}{\partial x} + \frac{\partial v^{\mathrm{II}}}{\partial y} = 0 \qquad (8.2.3c)$$

This step is performed implicitly in time and in spectral space because this step is linear. It is necessary to determine the pressure by solution of Eqs. (8.2.3a–8.2.3c) as a coupled set or by direct solution of the Poisson equation, which requires for compatibility that a coupled velocity–pressure boundary condition be applied at the surface. The condition is linear but complicates the solution for velocity. Deville and Orszag simplified this step by utilizing an inviscid pressure boundary condition $\nabla p \cdot \mathbf{N} = 0$.

Step III: (pseudospectral or spectral)

Correct the velocity field for viscosity

$$u(t + \Delta t) - u^{\mathrm{II}}(t + \Delta t) = \int_{t}^{t+\Delta t} \frac{\nabla^2 u}{\mathrm{Re}} \, dt \qquad (8.2.4a)$$

$$v(t + \Delta t) - v^{\mathrm{II}}(t + \Delta t) = \int_{t}^{t+\Delta t} \frac{\nabla^2 v}{\mathrm{Re}} \, dt \qquad (8.2.4b)$$

This correction can be made implicitly or explicitly since the step is linear.

As can be seen, this procedure does not avoid the boundary-condition questions for the pressure field and, at this time, it is not clear that it offers an advantage over the other approaches. The spectral and pseudospectral procedures outlined have suggested the use of simplified time integration. However, more complex time integration can be utilized by employing Chebyshev expansions in time (Morchoisne, 1981) or high-order Taylor series time integrations (Gazdag, 1973; Roy, 1980) which are generated by the rule

$$f(t + \Delta t) = f + f_t\, \Delta t + f_{tt}\, \frac{\Delta t^2}{2} + \cdots$$

The values of the time derivatives of the functions were generated by successive time derivatives of the Navier–Stokes equations, noting that one can cascade the process by the rule

$$f_t = L(f)$$
$$f_{tt} = L(f_t) = L(Lf)$$

where L is here a spatial differential operator commuting with the time derivative. Note that there is a trade-off made in loss of speed for increase in accuracy. As a result, for speed simple time-integration approaches are preferable, but if one needs accuracy, then either a Chebyshev or Taylor series time-integration approach will provide the desired results.

As mentioned earlier in this section, the spectral approaches have been applied primarily to study the possibility of transition from laminar to turbulent flows. In each of the studies the procedure employed was to select a flat plate or channel geometry to solve the primitive-variable laminar-flow equations. A perturbation velocity was imposed on the laminar flow and the problem was solved again and the behavior of perturbation studied to determine if any amplification or decay of the imposed disturbance occurs. Results obtained by Taylor and Murdock (1980) from such a study are discussed next.

In the Taylor–Murdock study the problem solution was developed by employing the pseudospectral collocation method and the velocity–pressure formulation of the form

$$v = -\int_0^y \frac{\partial u}{\partial x}\, dy \tag{8.2.5}$$

$$\frac{\partial u}{\partial t} + a = -\frac{\partial p}{\partial x} + \frac{\nabla^2 u}{Re}, \qquad a = \frac{\partial(u^2)}{\partial x} + \frac{\partial(uv)}{\partial y} \tag{8.2.6}$$

$$\nabla^2 p = -g = -\left[\left(\frac{\partial u}{\partial x}\right)^2 + 2\frac{\partial u}{\partial y}\frac{\partial v}{\partial x} + \left(\frac{\partial v}{\partial y}\right)^2\right] \tag{8.2.7}$$

The equations were time differenced in the following form

$$\frac{u(t+\Delta t) - u(t)}{\Delta t} + \frac{3}{2}a(t) - \frac{1}{2}a(t-\Delta t)$$

$$= -\frac{\partial p}{\partial x}(\bar{t}) + \frac{1}{Re}\left[\frac{\nabla^2 u(t+\Delta t)}{2} + \frac{\nabla^2 u(t)}{2}\right] \tag{8.2.8}$$

$$\nabla^2 p(\bar{t}) = -\tfrac{3}{2}g(t) + \tfrac{1}{2}g(t - \Delta t) \tag{8.2.9}$$

v is computed by numerical integration of Eq. (8.2.5) while the Poisson equation for p was solved by the tensor product method (Murdock, 1977). For the solution, each function was expanded in the form

$$f = \sum_{n=0}^{N} \sum_{m=0}^{M} f_{n,m}(t)\, T_n^*\!\left(\frac{x-x_1}{x_2-x_1}\right) T_m^*(Y) \qquad (8.2.10)$$

where

$$Y = \frac{\exp\,(-y/y_e) - \exp\,(y_{\max}/y_e)}{1 - \exp\,(y_{\max}/y_e)}$$

with $y_{\max}$ as the outer edge of the grid and y_e as a scaling parameter selected to give reasonable resolution of the calculation. Typically, y_e is of $O(1)$. $T_n^*(x)$ are the Chebyshev polynomials with the range $0 \le x \le 1$.

These expansions are used basically as interpolation formulas to accurately calculate function derivatives and integrals. This is accomplished by fitting the function f to a set of values of u, v, or p. This fit is accomplished by evaluating the series for $(N + 1)(M + 1)$ points at which at values of u, v, or p are known. This yields $(N + 1)(M + 1)$ equations for the unknown $f_{n,m}$ terms. This matrix of equations could be inverted by any matrix inversion routine but the fast Fourier transform routine was employed because of its speed. The series were employed to compute $a(t)$, $g(t)$, and $\int_0^y (\partial u/\partial x)\, dy$ in the calculations. The equations were solved for flow over a flat plate with the boundaries shown in Fig. 8.2.1. The boundary conditions were specified as follows. For the upstream boundary the flow was prescribed to be Blasius plus a time periodic solution of the Orr–Sommerfeld equation; i.e.,

$$u = u_{\text{Blasius}} + A\ \text{Real}\ [\phi_y(x,y)\,\exp(-i\omega\tau)] \qquad (8.2.11)$$

$$p = p_{\text{Blasius}} + p_{\text{Orr–Sommerfeld}}$$

where ϕ is the Orr–Sommerfeld solution for a wave number α and a frequency ω. τ is a dimensionless time defined by $\tau = tV_\infty/l$ where V_∞ is the characteristic velocity and l is the length from the leading edge of the flat plate to the boundary of the computation. The other boundary conditions employed were

$$u = v = 0, \qquad \frac{\partial p}{\partial y} = -\frac{1}{\text{Re}}\,\frac{\partial^2 u}{\partial x\,\partial y} \qquad \text{at } y = 0 \qquad (8.2.12)$$

$$u = 1, \qquad p = p_{\text{Blasius}} \qquad \text{at } y = \infty \qquad (8.2.13)$$

For the study, the downstream boundary condition was avoided by neglecting $\partial^2 u/\partial x^2$ in the u momentum equations so that a downstream boundary condition was not required on u. For the pressure downstream, the authors utilized the condition

$$\frac{\partial^2 p}{\partial x^2} = 0 \qquad \text{at } x = x_2 \qquad (8.2.14)$$

The solution of the specified problem is straightforward except for coupling between the pressure and velocity which occurs because of the surface boundary conditions. The significance of this coupling was investigated by employing the correct surface-pressure boundary condition as well as the condition

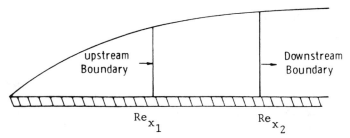

Fig. 8.2.1 Flow over a flat plate.

$$\frac{\partial p}{\partial y} = 0 \tag{8.2.15}$$

The procedure was used to compute two-dimensional flow over a flat plate for Reynolds numbers in the range $1.2 \times 10^5 \le \mathrm{Re}_x \le 3.8 \times 10^5$. Disturbances were introduced as upstream inputs to the flow. The disturbances were perturbation about the Blasius profile shown in Fig. 8.2.2. The total velocity is given by

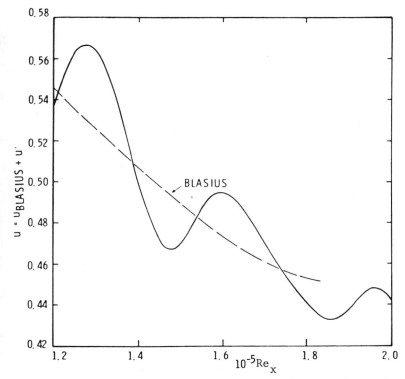

Fig. 8.2.2 Calculated total velocity u as a function of x for a fixed value of y, $\mathrm{Re}_y = 784$, $\partial p / \partial y_s = 0$, $t V_\infty / x = 4.0$.

$$u = u_{\text{Blasius}} + u' \qquad\qquad\qquad\qquad (8.2.16)$$

The disturbances was 5% of the free stream. The frequency of the initial wave for the calculation was $\omega\nu/V_\infty^2 = 56 \times 10^{-6}$.

Figure 8.2.2 displays the result obtained for the total velocity u as a function of x. Also shown is the Blasius velocity. The amplitude of the wave above the Blasius profile appears to decay as the wave propagates downstream, but no significant distortions are observed due to the nonlinear effects. The results computed from the v–p formulation for the Fourier amplitude of the first mode of u' are shown in Figs. 8.2.3 and 8.2.4. Also shown in Figs. 8.2.3 and 8.2.4 are results for the $\partial p/\partial y = 0$ boundary condition at the surface. There is some change in the results, but not anything significant. It is important to note, however, that the computer cost was reduced by approximately a factor of 2 when this boundary condition was applied.

In the course of this study, Taylor and Murdock also found that solving the equations and satisfying boundary conditions in real space was much easier

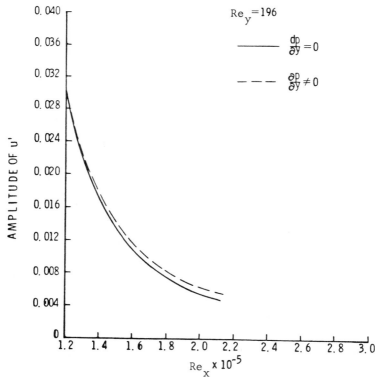

Fig. 8.2.3 Fourier amplitude of the velocity fluctuation u' about the Blasius as a function of x $\text{Re}_y = 196$ (first mode only).

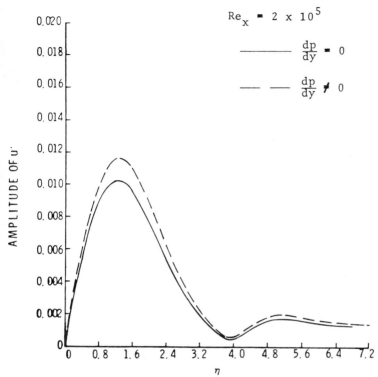

Fig. 8.2.4 Fourier amplitude of the velocity fluctuation about the Blasius as a function of the boundary-layer variable $\eta = y(\text{Re}_x/2)^{1/2}/x$ (first mode only).

than attempting a solution in spectral space and applying the tau method to satisfy boundary conditions.

Another example of the application of the pseudospectral approach is the combined use of finite-difference approximations and spectral expansions by Morchoisne (1981) to compute flow in a cavity and around a two-dimensional airfoil. Morchoisne applied the approach to both stream-function only and velocity–pressure formulations. We describe the velocity–pressure solution approach here and give sample results for a cavity flow. Morchoisne's approach is as follows:

Step 1:
  Write the two-dimensional Navier–Stokes equations in the form

$$\frac{\partial u}{\partial t} + \frac{\partial u^2}{\partial x} + \frac{\partial uv}{\partial y} + \frac{\partial p}{\partial x} - \frac{\nabla^2 u}{\text{Re}} = g_1 = 0 \qquad (8.2.17a)$$

$$\frac{\partial v}{\partial t} + \frac{\partial uv}{\partial x} + \frac{\partial v^2}{\partial y} + \frac{\partial p}{\partial y} - \frac{\nabla^2 v}{\text{Re}} = g_2 = 0 \qquad (8.2.17b)$$

$$\frac{\partial u}{\partial x} + \frac{\partial v}{\partial y} = g_3 = 0 \tag{8.2.17c}$$

Step II:

Introduce the iteration variable for all space and time values computed globally

$$U = u^{m+1} - u^m \tag{8.2.18a}$$

$$V = v^{m+1} - v^m \tag{8.2.18b}$$

$$P = p^{m+1} - p^m \tag{8.2.18c}$$

where m denotes the "global" iteration step with $m+1$ being the new result generated from the m result. Using these new variables, the flow equations can then be written

$$\frac{\partial U}{\partial t} + \frac{\partial}{\partial x}(U^2 + 2\,u^m U) + \frac{\partial}{\partial y}(UV + Uv^m + Vu^m) + \frac{\partial P}{\partial x} - \frac{\nabla^2 U}{\mathrm{Re}}$$

$$= -\left(\frac{\partial u^m}{\partial t} + \frac{\partial}{\partial x}(u^m)^2 + \frac{\partial}{\partial y}(u^m v^m) + \frac{\partial p^m}{\partial x} - \frac{\nabla^2 u^m}{\mathrm{Re}}\right) = g_1^*(u^m, v^m, p^m) \tag{8.2.19a}$$

and similarly

$$\frac{\partial V}{\partial t} + \frac{\partial}{\partial x}(UV + Uv^m + Vu^m) + \frac{\partial}{\partial y}(V^2 + 2\,v^m V) + \frac{\partial P}{\partial y} - \frac{\nabla^2 V}{\mathrm{Re}}$$

$$= g_2^*(u^m, v^m, p^m) \tag{8.2.19b}$$

$$\frac{\partial U}{\partial x} + \frac{\partial V}{\partial y} = g_3^*(u^m, v^m,) \tag{8.2.19c}$$

Step III:

The continuity equation can, however, be replaced by a Poisson equation for the pressure by combining Eqs. (8.2.19a) and (8.2.19b) with (8.2.19c) to obtain

$$\frac{\partial g_3^*}{\partial t} + \frac{\partial A}{\partial x} + \frac{\partial B}{\partial y} + \nabla^2 P - \frac{\nabla^2 g_3^*}{\mathrm{Re}} = \frac{\partial g_1^*}{\partial x} + \frac{\partial g_2^*}{\partial y} \tag{8.2.20}$$

where A and B are the combined nonlinear terms in Eqs. (8.2.19a) and (8.2.19b).

This equation must be solved subject to the usual boundary conditions for the pressure gradient normal to the boundary

$$\frac{\partial P}{\partial N} = \phi_N \tag{8.2.21}$$

where ϕ_N is obtained from Eqs. (8.2.19a) and (8.2.19b).

Step IV:
Introduce the finite-difference approximations for the A, B, P, U and V derivatives. These approximations are expressed in the nonuniform grid defined by the collocation points (x_i, y_j, t_n) associated with the pseudospectral approximations for u, v and p. A two-level scheme is used for the discretization of the time-derivatives and central differences (Eqs. (6.4.6) and (6.4.7)) are used for the space-derivatives.

Step V:
Resolve the equations and develop the solution by the following procedure.

a. Compute the values of g_k^* (u^m, v^m, p^m) by expanding in both space and time in Chebyshev polynomials. Compute derivatives of these expansions and then evaluate the derivatives and nonlinear terms in real space.
b. Given the g_k^* value, introduce finite differences for the left-hand side making diffusion and pressure terms implicit in time.
c. Develop the solution for U, V, and P at each time step using an iteration method analogous to the method described in Section 6.3.1.3.
d. If $|U|$, $|V|$, and $|P|$ are not sufficiently small, compute u^{m+1}, v^{m+1}, and p^{m+1} by Eq. (8.2.18) with possible relaxation, and return to (a) and start again.

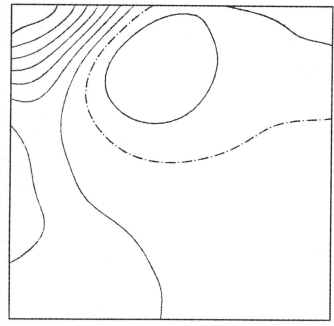

Fig. 8.2.5 Isobars computed by spectral-solution method, Re = 100. The isobars are spaced by $\Delta p = 0.025$. (Courtesy of Y. Morchoisne.)

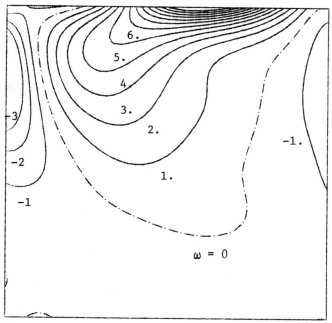

Fig. 8.2.6 Lines of constant vorticity computed by spectral-solution method, Re = 100. (Courtesy of Y. Morchoisne.)

Morchoisne used this procedure to calculate the flow in a square cavity (Section 6.6.1) using 17-term expansions in each space direction and 5 terms in time. The geometry and boundary conditions are shown in Fig. 6.6.1. The Reynolds number was 100 and the solution was converged to an accuracy of 10^{-4}. The results obtained for pressure and vorticity are shown in Figs. 8.2.5 and 8.2.6.

References

Cooley, J. W., and Tukey, J. W. *Math. Comput.* **19**, 297–301 (1965).

Deville, M., and Orszag, S. A. In: *Approximation Methods for Navier–Stokes Problems,* Lecture Notes in Mathematics, Vol. 771, pp. 159–176, Springer–Verlag, New York (1980).

Gazdag, J. *J. Comput. Phys.* **13**, 100–113 (1973).

Gottlieb, D., and Orszag, S. A. *Numerical Analysis of Spectral Methods,* SIAM, Philadelphia (1977).

Haidvogel, D. B., and Zang, T. *J. Comput. Phys.* **30**, 167–180 (1979).

Haidvogel, D. B., Robinson, A. R., and Schulman, E. E. *J. Comput. Phys.* **34**, 1–53 (1980)

Haidvogel, D. B. *J. Comput. Phys.* **33**, 313–324 (1979).

Hirsh, R. S. Taylor T. D., Nadworny, M. N., and Kerr, J. L. Techniques for efficient implementation of pseudo spectral methods and comparison with finite difference solutions of the Navier Stokes equations, Eighth International Conference on Numerical Methods in Fluid Dynamics, Aachen (1982).

Metcalfe, R. W., and Riley, J. J. *Proceedings of the 7th International Conference on Numerical Methods in Fluid Dynamics,* Lecture Notes in Physics, Vol. 141, pp. 279–284, Springer–Verlag, New York (1981).

Myers, R. B., Taylor, T. D., and Murdock, J. W. *J. Comput. Phys.,* **43,** 180–188 (1981).

Moin, P., and Kim, J. *J. Comput. Phys.* **35,** 381–392 (1980).

Morchoisne, Y. AIAA Paper 81-0109, January (1981).

Murdock, J. W. *AIAA J.* **8,** 1167–1173 (1977).

Murdock, J. W., and Taylor, T. D. AGARD Symposium on Laminar Turbulent Transition, Copenhagen, Denmark, May (1977).

Murdock, J. W. *IUTAM Proceedings of the Symposium on Laminar-Turbulent Transition.* Stuttgart, Germany, September (1979).

Orszag, S. A. *J. Comput. Phys.,* **37,** 70–92 (1980).

Orszag, S. A., and Kells, L. C. *J. Fluid Mech.* **96,** 159–205 (1980).

Orszag, S. A., and Gottlieb, D. In: *Approximation Methods for Navier–Stokes Problems,* Lecture Notes in Mathematics, Vol. 771, pp. 381–398, Springer–Verlag, New York (1980).

Riley, J. J., and Metcalfe, R. W. AIAA Paper No. 80-0274, January (1980).

Roy, P. *Rech. Aerosp.* 1980-6, 373–385 (1980).

Taylor, T. D., and Murdock, J. W. In: *Approximation Methods for Navier–Stokes Problems,* Lecture Notes in Mathematics, Vol. 771, pp. 519–537, Springer–Verlag, New York, (1980).

Taylor T. D., Hirsh, R. S., and Nadworny, M. N. FFT vs conjugate gradient method for solutions of flow equations by pseudo spectral methods, Fourth GAMM Conference on Numerical Methods in Fluid Mechanics, Paris (1981).

CHAPTER 9

Turbulent-Flow Models and Calculations

The Navier–Stokes equations apply equally to turbulent or laminar flows. Their completeness for turbulence computation have yet to be fully tested, however. For the present we will assume that the equations are adequate. The computation of turbulent flows with the above equations is a difficult task since the fully unsteady nature of turbulence must be calculated even if one is interested only in the time-averaged quantities. As a result one is faced with the solution of the full equations for a range of spatial length scales. Calculations using the full equations are just beginning to be implemented on the computer and remain limited in application due to cost. In the next few years this should change, however. In the following pages some of the recent work in this area is outlined.

In place of the full unsteady solutions much has been done to develop an alternate approach for solving applied turbulent problems. The approach is somewhat empirical but works well for certain classes of flows. This approach and the resulting equations will now be described.

9.1 Turbulence Closure Equations

The development of equations to describe turbulent flows for applied problems is accomplished by time averaging the turbulent fluctuations about a mean flow field. The logic of this process is available in a wide variety of books (see Rodi 1980) on turbulent flows and consequently will be abbreviated here.

Assuming that all the flow variables can be expanded in the form $f = \bar{f} + f'$, where $\bar{f}$ is a mean value of f and f' is a fluctuation about that mean, one can show that

$$\frac{\partial \bar{u}_i}{\partial x_i} = 0 \tag{9.1.1}$$

$$\frac{\partial \bar{u}_i}{\partial t} + \bar{u}_j \frac{\partial \bar{u}_i}{\partial x_j} = -\frac{1}{\rho} \frac{\partial \bar{p}}{\partial x_i} + \frac{\mu}{\rho} \nabla^2 \bar{u}_i - \frac{\partial R_{ij}}{\partial x_j} \tag{9.1.2}$$

where

$$R_{ij} = \overline{u_i' u_j'} \tag{9.1.3}$$

The quantity ρR_{ij} can clearly be viewed as a stress term in this equation. In order to establish a turbulence model that is less complicated than the full equations, it is necessary to establish an estimate of this stress in terms of the variables $\bar{u}_i$ and possibly some constants which are "tuned" to solve a given problem. Establishing such an estimate is the subject of an extensive number of publications and we can never hope to review all of the possible approaches. Instead, we will outline the basic principles. The first and foremost point in this approach is that somewhere in the progress of the model development one must assume the form of a transport term in the closure. A first closure is to assume the form of R_{ij}. By analogy with laminar flow one can take

$$-R_{ij} = \frac{\mu_t}{\rho}\left(\frac{\partial \bar{u}_i}{\partial x_j} + \frac{\partial \overline{u}_j}{\partial x_i}\right) - \frac{2\mu_t}{3\rho}\frac{\partial \bar{u}_i}{\partial x_i}\delta_{ij} \qquad (9.1.4)$$

where μ_t is a turbulent viscosity and δ_{ij} is the Kronecker symbol. When substituted into the right-hand term of Eq. (9.1.2) with μ_t constant, the result is

$$-\frac{\partial R_{ij}}{\partial x_j} = \frac{\mu_t}{\rho}\nabla^2 \bar{u}_i \qquad (9.1.5)$$

This is the simple approach which effectively adds an eddy viscosity to the laminar viscosity. Unfortunately, the constant-μ_t approach is not usually accurate, but it can give some indication of the flow behavior. For more accurate predictions, it is necessary to develop a better model to obtain estimates of R_{ij}. The models which are employed vary in the level of complexity from the so-called two-equation models to the complete Reynolds stress models (see Bradshaw, 1967 through Smagorinski, 1963). From the user point of view, it is not clear that the most complex models are of more value than the two-equation models, since all the models are built on empirical constants and postulates. As a consequence, the use of closure models in flow calculations remains a subject of continual controversy. At this time, the models which appear to be used most for engineering calculations are the two-equation boundary-layer-type models. As a result, these types of models will be outlined. To form such a model one must postulate some relationship for the R_{ij}'s in terms of mean flow quantities. This can vary in form but one of the least-complicated definitions is

$$-R_{ij} = \frac{\mu_t}{\rho}\left(\frac{\partial \bar{u}_i}{\partial x_j} + \frac{\partial \bar{u}_j}{\partial x_i}\right) - \frac{2\delta_{ij}k}{3} \qquad (9.1.6)$$

where μ_t is a turbulent viscosity and k is the turbulent kinetic energy $\frac{1}{2}\overline{u_i' u_i'}$. This definition introduces two unknowns, μ_t and k, which require two equations for closure. It is difficult, however, to derive an equation directly for μ_t. The modelers reasoned by dimensional analysis that for high-Reynolds-number flows

$$\frac{\mu_t}{\rho} \sim \frac{k^2}{\epsilon} \tag{9.1.7}$$

where ϵ is the dissipation rate of energy. As a result they assumed that

$$\frac{\mu_t}{\rho} = c_1 \frac{k^2}{\epsilon} \tag{9.1.8}$$

where c_1 may be a function of Reynolds number. Equations for k and ϵ were then derived using postulates, logic, and empiricism.

The equations follow an analogy with the classical energy-transport equations and take the form

$$\frac{Dk}{Dt} = \mathbf{\nabla} \cdot \left(c_2 \frac{\mu_t}{\rho} \mathbf{\nabla} k \right) + G - \epsilon \tag{9.1.9}$$

$$\frac{D\epsilon}{Dt} = \mathbf{\nabla} \cdot \left(c_3 \frac{\mu_t}{\rho} \mathbf{\nabla} \epsilon \right) + c_4 \frac{\epsilon}{k} G - c_5 \frac{\epsilon^2}{k} \tag{9.1.10}$$

where the left-hand side is the convection and the right-hand side is the diffusion along with source and sink terms. G in these equations is the source term of turbulent kinetic energy, and in its most general form can be found by analogy with the stress generation of energy in the conservation-of-energy equation. Due to its complexity, we will not write out the general form. A two-dimensional form of these equations which is frequently used is

$$\frac{Dk}{Dt} = \frac{\partial}{\partial x} \left(c_2 \frac{\mu_t}{\rho} \frac{\partial k}{\partial x} \right) + \frac{\partial}{\partial y} \left(c_2 \frac{\mu_t}{\rho} \frac{\partial k}{\partial y} \right) + G - \epsilon \tag{9.1.11}$$

$$\frac{D\epsilon}{Dt} = \frac{\partial}{\partial x} \left(c_3 \frac{\mu_t}{\rho} \frac{\partial \epsilon}{\partial x} \right) + \frac{\partial}{\partial y} \left(c_3 \frac{\mu_t}{\rho} \frac{\partial \epsilon}{\partial x} \right) + c_4 \frac{\epsilon}{k} G - c_5 \frac{\epsilon^2}{k} \tag{9.1.12}$$

where

$$G = \frac{\mu_t}{\rho} \left[2 \left(\frac{\partial \overline{v}}{\partial y} \right)^2 + 2 \left(\frac{\partial \overline{u}}{\partial x} \right)^2 + \left(\frac{\partial \overline{v}}{\partial x} + \frac{\partial \overline{u}}{\partial y} \right)^2 \right] \tag{9.1.13}$$

The c_n terms in these equations are empirical functions which in the simplest case are constants. For the standard model as defined by Launder and Spalding (1974) the constants have the dimensionless values

$$c_1 = 0.09, \qquad c_3 = 0.769, \qquad c_5 = 1.92$$
$$c_2 = 1.0, \qquad c_4 = 1.44$$

Each user is free to adjust (tune) these constants or even make them functions of dependent or independent variables. As a result, the accuracy of the model is problem dependent and is reported (Launder and Morse, 1979) to vary from 10% to 50% depending on the flow.

The numerical integration of the equations of the (k, ϵ) model also can contribute to the difficulties. This is because the last two terms in the equation for ϵ exhibit a singular nature near a wall since the turbulent kinetic energy k tends to zero. As a result, one is faced with some type of special consideration as the calculation approaches a solid wall. One additional problem with this type of model is specification of initial or boundary conditions on k and ϵ. The details, however, depend on the problem and no attempt will be made here to specify a guideline.

The next level of closure is to develop equations directly for the Reynolds stress terms R_{ij}. When one does this the equation set becomes more complicated because there is a differential equation for each R_{ij} to solve along with the mean-flow equations. The complexity of these equations depends on the level and technique of closure. The development of these equations is a lengthy subject and details are available in publications on the subject by Mellor and Yamada (1974), Donaldson (1973), Rodi (1980), Mellor and Herring (1971), and Reynolds (1976).

The problems with numerical solutions of the closure equations are very problem dependent and as a result cannot be easily set forth. Before ending the discussion on closure models, it is important to note that Rodi (1980) has pointed out the value of introducing algebraic stress models to reduce closure complications. This approach basically employs a (k, ϵ) model with algebraic relationships for the velocity correlations.

Beyond the full closure method for calculating turbulence, the next level of complexity in solving turbulence problems is to utilize the full set of flow equations in combination with a closure assumption for only the small scales. This approach avoids resolution problems and saves computer time by cutting off the scale length of the complete equation solution. As a result, one sets a minimum scale for the complete equation solution. Below this length no attempt is made to resolve the flow exactly, and the closure model is introduced to account for the flow at scale lengths below the cutoff. Such models are termed subgrid scale approximations in a large-eddy simulation. This approach is addressed next.

9.2 Large-Eddy Simulation Model

The model for large-eddy simulation of turbulent flow uses the full equations to compute the flow, but recognizes that it is very costly to compute the details of the very small-scale turbulence. As a result, one derives a closure model for only the scales below the selected cutoff length l. In order to accomplish this, one applies a filter to the flow equations which introduces a Reynolds stress for the flow with scale lengths less than the length l. There are a variety of filters that can be applied ranging from step functions to Gaussians. Here we present the results obtained using a Gaussian filter as described by Ferziger (1977). The filter is applied by defining the large-scale field by the general averaging formula

$$\bar{f}(\mathbf{x}) = \int_v H(\mathbf{x}, \mathbf{x}')\, f(\mathbf{x}')\, d\mathbf{x}'$$

where H is the filter taken to be

$$H = \left(\frac{6}{\pi l^2}\right)^{3/2} \exp\left[\frac{-6(\mathbf{x} - \mathbf{x}')^2}{l^2}\right]$$

and l is the scale length and $\mathbf{x} = (x_1, x_2, x_3)$.

If the momentum flow equations are averaged, one obtains

$$\frac{\partial \bar{u}_i}{\partial t} + \frac{\partial \overline{u_i u_j}}{\partial x_j} = -\frac{1}{\rho}\frac{\partial \bar{p}}{\partial x_i} + \frac{\mu}{\rho}\nabla^2 \bar{u}_i$$

if $u_i = \bar{u}_i + u_i'$. Then we find that

$$\overline{u_i u_j} = \overline{\bar{u}_i \bar{u}_j} + \overline{u_i' \bar{u}_j} + \overline{\bar{u}_i u_j'} + \overline{u_i' u_j'}$$

As Ferziger points out, these averages do not simplify as in the full closure case. As a result, the value of $\overline{u_i u_j}$ for the least-complicated closure has the form

$$\overline{u_i u_j} = \bar{u}_i \bar{u}_j + \frac{l^2}{24}\nabla^2 \bar{u}_i \bar{u}_j + \frac{l^2}{24}\bar{u}_i \nabla^2 \bar{u}_j + \frac{l^2}{24}\bar{u}_j \nabla^2 \bar{u}_i + \frac{\mu_t}{\rho}\bar{S}_{ij}$$

where μ_t is the turbulent viscosity. For many of the models used thus far, it has been assumed that the turbulent viscosity could be represented in the form (Smagorinsky, 1963)

$$\frac{\mu_t}{\rho} = c l^2 (\bar{S}_{ij} \bar{S}_{ij})^{1/2}$$

where

$$\bar{S}_{ij} = \frac{1}{2}\left(\frac{\partial \bar{u}_i}{\partial x_j} + \frac{\partial \bar{u}_j}{\partial x_i}\right)$$

and c is an empirical constant.

Recently, as interest in subgrid scale models has grown, new postulates for the turbulent viscosity have appeared in the literature (Clark et al., 1979; Leslie and Quarini, 1979; Love and Leslie, 1979). These vary from energy- to vorticity-based relationships and no attempt will be made here to evaluate the optimum approach since it is clearly an area for future research. As in the full closure, the subgrid closure models can be extended to higher order. This would follow along the lines already discussed.

Computations with a large-eddy simulation model will not be significantly different in complexity from the full closure model except that one can expect to obtain more of the true turbulent fluctuations in the time variation. As a result one will be more concerned with accuracy of initial conditions and integration scheme. Also computer time and storage become a problem as the cutoff length l becomes small compared to the characteristic length of the geometry. Some of these issues will be addressed in the examples presented later in this chapter.

9.3 Turbulent-Flow Calculations with Closure Model

In the beginning of this chapter the types of turbulence models that can be employed in incompressible-flow calculations were discussed. In this section we will present some of the more recent results obtained with a closure model.

The numerical methods employed in calculations with the turbulence models generally fall into two categories. Finite differences, typically employed with the closure calculations, and the pseudospectral approach have been used to perform the large-eddy and full simulation approach. The finite-element technique has not been employed as much in turbulent-flow predictions. In the future, one is likely to see the pseudospectral approach replace finite differences in the closure approach due to improvement in efficiency. The results reported here employ finite differences for the closure calculations. The numerical difficulties encountered when solving turbulence closure problems by finite differences center principally on the difference approximations used for the inertia term and whether the viscous terms for the unsteady case are implicitly or explicitly introduced. The problem basically is that to maintain stability at the high Reynolds numbers explicit central differences require very small grid sizes. As a result, there has been a tendency to introduce one-sided differences for the spatial derivatives in the convective terms. This is exactly equivalent to adding an artificial viscosity to damp oscillations and is equivalent to another closure model.

The use of such an approach improves stability but not accuracy. The investigation of the use of implicit viscous terms with time integrations to avoid stability problems seems limited for turbulence studies at this time.

Most other numerical difficulties with turbulence closure calculations arise due to the inadequacy of the model and not the numerics. For example, in a two-equation (k, ϵ) model, the terms with ϵ/k can be singular in poorly modeled regions of flow and the numerics simply will not account for this problem. In this section we will not attempt to pursue these types of difficulties with models. Recent publications on turbulent shear-flow predictions (Durst et al., 1979; Rodi, 1980) will serve as a useful guide to evaluating models. Numerical aspects are not strongly emphasized, however.

Two good examples of current progress in turbulence closure-model computations of flows are the work of Ha Minh and Chassaing (1979) along with the studies of Grant et al. (1980). Ha Minh and Chassaing used a hierarchy of closure models to compute turbulent axisymmetric flow. The modeling ranges

from the two-equation (k, ϵ) model, which basically was described earlier, to
a generalized Reynolds stress transport model. Beyond the two-equation model
the authors employed the three-equation model described by Launder et al.
(1973) in which one assumes

$$\frac{D\,\overline{uv}}{Dt} = \frac{c_1}{r}\frac{\partial}{\partial r}\left(r\,\frac{k}{\epsilon}\,\overline{v^2}\,\frac{\partial\,\overline{uv}}{\partial r}\right) - c_1\,\frac{k\,\overline{w^2}}{\epsilon}\,\frac{\overline{uv}}{r^2} - (1-\alpha)\,\overline{v^2}\,\frac{\partial U}{\partial r} - \frac{\beta\epsilon\overline{uv}}{k}$$

$$(9.3.1)$$

with $\overline{u^2} = a_1 k$, $\overline{v^2} = a_2 k$, $\overline{w^2} = a_3 k$ so that

$$\frac{D\,\overline{uv}}{Dt} = c_1 a_2 \frac{1}{r}\frac{\partial}{\partial r}\left(r\,\frac{k^2}{\epsilon}\,\frac{\partial\,\overline{uv}}{\partial r}\right) - c_1 a_3\,\frac{k^2}{\epsilon}\,\frac{\overline{uv}}{r^2} - (1-\alpha)\,a_2 k\,\frac{\partial U}{\partial r} - \frac{\beta\epsilon\overline{uv}}{k}$$

$$(9.3.2)$$

where the constants are $a_1 = 1.0$, $a_2 = 0.5$, $a_3 = 0.5$, $c_1 = 0.25$, $\alpha = 0.4$,
$\beta = 2.5$, U and V are the mean velocity components in x and r directions; u,
v and w are the fluctuating velocity components in x, r and θ.

A more general form of this model given by Rodi (1972) requires five
equations and was also applied in the form

$$\frac{D\overline{u^2}}{Dt} = c_2\frac{1}{r}\frac{\partial}{\partial r}\left(rk\,\frac{\overline{v^2}}{\epsilon}\,\frac{\partial\,\overline{u^2}}{\partial r}\right) - 2\,\overline{uv}\,\frac{\partial U}{\partial r} - c_3\frac{\epsilon}{k}\left(\overline{u^2} - \frac{2k}{3}\right)$$

$$+ \frac{4c_4\overline{uv}}{3}\frac{\partial U}{\partial r} - \frac{2\epsilon}{3} - 2\,\overline{u^2}\,\frac{\partial U}{\partial x} \qquad (9.3.3)$$

$$\frac{D\overline{v^2}}{Dt} = c_2\frac{1}{r}\frac{\partial}{\partial r}\left(rk\,\frac{\overline{v^2}}{\epsilon}\,\frac{\partial\,\overline{v^2}}{\partial r}\right) - c_3\frac{\epsilon}{k}\left(\overline{v^2} - \frac{2k}{3}\right) - 2\,\overline{v^2}\,\frac{\partial V}{\partial r}$$

$$- \frac{2c_4}{3}\,\overline{uv}\,\frac{\partial U}{\partial r} - \frac{2\epsilon}{3} \qquad (9.3.4)$$

$$\frac{D\overline{w^2}}{Dt} = c_2\frac{1}{r}\frac{\partial}{\partial r}\left(rk\,\frac{\overline{v^2}}{\epsilon}\,\frac{\partial\overline{w^2}}{\partial r}\right) - c_3\frac{\epsilon}{k}\left(\overline{w^2} - \frac{2k}{3}\right)$$

$$- \frac{2c_4}{3}\,\overline{uv}\,\frac{\partial U}{\partial r} - \frac{2\epsilon}{3} \qquad (9.3.5)$$

where the constants are given by $c_2 = 0.25$, $c_3 = 2.75$, and $c_4 = 0.42$.

The solution to the elliptic equations in the set was developed by using a
pointwise iteration method, similar to a relaxation procedure, as set forth by
Gosman et al. (1969)(see Sections 6.5.4 and 6.5.5). This methodology can be
improved upon by employing more-efficient methods such as block cycle
reduction or the conjugate-gradient method. For the parabolic equations, centered finite differences were employed for the first seven steps and then a switch
over to the noncentered DuFort–Frankel method was made.

The models were applied to three separate flow situations. The first was flow over a step; the second was flow past an expansion in a circular duct; and the third was flow in a circular jet. The authors also made experimental measurements for comparison. Figures 9.3.1 and 9.3.2 show comparisons of the calculations with experiments for the step in a pipe and for the circular jet case. For the step case two forms of a (k, ϵ) model were used. The solid line is the model presented and the dotted line is using $\epsilon = 0.09k^{3/2}/L$ where L is prescribed. The five equation model was used for the jet. For the step case

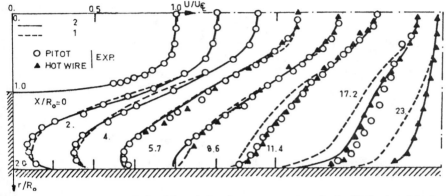

Fig. 9.3.1 Mean velocities calculated for turbulent flow over a step in a pipe for (1) (k, ϵ) model using k equation $\mu_t/\rho = L\sqrt{k}$ and $\epsilon = 0.09k^{3/2}L$, and (2) full (k, ϵ) model. (Courtesy of H. Ha Minh and P. Chassaing.)

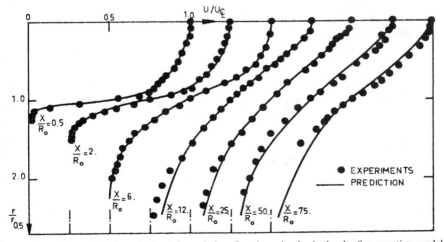

Fig. 9.3.2 Mean velocities calculated for turbulent flow in a circular jet by the five-equation model. (Courtesy of H. Ha Minh and P. Chassaing.)

$Re_0 = 7.2 \times 10^4$ and $\delta/R_0 = 0.3$.* Where δ is the boundary-layer thickness at the step and R_0 is the tube radius before the step. The step height was also R_0. For the jet case $Re_0 = 11 \times 10^4$ and $\delta/R_0 = 1.0$. Note that, in general, the predictions are in good agreement with experiment. These results demonstrate the level of turbulence simulations that can be attained by closure models. Furthermore, detailed testing of computational simulations have been conducted by Grant et al. (1980). These authors studied the refinement of the five-equation-type model for stratified flows. The authors "tuned" the model to compute the case of a collapsing mixed region in a stratified flow. No particular numerical innovations were developed in the study and the results demonstrated a similar level of accuracy as Ha Minh and Chassaing. Introduction of the equations and results would require extensive discussion. As a result, the interested reader should refer to the reference for the details of the model.

In addition to closure models, direct simulations of turbulence are beginning to appear. The progress in these types of simulations is limited due to the computer requirements. Some advances are being made, however, and these are outlined next.

9.4 Direct Simulations of Turbulence

The numerical solution of the Navier–Stokes equations without closure at high Reynolds numbers, in theory, will result in the simulation of a fully turbulent flow. The difficulty with this approach is the large grid-point requirement, which in turn leads to extreme computer costs. As computing has decreased in cost and numerical methods have increased in efficiency, full simulations are being attempted. Recent progress in this area has been made by Riley and Metcalfe (1980), who calculated a turbulent free shear layer flow. Previous studies by Schumann and Patterson (1978a,b), Leslie and Quarini (1979), Rogallo (1977), Orszag et al. (1969, 1972, 1974) and Patterson (1971) used direct simulations of the decay of turbulence with periodic boundary conditions. These studies employed the pseudospectral approach to solve the flow equations. A number of other studies have been conducted using finite differences and subgrid scale closure approximations (Clark et al., 1979; McMillan and Ferziger, 1979; Deardoff, 1970, 1972, 1974a,b) for simplified flows.

From the studies that have been conducted, it appears that the pseudospectral approach offers the most promise for direct simulation without closure. This is because of its improved efficiency compared to finite differences. This optimism is based on periodic-boundary-condition results, however, and additional research is needed to display this advance for flows with mean convection and nonperiodic boundary conditions.

---

*Re_0 is the Reynolds number at the entrance based on R_0.

The application of the finite-difference and pseudospectral approaches to solution of the flow equations in most of the simulations is made using a primitive-variable formulation, and typically the Poisson equation is used to determine the pressure. For the periodic-boundary-condition cases, one does not anticipate any particular problem with either finite-differences or the pseudospectral method, provided the standard time step and cell Reynolds number restrictions are observed. Once nonperiodic boundary conditions are introduced, then numerical difficulties due to strong gradients near boundaries may arise. The experience in this area is limited at this time and well-founded information on the subject is not yet available. Because of this, we are unable to present a full simulation of a turbulent boundary layer on a wall. The closest example is the calculation of a free shear layer by Riley and Metcalfe. The geometry of the flow is shown in Fig. 9.4.1. The problem is, given an initial state, to compute the late time turbulent behavior. The initial state for the flow was obtained by a random process (Orszag and Pao, 1974) associated with

$$\frac{\bar{u}(z)}{\Delta u} = \frac{1}{2} \tanh \left(\frac{0.55\, z}{z_{1/2}} \right) \qquad \text{(mean velocity)}$$

$$\frac{u'(z)}{\Delta u} = 0.18 \exp \left(\frac{-0.147\, z^2}{z_{1/2}^2} \right) \qquad \text{(turbulent intensity)}$$

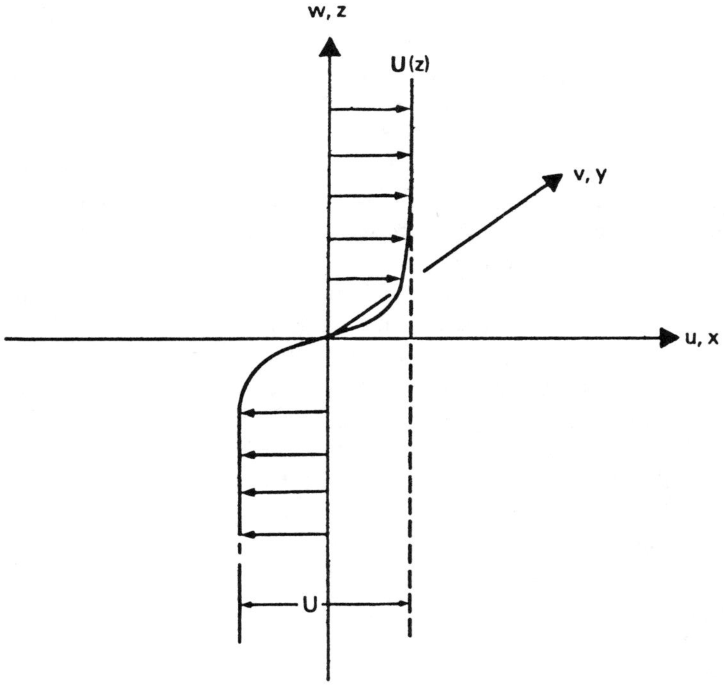

Fig. 9.4.1 Geometry of shear-layer flow.

with other velocity components zero. Δu is the mean velocity difference across the layer and $z_{1/2}$ is the distance from $z=0$ to where $\bar{u}$ takes on $1/2$ the free-stream value. The initial Reynolds number based on the Taylor microscale is about 50.

The computation domain was chosen to be cubic with dimensions

$$L = 22.8\, z_{1/2}$$

In the study of the free shear layer, Riley and Metcalfe employed a $32 \times 32 \times 33$ pseudospectral expansion to develop the solution. The dependent variables in the x (longitudinal) and y (lateral) directions were expanded in Fourier series.* In the z (vertical) direction, the solutions were expanded in sine and cosine series. The transformations from real space to spectral space were accomplished by employing fast Fourier transforms. Leapfrog time differencing was employed for the nonlinear terms and Crank-Nicolson (implicit) time differencing was employed for the viscous terms. At this point, it is important to note that employment of the implicit scheme on the viscous terms requires that the time integration be accomplished in spectral space. Also, this approach is optimum when Fourier series are employed since the second-order operators generate diagonal matrices in spectral space because $\partial^2 e^{ikx}/\partial x^2 = -k^2 e^{ikx}$ and hence there are no off-diagonal terms generated. This is not the case when Chebyshev polynomials are employed. As a result, the introduction of an implicit time-integration scheme may yield complications for such functions. In the calculations, the nonlinear terms are calculated by using the expansions to compute the values of the terms in real space. Then the real-space values of the nonlinear terms are transformed to spectral space for the numerical time integration. It is important to note that the time integration could also be accomplished in real space if a predictor-corrector was used for the viscous terms. Periodic boundary conditions were used in the x and y directions and are automatically satisfied by the Fourier series. In the z direction the net velocity difference across the shear layer was satisfied by employing sine and cosine expansions. An equal mesh spacing was employed, and this is again consistent with the Fourier expansions and the use of fast Fourier transforms. The calculations required 4 s per time step and the results obtained for vorticity at three different times are shown in Fig. 9.4.2. In these results the time is scaled by $0.275\, \Delta u/z_{1/2}$. Riley and Metcalfe have shown these results to be in reasonable agreement with experiments and theory.

Two-dimensional results were also calculated by Riley and Metcalfe using 64×64 and 128×128 grids. They initialized the calculations as in the three-dimensional case but superimposed on the initial conditions the most unstable mode with wave length λ_F velocity solution obtained from linear theory. Results obtained for the vorticity at three times are the results shown

---

*They use the form e^{ikx} and e^{ily} here.

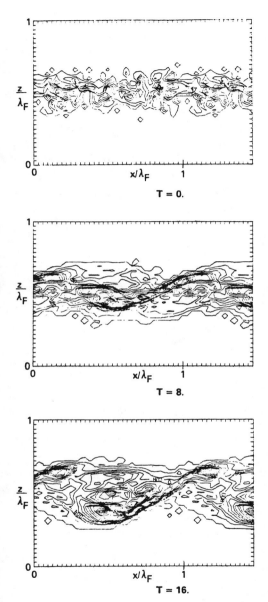

Fig. 9.4.2 Lateral (y) vorticity in a (x, z) plane at various times T. (Courtesy of J. Riley and R. Metcalfe.)

in Fig. 1.2 of Chapter 1. Note that the initial disturbance grows and rolls up into vortices. Needless to say these types of simulations, although somewhat crude, are beginning to offer meaningful simulations of a turbulent field with a mean flow.

Progress in direct simulations of turbulence can be expected to accelerate in the next few years as computers and memory cost decrease and it should be a significant period of advancement in the area.

References

Bradshaw, P., Ferriss, D. H., and Atwell, N. P. *J. Fluid Mech.* **28**, 593–616 (1967).

Bradshaw, P. *Aero. J.* **76**, 403–418, (1972).

Clark, R. A., Ferziger, J. H., and Reynolds, W. C. *J. Fluid Mech.* **91**, 1–16 (1979).

Daly, B. J., and Harlow, F. H. *Phys. Fluid* **13**, 2634 (1970).

Deardorff, J. W. *Workshop on Micrometeorology*, pp. 271–311, American Meteorological Society, Boston(1973).

Deardorff, J. W. *J. Fluid Mech.* **41**, 453–480 (1970).

Deardorff, J. W. *Boundary Layer Meteor.* **7**, 81–106 (1974a).

Deardorff, J. W. *Boundary Layer Meteor.* **7**, 199–226 (1974b).

Deardorff, J. W. *J. Atm. Sci.* **29**, 91–115 (1972).

Donaldson, C. du P., and Rosenbaum, H. Calculation of Turbulent Shear Flows through Closure of the Reynolds Equations by Invariant Modeling, ARAP Inc. Rept. 127, Princeton, NJ (1968).

Donaldson, C. du P. *Workshop on Micrometeorology*, pp. 313–392, American Meteorological Society, Boston (1973).

Durst, F., Launder, B. E., Schmidt, F. W. and Whitelaw, J. H. (eds.) *Turbulent Shear Flows I,* Springer-Verlag, New York (1979).

Ferziger, J. H. *AIAA J.* **15**, 1261–1267 (1977).

Gosman, A. D., Pun, W. M., Runchal, A. K., Spalding, D. B., and Wolfshtein, M. Y. *Heat and Mass Transfer in Recirculating Flows,* Academic Press, London (1969).

Grant, J. R., Innis, G. E., and Shaw, W. Y. A Computational Model for Turbulent Wake Flows in a Stratified Fluid, SAI Rept. 200-80-483-LJ, La Jolla, CA, August (1980).

Ha Minh, H., and Chassaing, P. *Turbulent Shear Flows I,* pp. 178–197, Springer-Verlag, Berlin, Heidelberg, New York (1979).

Hanjalic, K., and Launder, B. E. *J. Fluid Mech.* **52**, 609–638 (1972).

Harlow, F. H., and Nakayama, P. I. *Phys. Fluids* **10**, 2323–2332 (1967).

Jones, W. P., and Launder, B. E. *Int. J. Heat Mass Trans.* **15**, 301–314 (1972).

Launder, B. E., and Spalding, D. B. *Comp. Methods Appl. Mech. Eng.* **3**, 269 (1974).

Launder, B. E., Reece, G. J., and Rodi, W. Progress in Development of a Reynolds Stress Turbulence Closure, Rept. HTS/73/31, Imperial College of London, December (1973).

Launder, B. E., Reece, G. J., and Rodi, W. *J. Fluid Mech.* **68**, 537 (1975).

Launder, B. E., and Morse, A. *Turbulent Shear Flows I,* pp. 279–294, Springer–Verlag, Berlin, Heidelberg, New York (1979).

Leslie, D. C., and Quarini, G. L. *J. Fluid Mech.* **91**, 65–91 (1979).

Love, M. D., and Leslie, D. C. *Turbulent Shear Flows I,* pp. 353–369, Springer–Verlag, New York (1979).

Lumley, J. L., and Khajeh-Nouri, B. *Proceedings of the 2nd IUGG-IUTAM Symposium on Atmospheric Diffusion,* Academic Press, New York (1974); *Adv. Geophys.* **18,** 169–192 (1974).

McMillan, O. J., and Ferziger, J. H. AIAA Paper 79-0072, New Orleans (1979).

Mellor, G. L., and Herring, H. J. A Study of Turbulent Boundary Layer Models Part II, Sandia Corp., Rept. SC-CR-70-6125B, January (1971).

Mellor, G. L., and Yamada, T. *J. Atm. Sci.* **31,** 1791–1806 (1974).

Metcalfe, R. W., and Riley, J. J. In: *Proceedings of the 7th International Conference on Numerical Methods in Fluid Dynamics,* Lecture Notes in Physics No. 141, pp. 279–284, Springer–Verlag, New York (1981).

Orszag, S., and Pao, Y. J. *Adv. Geophys.* **18A,** 225–236 (1974).

Orszag, S. A., and Patterson, G. S. *In Statistical Models and Turbulence,* Lecture Notes in Physics No. 12, pp. 127–147, Springer–Verglag, Berlin, Heidelberg, New York (1972).

Orszag, S. A. *Phys. Fluids Suppl.* **12,** 250–257 (1969).

Patterson, G. S., and Orszag, S. A. *Phys. Fluids* **14,** 2538–2541 (1971).

Reynolds, W. C. *Ann. Rev. Fluid Mech.* **8,** 183–208 (1976).

Riley, J. J. and Metcalfe, R. W. AIAA Paper 80-0274, Pasadena (1980).

Rodi, W. The Prediction of Free Turbulent Boundary Layers by Use of a Two-Equation Model of Turbulence, PhD Thesis, University of London (1972).

Rodi, W. *Turbulence Models and Their Application in Hydraulics, International Association for Hydraulic Research,* Delft (1980). (Contact W. Rodi, Univ. of Karlsruhe, Karlsruhe, West Germany.)

Rodi, W., and Spalding, D. B. *Wärme und Stoffübertrag* **3,** 85 (1970).

Rodi, W. On the Equation Governing the Rate of Energy Dissipation Imperial College Rept. No. TM/TN/A/14, London (1971).

Rogallo, R. S. An ILLIAC Program for the Numerical Simulation of Homogenous Incompressible Turbulence NASA Document No. NASA TM-73, 203 (1977).

Rotta, J. C. *Z. Phys.* **129,** 547–572 (1951).

Saffman, P. G. *Proc. R. Soc. Lond.* A **317,** 417–433 (1970).

Schumann, V., and Patterson, G. S. *J. Fluid Mech.* **88,** 711–735 (1978a).

Schumann, V., and Patterson, G. S. *J. Fluid Mech.* **88,** 685–709 (1978b).

Smagorinski, J. *Mon. Weather Rev.* **91,** 94–165 (1963).

PART III
COMPRESSIBLE FLOWS

General Comments on Compressible-Flow Calculations

In the following two chapters the subject of compressible flow calculations is addressed. Chapters on viscous and inviscid flows are included along with a brief discussion on the special features of compressible flow. The amount of material in this field of computational fluid mechanics is large and we have attempted to abstract as much of the important work as possible. As a result, we have limited the details and the reader may find it necessary to refer to the reference material for the complete coverage of a specific calculation.

Inviscid regions

The calculation of compressible flows differs considerably from incompressible flows. The first and most apparent change is the addition of a time derivative to the continuity equation as well as an additional unknown—the density. This addition simplifies the numerics of the problem by permitting direct solution for all flow variables.

There is also the introduction of an additional parameter—termed Mach number—which characterizes the flow as subsonic, $M < 1$, and supersonic, $M > 1$. As soon as the Mach number exceeds a magnitude of "one," compressible flows can produce serious difficulty for the computational fluid dynamicist. The reason is that shock waves, strong rarefactions, sonic lines, and contact discontinuities begin to appear, and each of these quantities can introduce a special numerical requirement. For example, the shock wave requires some type of damping to stabilize the calculation, and a sonic line in a steady flow introduces a saddle-point-type singularity in the flow equations which requires a special integration technique to obtain a solution (see Holt, 1977). Strong rarefactions tend to produce significant oscillations in finite-difference methods and contact surfaces (density jumps) can introduce similar phenomena in computations.

The sonic line denotes another significant feature of inviscid compressible flow—the change from a subsonic elliptic-type flow to a supersonic marching or hyperbolic-type flow. The different nature of each of these types of flow regimes can in some cases make it necessary to consider methods for each in order to perform efficient computations.

Viscous regions

Compressible flows also exhibit another feature which can make computations difficult. This feature is that most compressible flows, with the exception of the very-low-density problems, exhibit large inviscid regions and limited viscous regions. Typically, however, the viscous-region computation requirements dominate the overall calculation due to grid resolution and time-step requirements. The result is that in complete calculations the inviscid regions may be computed with viscous time-step requirements, and hence large computer costs result. The alternative is to split the flow into an inviscid flow and a viscous boundary layer. This, however, is not satisfactory in separated-flow regions or regions of strong inviscid viscous interaction. Typically, these types of flow situations have lead to "patching" of the viscous and inviscid computations. This art needs improvement and a natural technique for this is to employ numerical inner and outer expansion concepts. Such an approach would be well suited for spectral solution methods where functional expansions are employed. The concept is to employ two separate solutions that overlap but which are computed for different scales. For example, the inner solution could include the full viscous equations and the outer solution could be for the inviscid flow. The solution of the outer and inner solutions could then be relaxed simultaneously with the inner solution serving as a boundary condition for the outer solution and the outer solution also serving as the inner-solution boundary condition. Note that these conditions are not necessarily applied at the same points in the inner and outer regions. Further discussions on these points of coupling can be found in LeBalleur et al. (1980).

Noncontinuum problems

Another and sometimes troublesome aspect of compressible flows is the fact that low-pressure rarefaction effects can enter into the calculations. The effects can lead to the solution of noncontinuum equations which can require lengthy and difficult calculation procedures. In this book, we have not included this area of research, but the interested reader is referred to Bird (1978) for a description of the state of rarefied-flow calculations.

Artificial viscosity in compressible-flow solutions

In the calculation of inviscid compressible flows, it can become necessary to include some type of damping to stabilize a calculation and to compute shock waves. Various approaches have been proposed ranging from the sophisticated "flux-correction" approach of Boris and Book (1976) to a straightforward artificial viscosity term of the form $\alpha \nabla^2 f$ in the momentum equations. Most of the unsteady methods employed to integrate inviscid equations have some built-in damping and do not require the introduction of an additional term. Within the approaches available for finite difference, the flux-correction approach is interesting since it only applies damping in the region where it is required. Other methods apply damping in all regions and hence the artificial terms gradually impact the solution. In a recent publication Zalesak (1979) presented a generalized form of the original Boris and Book flux-correction method, and this generalization makes the approach practical for multidimensional application. However, it will increase the computer time of a calculation when the approach is used. The approach is simple in principle and states that one can compute a solution with damping or artificial viscosity and then subtract it out in the regions where it is not required for stability. This rule of course can be implemented in many different ways as the numerous publications of Boris and Book indicate. The problem is that there is no clear "best" approach. As a result, the problem solver must rely on his or her own decision as to what is best. Clearly, the minimum use of artificial viscosity terms is best.

Turbulence terms

In the incompressible-flow discussion, the various models and techniques for turbulent-flow calculations were discussed. The same approaches, in general, apply to compressible flow with the exception that one must also introduce a density fluctuation. This leads to the introduction of a number of new correlation terms which can be kept to a minimum by using mass-averaged variables (Favre, 1965). One can close the equations at different levels of approximation or proceed to complete simulation. The difficulty in this, however, is that with the exception of some boundary-layer models which limit analysis to one or two Reynolds stress terms, the closure models for compressible flows are not well developed. The result is that it is not possible to present a meaningful hierarchy of turbulence models for flows where density variations are important. Bradshaw (1977) has pointed out that for Mach numbers less than 5, the density-fluctuation terms are probably not important for most flows and the incompressible models could be employed. This is a general statement, and one should take care in applying the incompressible model. Unfortunately, right or wrong, if it is necessary to solve a nonboundary-layer engineering problem, one will have to rely heavily on the incompressible results.

In addition to the closure models, there is, of course, the possibility of complete simulations. At this time, no such simulations are known to exist, but this would be a fruitful area of research.

Due to the fact that the state of turbulence modeling in compressible flow is so highly concentrated on boundary-layer theory and the boundary-layer approximations are adequately covered in the text of Cebeci and Smith (1974), we will not deliberate on the the details in this book.

References

Bird, G. A. *Ann. Rev. Fluid Mech.* **10**, 11–31 (1978).

Boris, J. P., and Book, D. L. *J. Comput. Phys.* **20**, 397–431 (1976).

Bradshaw, P. *Ann. Rev. Fluid Mech.* **9**, 33–54 (1977).

Cebeci, T., and Smith, A. M. O. *Analysis of Turbulent Boundary Layers,* Academic Press, New York (1974).

Favre, A. J. *J. Méc.* **4**(3), 361–390 (1965).

Holt, M. *Numerical Methods in Fluid Dynamics,* Springer–Verlag, New York (1977).

LeBalleur, J. C., Peyret, R., and Viviand, H. *Comp. Fluids* **8**, 1–30 (1980).

Zalesak, S. T. *J. Comput. Phys.* **31**, 335–362 (1979).

CHAPTER 10

Inviscid Compressible Flows

In the area of inviscid compressible flows there are a variety of practical problems that arise in everyday engineering applications. These include rocket nozzle flows, aircraft and missile engine inlet flows, reentry vehicle and rocket aerodynamics, blast fields generated by different types of energy release, and aircraft flow fields. One can, of course, continue the list, but these serve as examples whose discussion will adequately display the techniques for solving compressible-flow problems.

The methods that have evolved and are used currently (for history see Taylor, 1974) for the solution of compressible flows fall roughly into three categories. The first is the unsteady method which is used to time integrate the flow equations both for unsteady and steady flows. This approach has been successful in overcoming many of the difficulties associated with shock waves and transonic regions. The second approach is the relaxation solution of the potential flow equations for transonic flows and the third approach is marching techniques for integration of the steady equations of supersonic flow. At this time, these approaches are based principally on finite-difference methods. Some work on finite elements has been accomplished (Bristeau et al., 1978; Bristeau, 1977; Periaux, 1975; Chan et al., 1975; Fletcher, 1977, 1979) but these approaches have not been developed to the extent of the other approaches. Due to the nature of transonic flows, it is not at all clear that the finite-element approach will be optimum for calculating cases when a flow has mixed subsonic–supersonic regions. The reason is that the finite-element methods tend to give difficulty in computing both hyperbolic and elliptic flows in one problem because of the inherent central-difference approximation of derivatives. This can be controlled by introducing artificial viscosity—an approach taken by Bristeau et al. (1978)—but more study is required to arrive at an optimum scheme. Spectral methods have not yet been applied because of difficulties with shock waves. In Chapter 3, we pointed out at least one approach to overcome the problems for the pseudospectral approach. As a consequence, one can expect to observe spectral-method applications in compressible flows to appear in the near future.

In the solution of inviscid problems there are two main types of difficulties. The first is the subsonic free-stream case where the flow initially is subsonic and is accelerated to a supersonic condition by some flow disturbance. The second is a supersonic free stream that is slowed or shocked down by some type of body. Both of these problems are transonic in principle, but a significant difference between the two is the elliptic nature exhibited by the subsonic free-stream case at steady state. As a consequence, subsonic flows for steady situations can become sensitive to the prescribed boundary conditions and yield difficulty in their solution. This fact has seemed to limit the acceptance of unsteady finite-difference methods for solution of such problems since these methods tend to generate waves that propagate outward and reflect from the boundaries into the calculated region. Therefore, the user of an unsteady method for $M_\infty < 1$ must be aware of this phenomena in order to eliminate uncertainty in the results at long times. Magnus and Yoshihara (1970, 1975), Grossman and Moretti (1970), and Masson and Friedman (1972) noted this problem in transonic studies.

One other area that is important in the solution of inviscid problems is the use of mapping to generate grid systems for nonregular geometries. This approach has grown in importance and has made the previous arguments for the use of finite elements to compute complex geometries less appropriate. In the examples which follow, mapping techniques have played a strong role in making the calculations practical. The improvements have mainly been with finite-difference applications, but the use with spectral-type approaches will greatly enhance the usefulness of the spectral approach in complex geometries. Having presented comments on boundary conditions and mapping methods, we next proceed to the discussion of the application of the steady and unsteady methods and their applications.

10.1 Application of Unsteady Methods

The general procedure employed in solving inviscid problems by unsteady methods advances in the following fashion.

Step I:

Write the inviscid nonlinear equations in the form

$$\frac{\partial f}{\partial t} + \frac{\partial F}{\partial x} + \frac{\partial G}{\partial y} = 0 \qquad (10.1.1)$$

as defined in Chapter 1.

Step II:

If the geometry is irregular, develop a numerical or analytical mapping of the geometry to a regular condition such as a cylinder, rectangle, or flat plate. In two dimensions this can be accomplished by construction of the potential solution for the geometry at hand. Such a solution can be constructed by a variety of methods ranging from the panel methods discussed earlier to the well-known Schwarz–Christoffel transformation. A rather general computer program for boundary-fitted mesh generation has been developed by Thompson et al. (1974) and the technique can be implemented by following their reports.

Step III:

Transform the original equations of Step I into the new coordinate system either analytically or numerically as you perform the computation, or use a discretization technique based on the equations kept in Cartesian coordinates (for example, finite volume or finite-difference scheme of Section 11.4.2).

Step IV:

Apply the unsteady method of choice to integrate the equations forward in time subject to the inviscid flow-boundary conditions of no flow normal to the surface and a prescription of the appropriate free-stream conditions.

For an explicit scheme, the time integration proceeds in the same manner as solution of a first-order differential equation. However, if one decides to employ an implicit scheme, it will be useful to split the equations into two unsteady problems; i.e., one in each direction as employed in the example of Section 6.4. The solution of the split equations will depend on the nature of the method employed. If the unknown in each equation forms a set of the points $i+1$, i, and $i-1$, then the factorization method described in Chapter 2 can be used to directly solve the equation set. However, if the set extends pointwise beyond $i+1$ or $i-1$, a technique such as the conjugate-gradient method (Kershaw, 1978) should be employed. Other solution approaches can be used but indications at this time are that a preconditional conjugate-gradient approach is exceptionally efficient (Koshla and Rubin, 1981).

In performing calculations, one can fit shock waves as proposed by Richtmyer and Morton (1967), and by Moretti (1974) but this seems very time consuming and tedious for complicated flows. Of course, this will yield the most accurate solution, but for many engineering applications shock capturing (or smearing) by addition of some artificial viscosity appears preferable. If accuracy is required, shock capturing as the first step can be employed.

Before displaying some examples of the unsteady approach, it is important to provide comment on integration of the energy equation for those interested only in steady-state cases. For such problems one can use the constant-total enthalpy assumption (valid at steady state only). For shockless flow or with weak shock one can employ the constant-entropy assumption—i.e., $p \propto \rho^\gamma$ (valid for both steady and unsteady flow). Both of these assumptions avoid time integration of the energy equation and lead to converged correct solutions. A complete study of these approaches was made by Viviand and Veuillot (1978).

For external flows, one may also wish to employ inner and outer expansion concepts to simplify the boundary-condition application and reduce the solution time. For those cases when such concepts can be utilized one must look for either an analytical or simplified numerical solution for the far field of the flow (see Murman and Cole, 1971; Euvrard and Tournemine, 1973). This result can be used as the boundary condition of the nonlinear calculation instead of the free-stream conditions. This approach frequently will reduce the computer time by reducing the size of the region that must be covered by a grid.

Holt and Masson (1971) used this type of approach in their studies of transonic flow by integral methods.

In addition to the boundary condition, one must also take care to apply reasonable initial conditions to start the calculations. Grossman and Moretti (1970), for example, found that an impulsive start tended to make transonic calculations unstable while a ramped or gradual start worked reasonably well.

10.1.1 Finite-difference solutions

10.1.1.1 *Examples for $M_\infty < 1$:* There are a number of examples of the application of unsteady methods[†] to solve problems with $M_\infty < 1.0$. Obviously it is not possible to present them all, but three examples which display the general capabilities of the approach are shown. The first example is a study of a nozzle flow performed by Chang (1980). Chang employed the MacCormack method with fourth-order damping to integrate the nonlinear inviscid equations to a steady state. The damping is added to the corrector step of the scheme and has a form

$$D_{i,j}^n = -\alpha_1 [f_{i+2,j}^n + f_{i-2,j}^n - 4(f_{i+1,j}^n + f_{i-1,j}^n) + 6 f_{i,j}^n]$$
$$- \alpha_2 [f_{i,j+2}^n + f_{i,j-2}^n - 4(f_{i,j+1}^n + f_{i,j-1}^n) + 6 f_{i,j}^n] \qquad (10.1.2)$$

where α_1 and α_2 are damping coefficients taken to be 0.01. In addition, the mapping approach of Thompson et al. (1974) was used to generate the grid system. Chang studied the three geometries shown in Fig. 10.1.1 and employed the grid shown for each figure. The upstream conditions were obtained from rocket-engine combustion calculations and the downstream condition was supersonic extrapolation. Chang was able to successfully compute all of the flows for both gas and gas-particle cases. The calculations were for direct application to engineering evaluations and proved very useful.

The nozzle flow represents a classic internal flow. A classic external flow is the two-dimensional airfoil problem. The literature has a number of unsteady-method solutions to airfoil problems and we make no attempt to recount these. Instead, we have selected an example that demonstrates the type of result that can be expected. The result is for an NLR 7301, 16.5% thick airfoil computed by Magnus (1978) using a two-step method (Thommen, 1966) to solve the complete nonlinear Euler equations in conservative form. An artificial viscosity was also employed. The calculation used 5000 nodes arranged with different grid densities. A fine mesh was used around the nose. A regular square mesh of 0.04 chord thickness was used near the remainder of the airfoil and the mesh was made coarse to extend the calculation several chord dimensions away from the body. The tangency body conditions and upper and lower pressures, as well

---

[†]See Sections 2.7 and 2.8.

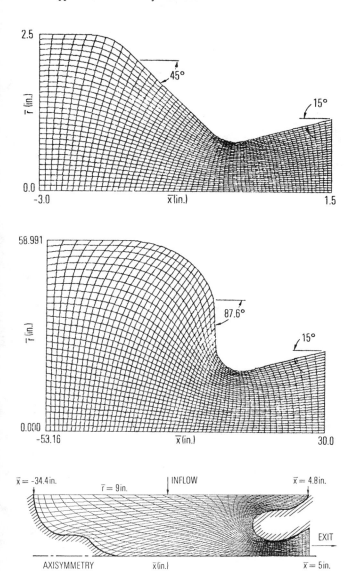

Fig. 10.1.1 Geometries and grid system for finite-difference solutions of the Euler equations by Chang (Courtesy of I. Chang.)

as flow directions, were satisfied using a Euler predictor, simple-wave-corrector condition given by Abbett (1973). The outer boundaries were located 7.0 chords upstream and downstream and 10.4 chords in the vertical. On these boundaries a doublet and vortex condition along with the free stream were imposed. The calculations took about 2400 cycles or 580 s of CDC 7600 time to start achieving stable solutions, and steady state was reached after about

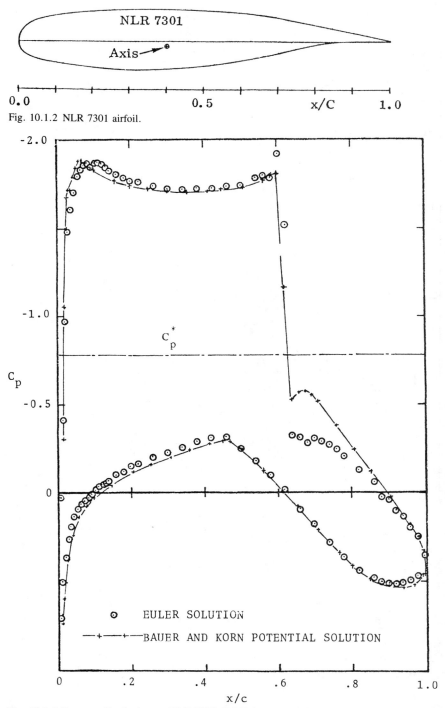

NLR 7301

Axis

0.0   0.5   x/C   1.0

Fig. 10.1.2 NLR 7301 airfoil.

- 2.0

-1.0

C_p^*

C_p

-0.5

0

⊙   EULER SOLUTION

—+——+—BAUER AND KORN POTENTIAL SOLUTION

0   .2   .4   .6   .8   1.0

x/c

Fig. 10.1.3 Pressure distribution on NLR 7301 airfoil for $M_\infty = 0.7$ and angle of attack $\alpha = 3.0°$. Computed by R. Magnus (1978).

2200 s of computing. The airfoil shape and the resulting pressure distribution are shown in Figs. 10.1.2 and 10.1.3. Note that the free-stream Mach number was $M_\infty = 0.7$. The results shown are also compared with a potential solution by Bauer et al. (1975) developed by steady methods. The results agree closely except near the base of the shock wave. In concluding discussion of this example, it is important to note that in these types of calculations, at long times, the reflection of small disturbances due to flow relaxation can create problems with the calculations.

The last $M_\infty < 1$ example of unsteady transonic applications to be discussed is the use of the constant-total enthalpy and constant-entropy assumptions to reduce the calculation complexity for steady flows. These two approaches were used to compute flow at various Mach numbers for a variety of shapes by Viviand and Veuillot (1978) (see also Section 11.4.2). For the $M_\infty < 1$ application they examined the flow over a circular profile placed in a channel. The profile is a semicircular profile of 8.4% thickness and the distance between the centerline of the profile and the channel wall was taken to be 2.073 times the chord dimension. A grid system that conformed to the wall shape of 81 × 26 points was employed. For the boundary conditions, the upstream was assumed parallel at the wall with a Mach number of 0.85. The pressure for this condition was imposed as the upstream condition. For the downstream, it was assumed that $p_2/p_\infty = 0.6235$. Figure 10.1.4 shows the results obtained for lines of constant Mach number. The constant-total enthalpy results utilize the steady state $h + \frac{1}{2} q^2 = h_0$ for an energy equation. The constant-total enthalpy–entropy results also use $\rho/\rho^\gamma = $ const. Using both an enthalpy and entropy equation will determine two unknowns but two other equations are required to determine the other unknowns. These are continuity and some form of a momentum equation. Viviand and Veuillot used the continuity equation and a momentum equation written in terms of a flow angle θ in the form

$$\frac{\partial \theta}{\partial t} = -\omega = -\left(\frac{\partial v}{\partial x} - \frac{\partial u}{\partial y}\right)$$

when $\tan \theta = v/u$ and ω is the vorticity.

This equation was derivable because of the initial assumptions. It may be possible, however, to employ other forms for the momentum equations that work better. The figures show that the constant entropy–total enthalpy results tend to be oscillatory near the shock. It may be that changing the form of the momentum equation employed could help the results.

To this point we have briefly outlined some two-dimensional solutions with free-stream Mach numbers less than 1 obtained by using the unsteady finite-differences approach. We next consider solutions for a supersonic free stream.

10.1.1.2 *Examples for $M_\infty > 1$:* The use of unsteady methods for supersonic free streams is primarily for the computation of unsteady flow or steady flows

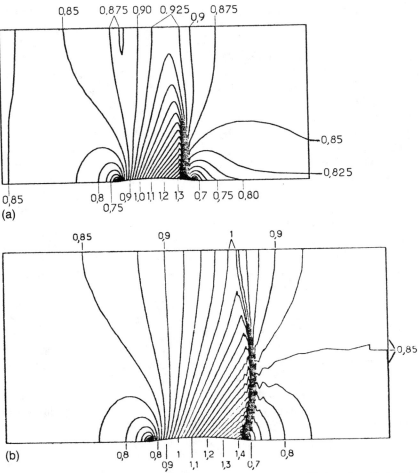

Fig. 10.1.4 Flow lines of constant mach number for flow past a profile in a channel at $M_\infty = 0.85$. (Courtesy of H. Viviand.) $\Delta M = 0.25$, $e/c = 8.4\%$ and $h/c = 2.073$. (a) Results from constant total enthalpy method; (b) results from constant total enthalpy and entropy.

with embedded subsonic regions. For supersonic flows with no subsonic region, steady marching methods should be employed. These are discussed in the next section.

A significant advance in the computation of compressible flows occurred when the application of unsteady finite-difference methods to $M_\infty > 1$ flows occurred. The work of Bohachevsky and Rubin (1966) was a significant milestone in this area. These authors employed the highly damped method of Lax (1954), which we have not discussed, to solve the inviscid Euler equations for flow past a blunt body. Next came solutions by Moretti and colleagues using schemes suggested by Lax and Wendroff (1960, 1964) and, successively, by the MacCormack method (1969). Simultaneously, solutions were developed by Rusanov (1968, 1970) using his own techniques.

The solution by the Lax–Wendroff, MacCormack, and Rusanov methods worked satisfactorily for flow past bodies with fairly regular flows such as a sphere cone, but they have a tendency to fail for complicated flow geometries such as the indented body shown in Fig. 10.1.5. Recently, however, Moretti (1978, 1979, 1980) has published a method called the λ scheme which seems to work very well for both steady and unsteady supersonic flow problems. This scheme relies strongly on the physics of the problem much like the Godunov cell method. Moretti, in his development, clearly points out why the schemes tend to fail in the supersonic problems. The reason simply is that such schemes use fixed rules for constructing finite differences which on occasion do not adequately account for the physics of signal propagation in a hyperbolic problem. Recognizing this fact, Moretti set up a method that examined characteristic directions and selected difference formulas to approximate the flow equations depending on the behavior of the characteristics. Although not stated

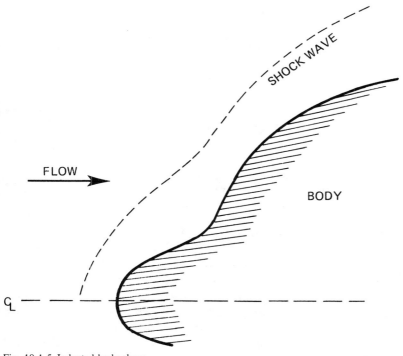

Fig. 10.1.5 Indented body shape.

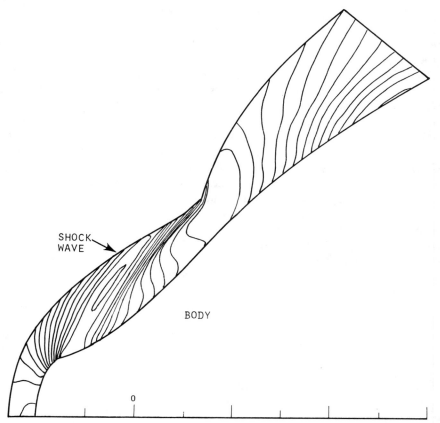

Fig. 10.1.6 Isobars computed by Moretti for flow past an indented axisymmetric body ($M_\infty = 12$). (Courtesy of G. Moretti.)

explicitly in publications, this also is exactly the procedure employed in the Godunov method and perhaps is the reason for its success.

Moretti's procedure was to write the inviscid flow equations in the form which in characteristic methods would be called the compatibility equations, and if they are integrated along characteristic directions they are reduced to ordinary differential equations. Moretti, however, did not impose that the equations hold only along characteristics. Instead, he employed the compatibility equations in their general form. He then introduced a splitting of the equations. The resulting equations were finite differenced and solved subject to differencing selection rules based on characteristic directions. Moretti's formulation makes incorporation of boundary conditions more efficient and tends to give better accuracy as shown by Abbett (1973). The significance of the method is that it will compute flows such as the one shown in Fig. 10.1.5, which before could only be computed by the Godunov approach. Moretti has applied the method to computation of flow past blunt bodies and flow past supersonic aircraft configurations. The isobars obtained for flow past an ablated shape are shown in Fig. 10.1.6. The flow Mach number was 12 and the grid employed 30 intervals along the body with 15 intervals between the shock and body. The grid was constructed by conformal mapping using a program developed by Moretti (1977). Moretti applied the scheme to compute steady supersonic flow past a complex aircraft configuration and the results are discussed in the section on steady $M_\infty > 1$ flows, Section 10.2.2.

For smooth flows one can employ the method proposed by Moretti or the well-tested MacCormack scheme to integrate the inviscid flow equations and expect to obtain good results. For steady flows one can also replace the unsteady energy equation by the steady constant-total enthalpy equation and expect good results. Viviand and Veuillot (1978) (see also Section 11.4.2) followed this approach for flow past a sphere at $M_\infty = 3.0$. The results obtained for pressure and Mach number profiles as well as the mesh employed are shown in Fig. 10.1.7. A mesh of 41×21 points was used, but it was later found that a 21×21 mesh was adequate.

The two displayed examples are what the authors consider as good finite-difference method prediction of supersonic flows by unsteady methods. This area unfortunately has not been fully developed for computation of arbitrary shapes. Most of the past approaches that have been demonstrated for sphere cone geometries unfortunately fail for more complicated shapes. Moretti, however, appears to have overcome the problem. More tests of his method would be desirable.

10.1.2 Cell and finite-volume solutions

In Chapter 3 we outlined the cell methods of Godunov and Glimm and the finite-volume approach. Recall that the Godunov method is basically a predictor-corrector unsteady method. The predictor step consists of solving local Riemann problems at each cell boundary to obtain the flow variables and hence fluxes. These fluxes are then combined with the finite-differenced flow equations to obtain the new time values in each cell center. The Glimm method is similar in that it uses the same predictor as the Godunov approach. However, it uses a random choice of the Riemann solution to estimate values at the cell boundary. The corrector step differs significantly because the mean flow equations are not employed. Instead, the Godunov predictor is reemployed to obtain the values of the flow variables at the cell centers. The Riemann problem is solved for a plane located at the cell center for the states previously calculated at each cell boundary. Again, random choice is applied to the Riemann problem to select the values to be applied at the cell center. In one dimension this approach appears to work well; however, in two and three dimensions it is not clear that the optimum procedure for application is fully developed. In the following discussion we will outline the progress made with Glimm's method

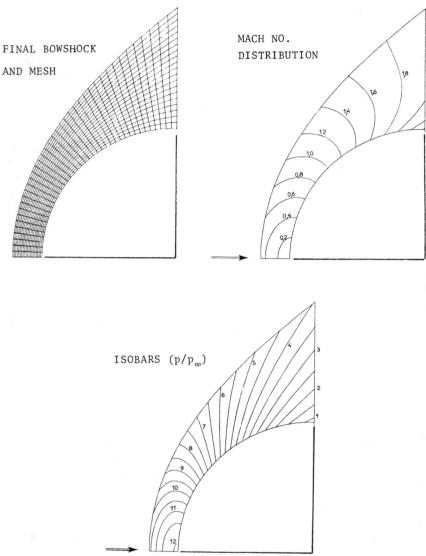

Fig. 10.1.7 Flow past a sphere, $M_\alpha = 3.00$. (Courtesy of H. Viviand.)

as well as with the Godunov approach for both steady and unsteady flows. Also we present some results obtained by a finite-volume approach.

In compressible-flow calculations one of the most successful approaches for computing inviscid flows with complex shock-wave structure or extreme gradients has been the cell method proposed by Godunov (1959). This method invariably succeeds in computing a flow when most methods based on a Taylor series development fail. The success of the method appears to be based on two principal facts. First, the method utilizes the local physics of the problem, i.e., the local characteristics. Secondly, the method appears to have sufficient damping to overcome the tendency of strong gradients to destabilize a calculation.

In Chapter 3 the principles of the Godunov method were given for one dimension. The application to two-dimensional problems uses the same principles combined with splitting. As a consequence, the basic flow to be computed is divided into cells which may or may not be rectangular. At each cell boundary plane the one-dimensional Godunov formulas are applied to determine the fluxes on the boundary. These fluxes are then used to approximate the spatial fluxes in the inviscid flow equations. The resulting equations are then integrated forward one time step. Note that this procedure applies to the original Godunov method. There are, however, higher-order Godunov approaches as discussed by Holt (1977) and VanLeer (1979). These improvements have met with mixed success. The best results seem to have come from a recent combination of the improved Godunov scheme, proposed by VanLeer, and Glimm method. The approach is being developed by Glaz and Colella (1980), and unpublished results indicate that the approach works well for strongly shocked flows. Unfortunately, complete details of the work are not available at this time. They should, however, appear in the literature in the near future. As a result of these facts, we will limit the discussion to the original method. It will be interesting in the future to compare the different Godunov scheme results. In the discussion which follows we discuss the application of the Godunov and Glimm methods to both $M_\infty < 1$ and $M_\infty > 1$ cases. We begin with the subsonic cases. Note that both unsteady- and steady-state cases are included.

10.1.2.1 *Examples for $M_\infty < 1$*: For subsonic free streams there are basically two classes of problems. These are flow past or within configurations and flows expanding into ambient conditions. An example of the first case is transonic flow past a body or airfoil. Masson and Friedman (1972) applied the Godunov method to compute transonic flow past axisymmetric shapes. These authors solved the axisymmetric equations using a body-oriented grid system and analytical far-field solutions for the free-stream boundary conditions. A typical result obtained for an axisymmetric ogive-cylinder flow is shown in Fig. 10.1.8. Also shown is the calculated pressure compared with experiment. As can be seen, the results appear reasonable, but the computations were rather lengthy and took a large number of time iterations to converge.

Other studies of subsonic free-stream problems include the solution by Taylor and Lin (1980) of the flow from the muzzle of a gun and the solution by Reddall (1980) of the flow produced by the launch of a Titan missile. In each of these studies the Godunov method was employed to solve the two-

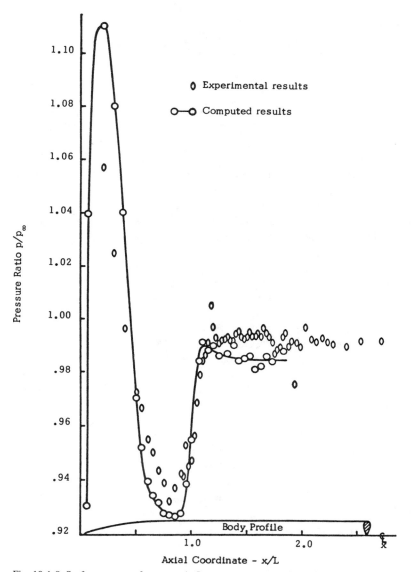

Fig. 10.1.8 Surface pressure for transonic flow past a tangent ogive-cylinder computed by Masson and Friedman (1972), $M_\infty = 0.9$.

dimensional unsteady inviscid equations. The muzzle blast problem which was solved is shown in Fig. 10.1.9. Here the flow from the gun barrel was prescribed along with the velocity of the shell leaving the muzzle. The flow resulting as the gun empties into the ambient atmosphere is computed. The principal difficulty with the calculation occurs because the pressure ratios

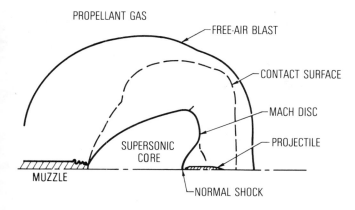

E = FREE STREAM UNDISTURBED PROPERTIES
C OR A = BOUNDARY CONDITIONS WITH $v = 0$
D = BOUNDARY CONDITION WITH u_s
F = SYMMETRY BOUNDARY CONDITION
B OR G = SPECIFIED BOUNDARY CONDITIONS AT MUZZLE EXIT

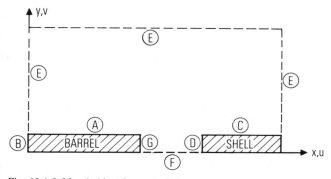

Fig. 10.1.9 Muzzle blast flow calculation geometry.

between cells can be larger than 100 to 1. This occurs as the flow expands near the axis of symmetry and near the edge of the gun muzzle. A typical flow result computed by Taylor and Lin for an M16 rifle with no shell in the flow was shown in the Introduction. The results obtained for the expanding blast-wave position and the comparison with experiment are shown in Fig. 10.1.10.

A calculation with less rigor, but with interest because it displays the flexibility of the unsteady approach for quick looks, is the simulation by Reddall of a Titan III missile launch. Reddall used the Godunov approach with a rectangular grid system and computed the flow exhausting from the missile into a launch duct. The simulation was two-dimensional and a visualization of the flow patterns computed was shown in the Introduction. We make no

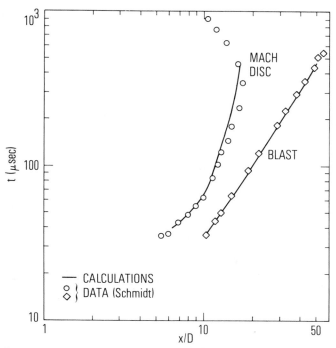

Fig. 10.1.10 Comparison of muzzle blast predictions with data of Schmidt and Shear (1974).

attempt to analyze this calculation, but it serves as a good example of how the unsteady approach can be applied to examine complex flow geometries where measurements are difficult to obtain.

So far only the Godunov cell method has been discussed, but is important to indicate the present state of Glimm's method which appears to have promise for the future. The method has been studied by Chorin (1976, 1977), Sod (1978, 1980) Collella (1978), and Glaz (1979). The findings at this time seem to leave some question on the general usefulness of the Glimm method for two-dimensional flows with strong shock waves present. Glaz attempted to apply the method to compute flow over a blunt body in a supersonic flow and found that the method did not perform well. Colella in his work alludes to a problem in handling shock waves and suggests introducing an artificial damping to eliminate the difficulty. Clearly more work is needed in this area before one can display definitive two-dimensional results. For weak waves, as in combustion engines, Sod (1980) has made the method work successfully.

In addition to the Godunov and Glimm methods, there has been another recent finite-volume approach successfully applied to compressible flows. The scheme was developed and applied by Lerat and Sides (1981) to compute transonic flow past an airfoil. The scheme is straightforward in computing the flux terms at the cell boundary but differs from the Godunov and Glimm approaches in that mathematical averages involving both space dimensions are employed at each boundary rather than only one direction. The general scheme is as described in Section 3.7. It is applied in predictor-corrector form. The predictor predicts the flux values at the original cell boundary and the corrector yields the values at the original cell center, Fig 3.7.1. The predictor is obtained by shifting the cell by $1/2$ a cell width so that the original cell boundary is at the center of the predictor cell. The method of Section 3.7 is applied to each predictor cell. The following flux terms are used for the right-shifted cell

$$H^n_{A'B'} = F^n_{i+1,j} \Delta y_{A'B'} - G^n_{i+1,j} \Delta x_{A'B'} \tag{10.1.3}$$

$$H^n_{B'C'} = +\tfrac{1}{4}(F^n_{i,j} + F^n_{i+1,j} + F^n_{i+1,j+1} + F^n_{i,j+1}) \Delta y_{B'C'} \tag{10.1.4}$$
$$-\tfrac{1}{4}(G^n_{i,j} + G^n_{i+1,j} + G^n_{i+1,j+1} + G^n_{i,j+1}) \Delta x_{B'C'}$$

Note that A'-B' and C'-B' are the coordinates (Fig. 3.7.2) of the shifted cell not the original cell. The original cell is centered at (i,j). The right predictor cell is centered at $(i+\tfrac{1}{2},j)$. The flux expressions for boundaries C'-D' and D'-A' are analogous except the indices are shifted. Note that in these expressions the fluxes in the direction of shift, i.e., i to $i+\tfrac{1}{2}$, are simple but in the other direction, i.e., $j+\tfrac{1}{2}$ and $j-\tfrac{1}{2}$, there are averages over four cells. Again note that if the cell is shifted in j rather than i the simple fluxes are then $H_{B'C'}$ and $H_{D'A'}$ and the average fluxes are $H_{A'B'}$ and $H_{C'D'}$. From these facts one can construct the predictor fluxes for the other predictor steps. The predictor fluxes are employed in the time-integration equation for each of the four shifted cells. For the $i+\tfrac{1}{2},j$ boundary, Lerat and Sides employ the relationship

$$\tfrac{1}{2}(S_{i+1,j} + S_{i,j}) f^{n+\alpha}_{i+1/2,j} = \tfrac{1}{2}(S_{i,j} f^n_{i,j} + S_{i+1,j} f^n_{i+1,j})$$
$$- \alpha \, \Delta t (H^n_{A'B'} + H^n_{B'C'} + H^n_{C'D'} + H^n_{D'A'}) \tag{10.1.5}$$

where α is the fraction of the time step employed for the prediction and $S_{i,j}$ is the area of the cell centered at (i,j).

For the corrector step, the cell is centered at (i,j) and the flux terms on the boundaries A-B and B-C are chosen to have the form

$$\tilde{H}_{AB} = \frac{1}{2\alpha}\left[(\alpha - \tfrac{1}{2})(F^n_{i+1,j} + F^n_{i,j}) + F^{n+\alpha}_{i+1/2,j}\right]\Delta y_{AB}$$
$$- \frac{1}{2\alpha}\left[(\alpha - \tfrac{1}{2})(G^n_{i+1,j} + G^n_{i,j}) + G^{n+\alpha}_{i+1/2,j}\right]\Delta x_{AB} \tag{10.1.6}$$

$$\tilde{H}_{BC} = \frac{1}{2\alpha}\left[(\alpha - \tfrac{1}{2})(F^n_{i,j+1} + F^n_{i,j}) + F^{n+\alpha}_{i,j+1/2}\right]\Delta y_{BC}$$
$$- \frac{1}{2\alpha}\left[(\alpha - \tfrac{1}{2})(G^n_{i,j+1} + G^n_{i,j}) + G^{n+\alpha}_{i,j+1/2}\right]\Delta x_{BC} \tag{10.1.7}$$

where $F^{n+\alpha}_{i+1/2,j} = F(f^{n+\alpha}_{i+1/2,j})$ and similarly for G. By shifting indices one can obtain the flux terms for the boundaries C-D and D-A. For the corrector time integration, note that the time step is Δt and not $\alpha\,\Delta t$; therefore the time integration takes the form

$$S_{i,j}f^{n+1}_{i,j} = S_{i,j}f^n_{i,j} - \Delta t[\tilde{H}_{AB} + \tilde{H}_{BC} + \tilde{H}_{CD} + \tilde{H}_{DA}] \qquad (10.1.8)$$

In the calculations, α was selected to be $1 + \sqrt{5}/2$ which corresponds to the optimum value for an equivalent finite-difference scheme (Section 2.7.1) in one dimension. Note also, for a uniform grid, the finite-volume approach reduces to the finite-difference scheme (2.8.9) with all parameters equal to $1/2$ except α (Thommen's type scheme). Lerat and Sides (1981) calculated the flow over an NACA 0012 airfoil at Mach 0.8 with an angle of attack of 1.25° as shown in Fig. 10.1.11; the computations were performed with 298×34 cells. The resulting pressure distribution is shown in Fig. 10.1.12. Figure 10.1.13 shows the lines of constant Mach number which were computed. Figure

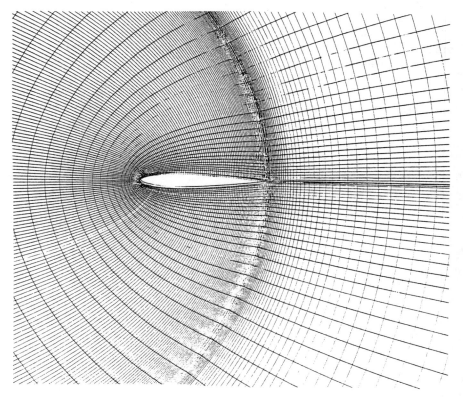

Fig. 10.1.11 NACA 0012 airfoil and grid system employed by Lerat and Sides (1981).

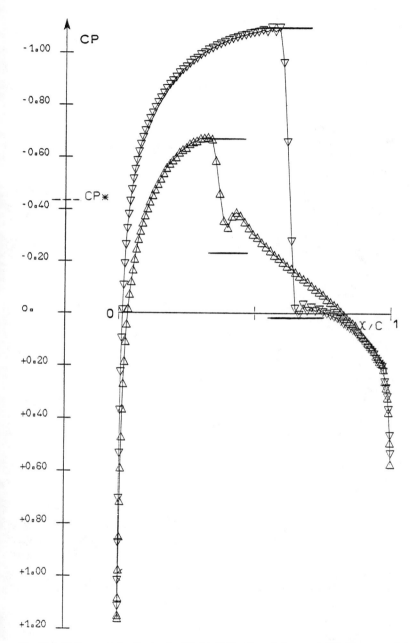

Fig. 10.1.12 Pressure distribution for NACA 0012, $M_\infty = 0.8$, $\alpha = 1.25°$. (Courtesy of A. Lerat and J. Sides.)

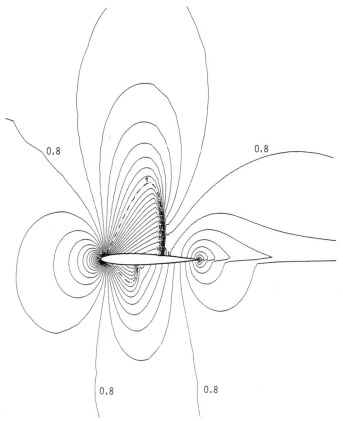

Fig. 10.1.13 Lines of constant mach number, $M_\infty = 0.8$, $\alpha = 1.25°$. (Courtesy of A. Lerat and J. Sides.)

10.1.14 shows the constant-entropy lines. The results obtained indicate that the method yields very satisfactory results.

10.1.2.2 *Examples for Cell Method* $M_\infty > 1$: The use of the unsteady approach to solve supersonic free-stream problems has met with great success. This approach in fact was a significant advance in the solution of problems with embedded subsonic regions. A paper by Bohachevsky and Rubin (1966) seemed to be the start of this advance.

    The application technique follows the same pattern as for the $M_\infty < 1$ cases except boundary conditions are not difficult to introduce. For efficiency, one may want to fit shock waves that bound the flow region being calculated. This approach was employed by Taylor and Masson (1970, 1974) and Masson et al. (1969, 1971) in a variety of blunt-body flow studies. Both finite-difference and cell methods have been successfully employed in the computation of supersonic flows.

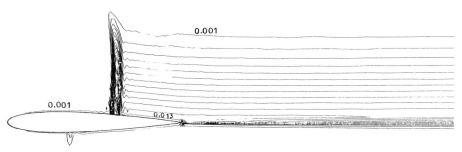

Fig. 10.1.14 Lines of constant entropy, $M_\infty = 0.8$, $\alpha = 1.25°$. (Courtesy of A. Lerat and J. Sides.)

For the calculation of flows with supersonic free streams and embedded subsonic regions, the cell methods have been very successful—the most successful being the Godunov method. This approach was applied by Taylor and Masson to compute a variety of flows past blunt bodies in a supersonic stream. To date, this approach is the only method that has successfully predicted, without difficulty, complicated flows with embedded shock waves, contact surfaces, and sonic lines. The λ scheme of Moretti, however, may become competitive. As a result we will outline briefly how the method can be applied to obtain the solutions. With possible improvements as suggested by VanLeer, the approach may become optimum for solving complicated supersonic flow problems. It is important to note that for steady supersonic flows without subsonic regions, it is not wise to employ an unsteady integration since the flow in the steady case is already hyperbolic and can be marched. The steady equivalent of the Godunov method has not been developed for compressible flows and, therefore, the unsteady Godunov approach at this time is not optimum for computing large steady domains that are totally supersonic. The extension of the Godunov method to steady supersonic flows could be a useful research effort.

The application of the Godunov approach to computations of flow past bodies can be accomplished by using a time-varying mesh that adapts to the body and the shock shape. For two dimensions or axisymmetric flows, such a system is shown in Fig. 10.1.15.

The cell coordinates are formed by first introducing a set of rays that are approximately normal to the body surface. Each ray is characterized by the angle, θ_i, which it forms with the symmetry axis. Each ray is then divided into equally spaced segments between the body and the external bounding shock. The distance along the ray, i, from the axis of symmetry to the cell coordinate j is denoted by $l_{i,j}$. The difference equations for each cell are derived by integrating the inviscid equations over the volume of the cell. The resulting expression can be written in the form

$$\frac{\Delta_t (f_{(i,j)} A)}{\Delta t} + \Delta_i (Hs)_{\langle j, t \rangle} + \Delta_j (Hs)_{\langle i, t \rangle} + \left(\frac{QA}{y} \right)_{\langle i, j, t \rangle} = 0 \qquad (10.1.9)$$

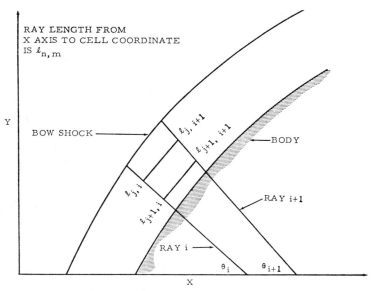

Fig. 10.1.15 Cell coordinate system for application of the Godunov method.

where

$$H = \begin{bmatrix} \rho(q_N - V_N) \\ \rho u(q_N - V_N) + p \sin \theta \\ \rho v(q_N - V_N) + p \cos \theta \\ \rho E + p \end{bmatrix}, \qquad \Delta_\nu f = f_{\nu+1} - f_\nu \qquad \text{for } \nu = i, j, t$$

$$Q = \begin{bmatrix} \rho v \\ \rho u v \\ \rho v^2 \\ \rho v(E + p/\rho) \end{bmatrix}$$

$$E = e + \tfrac{1}{2}(u^2 + v^2)$$

and the symbol $\langle i, j \rangle$ denotes the average over the interval i to $i+1$ and j to $j+1$ defined by $f_{\langle i,j \rangle} = \iint f\, dx\, dy / A$. Similarly, the symbol $\langle j, t \rangle$ denotes the average over the intervals j to $j+1$ and Δt. For the boundary under consideration,

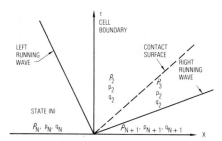

Fig. 10.1.16 Cell-boundary wave-pattern geometry for the Godunov method.

the quantity q_N denotes the velocity of flow normal to the boundary and V_N is the normal velocity at which the boundary is moving. A is the area of the mesh element and s is the arc length of the boundary under consideration.*

The finite-difference equations describe the behavior of the space-averaged flow variables $f_{\langle i,j \rangle}$, in each cell. In order to solve the difference equations, it is necessary to relate the values of the boundary fluxes, $H_{\langle j,t \rangle}$ and $H_{\langle i,t \rangle}$, to the averaged cell flow variables, $f_{\langle i,j \rangle}$. The basic concept for relating these quantities is the Godunov procedure. The procedure is to consider each boundary of a cell as a one-dimensional initial value (Riemann) problem and utilize the averaged-flow quantities of the cells on each side of the boundary for the initial states. The value of the flux $H_{\langle i,t \rangle}$ or $H_{\langle j,t \rangle}$ is then determined in the following manner.

The one-dimensional problem is posed so that the cell boundary is the reference point for all waves resulting from the initial discontinuity. The

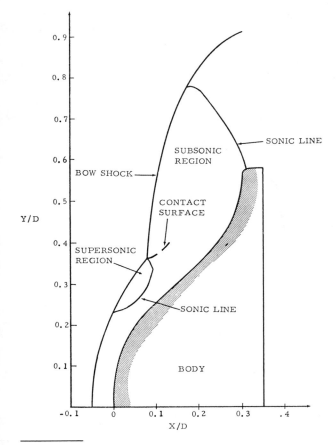

Fig. 10.1.17 Flow-pattern results from a 1200 time-step calculation by the God-unov method of inviscid flow past an indented body.

*Note $Q = 0$ for plane flows.

typical wave pattern is shown in Fig. 10.1.16. The strengths and velocities of the waves are calculated from the relationships given in Chapter 3.

The method described was employed to compute flow past a blunt bell shape with a nose radius $4/10$ the base diameter. The free-stream Mach number for the calculation was 8.0. Steady-state (1200 time-step) results are shown in Fig. 10.1.17. Note that the flow exhibits a number of significant singularities that will create extreme difficulty with most solution methods. For example, observe the embedded supersonic region, the triple point, and the contact surface. Each of these are known to be trouble spots in supersonic flow calculations. The surface-pressure distribution for the calculation is shown in Fig. 10.1.18, where the impact of the secondary shock on the surface can be seen.

A similar calculation to the blunt bell-shaped body was also conducted on a body for which the flow was known to oscillate. The result for the shock-wave position is shown in Fig. 10.1.19. The difficulty with this calculation was that the method tended to damp the oscillation at long times. The oscillation was a real flow condition and was confirmed experimentally by Holden (1975). The pressure distribution for the problem as computed by the inviscid calculation and a viscous solution to the same problem are shown in Fig. 10.1.20. The approach was also applied to compute flows past bodies with surface injection (Masson and Taylor, 1971).

The results displayed thus far for cell methods have considered only the use

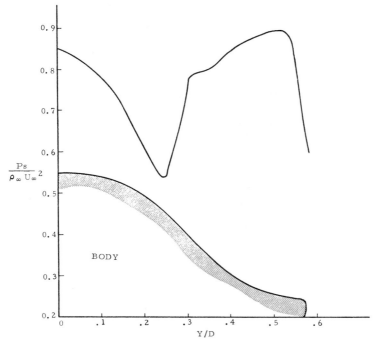

Fig. 10.1.18 Surface-pressure distribution from indented-body calculation.

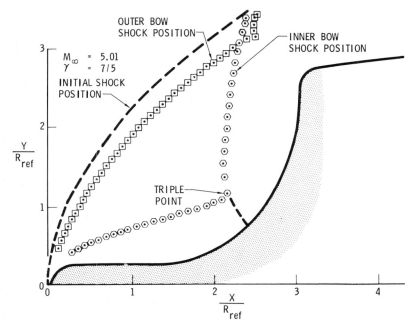

Fig. 10.1.19 Oscillating flow pattern for an indented body by Godunov's method.

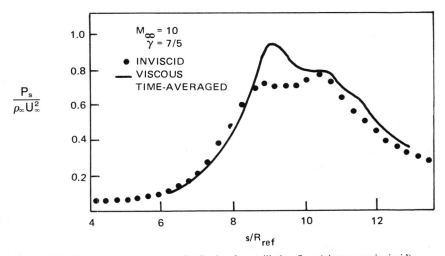

Fig. 10.1.20 Time-averaged pressure distribution for oscillating flow (viscous vs. inviscid).

of the Godunov approach. Other cell methods which yield equivalent two-dimensional results are not known to be available at this time. The method of Lerat and Sides (1981), however, may be useful. There are prospects of an improved Godunov method as indicated earlier but the results are not yet available for publication.

10.2 Steady-Flow Methods Using Finite-Difference Approaches

The computation of steady compressible flows by steady methods has advanced considerably in two areas. These are finite-difference potential flow solutions for $M_\infty < 1$ and finite-difference marching solutions for $M_\infty > 1$. The potential flow solutions have been developed principally for transonic airfoils and clearly have replaced unsteady solution methods in terms of efficiency. The limitation, of course, is in the irrotational flow and constant-entropy assumption which becomes less correct as M_∞ approaches 1. For $0.7 \leq M_\infty \leq 1$, the potential solutions at this time look like the best approach for solving steady-flow problems. Finite element or spectral methods could impact this in the future by yielding solutions to the nonlinear Euler equations, but at this time finite-element, spectral, and unsteady methods do not appear to be as efficient as the potential flow solutions. We now turn to the description of the potential solution approach.

10.2.1 Examples for $M_\infty < 1$

The research in solution of compressible flows for $M_\infty < 1$ has evolved toward solution of the potential formulation of flow by relaxation methods. The formulation is straightforward. One has the continuity, energy, and potential definition equations which take the form

$$\frac{\partial \rho u}{\partial x} + \frac{\partial \rho v}{\partial y} = 0 \tag{10.2.1}$$

$$\frac{\gamma}{\gamma-1} \frac{p}{\rho} + \frac{q^2}{2} = h_0 \tag{10.2.2}$$

$$u = \frac{\partial \phi}{\partial x}, \qquad v = \frac{\partial \phi}{\partial y} \tag{10.2.3}$$

where $q = \sqrt{u^2+v^2}$. Also required is the nondimensional state relationship which for isentropic flow is $p/p_\infty = \rho^\gamma$ where $p_\infty = 1/\gamma M_\infty^2$. The resulting equations in rectangular coordinates when combined are

$$\frac{\partial}{\partial x}(\rho \phi_x) + \frac{\partial}{\partial y}(\rho \phi_y) = 0 \tag{10.2.4}$$

$$c^2 = c_0^2 - \frac{(\gamma-1)}{2} q^2$$

$$\rho^{\gamma-1} = M_\infty^2 c^2$$

and c is the sound speed defined by $c = \sqrt{\gamma p/\rho}$.* In order for the solution of

---

*The subscripts x and y denote derivatives.

these equations to proceed properly in both the subsonic and supersonic regions, one must apply the proper finite differencing. If one uses central differences for subsonic regions these must be converted into forward or backward differences for stable supersonic flow calculations. This was pointed out early on by Murman and Cole (1971). An alternative to this concern, however, is to use central differences and an artificial viscosity term which automatically switches the differencing as the Mach number exceeds 1. This is the approach now employed by most researchers since it automates the calculation.

The artificial viscosity term that is usually added to the right-hand side of the continuity equation has the form

$$\frac{\partial}{\partial x}\left(\Delta x \mu |u| \frac{\partial \rho}{\partial x}\right) + \frac{\partial}{\partial y}\left(\Delta y \mu |v| \frac{\partial \rho}{\partial y}\right)$$

where μ is an artificial function that can be used to turn the viscosity on or off based on the Mach number. Jameson (1975) employed the form

$$\mu = \text{const} \times \max\left(0, 1 - \frac{1}{M^2}\right)$$

which turns the viscosity off for $M < 1$. It is important to note that there is no uniqueness to this approach and possibly another form would work equally well or better.

This approach stabilizes the flow both for switching from subsonic to supersonic regions as well as in the presence of shock waves. Once the differencing is established one then is faced with solving the resulting algebraic equations. At this point there are two approaches that one can take. First, one can attack the problem head-on and develop a solution to the complete equations. Second, one can reduce the complexity and employ the small disturbance equations in the classical manner for transonic flow. A solution to the reduced set is then developed. Both procedures have been followed. Jameson (1974, 1975, 1979), Caughey and Jameson (1977a,b, 1979a,b), and Chattot and Coulombeix (1978) have addressed the full nonlinear problem while Murman and Cole (1971), Boppe and Aidala (1980), Boppe and Stern (1980), Ballhaus (1972, 1978), and others (see Schmidt, 1978) addressed the reduced case. Among all the studies the struggle is to improve the speed of solution. At this time there appears to be no ideal approach to this. South and Jameson (1973) have employed a multiple-grid approach (Appendix B) to speed up the solution, and a variety of relaxation techniques have been employed to solve the algebraic equations. Interestingly, the conjugate-gradient method (Appendix C) with factorization seems not to have been exploited in this area. Because of the variety of methods employed to relax the solutions—successive relaxation, line relaxation, alternating directions implicit—we will not attempt to review the relaxation procedures in detail since a review by Lomax and Steger (1975) covers much of this.

For most problems equation set (10.2.4) is transformed into a curvilinear coordinate system that is consistent with the body geometry. An example is the use of the incompressible stream-function potential solution as the coordinate system for the transonic problem. In a generalized system ξ and η, the continuity equation will take the form (see Chattot and Coulombeix, 1978)

$$A \frac{\partial}{\partial \xi}\left(\rho \frac{\partial \phi}{\partial \xi}\right) + B \frac{\partial}{\partial \eta}\left(\rho \frac{\partial \phi}{\partial \eta}\right) + D \left[\frac{\partial}{\partial \xi}\left(\rho \frac{\partial \phi}{\partial \eta}\right) + \frac{\partial}{\partial \eta}\left(\rho \frac{\partial \phi}{\partial \xi}\right)\right] + G = 0$$

$$(10.2.5)$$

where

$$A = \left(\frac{\partial \xi}{\partial x}\right)^2 + \left(\frac{\partial \xi}{\partial y}\right)^2$$

$$B = \left(\frac{\partial \eta}{\partial x}\right)^2 + \left(\frac{\partial \eta}{\partial y}\right)^2$$

$$D = \frac{\partial \xi}{\partial x} \frac{\partial \eta}{\partial x} + \frac{\partial \xi}{\partial y} \frac{\partial \eta}{\partial y}$$

$$G = \left(\frac{\partial^2 \xi}{\partial x^2} + \frac{\partial^2 \xi}{\partial y^2}\right) \rho \frac{\partial \phi}{\partial \xi} + \left(\frac{\partial^2 \eta}{\partial x^2} + \frac{\partial^2 \eta}{\partial y^2}\right) \rho \frac{\partial \phi}{\partial \eta}$$

In some cases these coordinates are oriented along the flow direction so that the addition of aritificial viscosity will principally impact the streamwise convective terms. This potential equation must be solved, subject to the following boundary conditions:

1. At the surface $\partial \phi / \partial N = 0$, where N refers to the normal.
2. The Kutta condition holds, requiring that the tangential velocity at the trailing edge is bounded. Also the normal velocity across the slip line from the trailing edge must be continuous.
3. At the outer boundary the potential approaches the value of that of a vortex of prescribed circulation or lift plus the free-stream flow.

A variety of solutions have been developed for both the full potential and the small disturbance potential equations which satisfy these boundary conditions. As in other areas we cannot account for all of this work, but a forthcoming article by Caughey (1982) will review the complete field. We present here three examples that give a reasonable picture of the progress.

The first example is the work of Boppe and Aidala (1980) along with Boppe and Stern (1980). These authors solved the slender-body potential equations for a number of complicated flow problems. The geometry problems were overcome by using grids of different magnitude for different regions. The grid regions were modularized and overlapping much like an inner and outer numerical expansion. In the calculations one grid serves as the boundary condition of the adjoining grid system. A picture of this approach is shown in Fig. 10.2.1. Note that a fine grid is employed near the surface to incorporate the

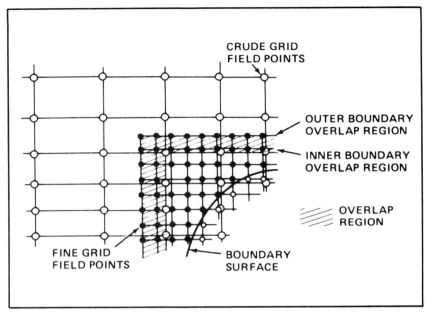

Fig. 10.2.1 Grid system of Boppe and Aidala. (Courtesy of C. Boppe.)

boundary conditions and a coarse grid is employed away from the boundary. Using these overlapping rectangular mesh systems, Boppe and colleagues solved the potential equation in the form

$$\left[1 - M_\infty^2 - (\gamma+1)M_\infty^2\,\phi_x - \frac{(\gamma+1)}{2}\,M_\infty^2\,\phi_x^2 \right]\phi_{xx} - 2\,M_\infty^2\,\phi_y\,\phi_{xy}$$

$$+ \left[1 - (\gamma-1)M_\infty^2\,\phi_x \right]\phi_{yy} + \phi_{zz} = 0 \qquad (10.2.6)$$

by a successive line relaxation procedure. The Murman–Cole scheme of switching differences from central for subsonic regions to forward differences for supersonic regions was employed. Boppe furnished the authors with a number of results of his group's studies and Fig. 10.2.2 was selected to display the practicality of the approach. The results shown are the pressures computed for KC135 wing sections compared with experiment. Note that the agreement is generally good, but some deviation from experiment occurs on the lower surface predictions. Boppe and Aidala also computed flows for a complex space shuttle configuration. A portion of the geometric model employed is shown in Fig. 10.2.3. A result obtained for the external tank-pressure distribution is shown in Fig. 10.2.4. For this complex configuration there is some variation from the data, but the results give a reasonable indication of the pressure level. The typical run times for these computation was 10–12 min on the CDC 7600.

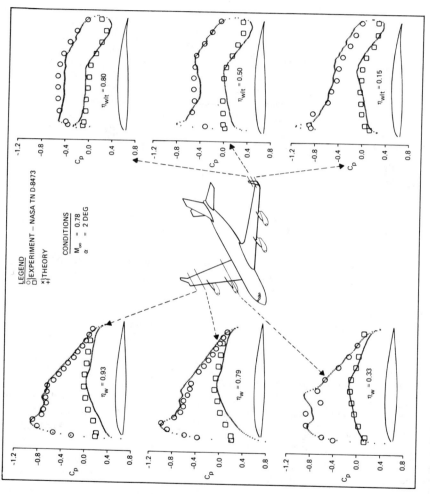

Fig. 10.2.2 KC 135 pressure distribution as computed by Boppe and Stern (1980). (Courtesy of C. Boppe.)

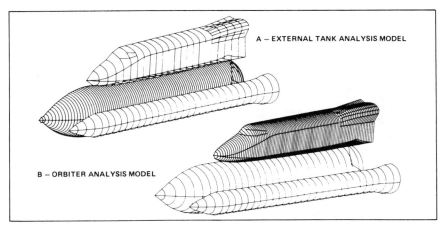

Fig. 10.2.3 Resolution of geometry utilized by Boppe and Aidala to compute flow past space shuttle and external tanks. (Courtesy of C. Boppe.)

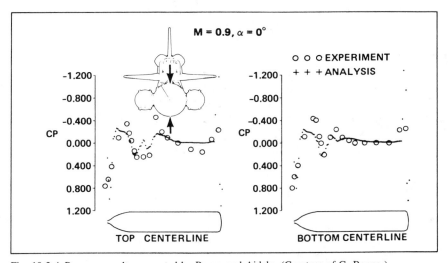

Fig. 10.2.4 Pressure results computed by Boppe and Aidala. (Courtesy of C. Boppe.)

The second example is the work of Chattot and Coulombeix (1978) who solved the complete nonlinear potential equation by both the Murman and Cole (1971) approach of difference switching and the Jameson (1974) artificial viscosity technique. For the solution of the difference equations, Chattot and Coulombeix employed over-relaxation for the Murman–Cole portion and operator splitting for the Jameson portion. For the calculation, a grid of $128 \times 32 \times 16$ points was required to converge the nonconservative Murman–Cole approach and a $96 \times 32 \times 16$ mesh was required to converge results for the Jameson conservative formulation. A typical run time was

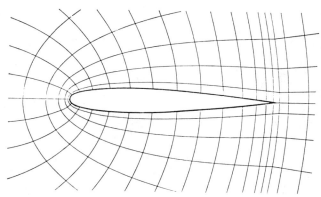

Fig. 10.2.5 Grid and geometry employed by Chattot and Coulombeix to compute transonic flow past NACA 0012 airfoil. (Courtesy of J. Chattot.)

15–30 min on a CDC 7600. Figure 10.2.5 displays the grid used for an NACA 0012 airfoil calculation. The results of calculations for the pressure on a finite wing using the NACA 0012 airfoil are shown in Fig. 10.2.6 along with experimental results. Note that the experiments fall between two predictions for the shock locations and that the results shown are essentially two dimensional due to the location. The Mach number was 0.85 and the angle of attack was zero. Chattot and Coulombeix also presented results for swept wings.

The last example solution of the complete potential equation is from the work of Caughey (1978) and Chen and Caughey (1980). The problem studied was flow past axisymmetric inlet cowls with blunt center bodies. The geometry of a case studied is shown in Fig. 10.2.7 along with the grid which was

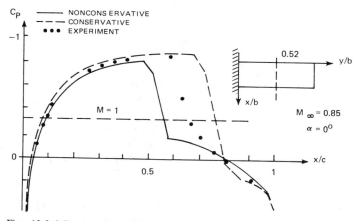

Fig. 10.2.6 Pressure distribution on airfoil computed by Chattot and Coulombeix employing Murman–Cole nonconservative and Jameson conservative schemes. (Courtesy of J. Chattot.)

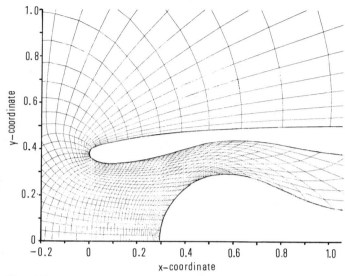

Fig. 10.2.7 Geometry and grid of engine inlet calculation of Caughey (Courtesy of D. Caughey.)

generated by conformal mapping with some type of shearing. The equation solved was of the form

$$c^2 \left[\frac{\partial(Du)}{\partial X} + \frac{\partial(Dv)}{\partial Y} \right] - D \left[\frac{u}{2} \frac{\partial q^2}{\partial X} + \frac{v}{2} \frac{\partial q^2}{\partial Y} \right] + \frac{Dc^2 v}{Y} = 0$$

where $q^2 = u^2 + v^2$ and D is the coefficient that arises in the equation when one transforms from an x, y coordinate system to mapping X, Y coordinate. This term was presented earlier. The value of D are generated from the transformations used to obtain the grid. This equation was approximated by central differences with the introduction of an artificial viscosity that is added in the supersonic region. The equation set was solved by successive line relaxation with constraints on the coefficients such that the relaxation corresponds to the solution by an unsteady equation. Results of the computation approach provided by Caughey are shown in Fig. 10.2.8 for a small free-stream Mach number.

The discussion thus far has given an indication of the problems that can be attacked by solution of the steady potential formulation. Also, we have pointed out the three principal points—geometry, differencing, and iteration techniques—that are the main issues that must be faced with this approach. For a variety of practical problems these issues have been resolved. The job ahead is to solve the full nonlinear Euler equations so that the potential and constant-entropy restrictions can be relaxed.

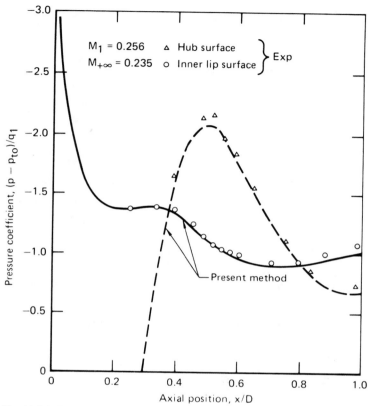

Fig. 10.2.8 Calculated pressure distribution on engine surface. (Courtesy of D. Caughey.)

10.2.2 Examples for $M_\infty > 1$

The solution of steady supersonic flows by steady methods is generally restricted to flows with totally supersonic regions. The procedure is to integrate the flow equations forward in the flow direction just as the flow direction is a time variable. As a result, any finite-difference scheme that is useful for unsteady flows can be adapted to steady flows by eliminating the time derivative and integrating the derivative in the flow direction as if it were a time derivative. The procedure for general curvilinear coordinates is first to rearrange the Euler equations of Chapter 1 into the form

$$E_\xi + F_\eta + G_\zeta + H = 0$$

where

$$E = h_2 h_3 \begin{pmatrix} \rho u \\ p + \rho u^2 \\ \rho u v \\ \rho u w \end{pmatrix}, \quad F = h_1 h_3 \begin{pmatrix} \rho v \\ \rho u v \\ p + \rho v^2 \\ \rho v w \end{pmatrix}, \quad G = h_1 h_2 \begin{pmatrix} \rho w \\ \rho u w \\ \rho v w \\ p + \rho w^2 \end{pmatrix}$$

$$H = \begin{pmatrix} 0 \\ \rho u v h_3 \dfrac{\partial h_1}{\partial \eta} + \rho u w h_2 \dfrac{\partial h_1}{\partial \zeta} - (p+\rho v^2) h_3 \dfrac{\partial h_2}{\partial \xi} - (p+\rho w^2) h_2 \dfrac{\partial h_3}{\partial \xi} \\ \rho u v h_3 \dfrac{\partial h_2}{\partial \xi} + \rho v w h_1 \dfrac{\partial h_2}{\partial \zeta} - (p+\rho u^2) h_3 \dfrac{\partial h_1}{\partial \eta} - (p+\rho w^2) h_1 \dfrac{\partial h_3}{\partial \eta} \\ \rho u w h_2 \dfrac{\partial h_3}{\partial \xi} + \rho v w h_1 \dfrac{\partial h_3}{\partial \eta} - (p+\rho u^2) h_2 \dfrac{\partial h_1}{\partial \zeta} - (p+\rho v^2) h_1 \dfrac{\partial h_2}{\partial \zeta} \end{pmatrix}$$

In these equations h_1, h_2, and h_3 are the metric coefficients and for cylindrical coordinates (x, r, ϕ) these take values $h_1 = 1$, $h_2 = 1$, and $h_3 = r$. These equations are integrated with ξ as the marching variable using a marching approach of one's choice which has been adequately tested to know that it will not fail upon encountering a shock wave. The approach has been employed extensively by Kutler and coworkers (1971, 1973, 1974, 1975, 1978) at NASA Ames, Thommen and D'Attorre (1965), Thomas et al. (1972), and Babenko et al. (1964). An alternate approach very similar to the approach of Moretti (1979) has been proposed by Walkden et al. (1978). Basically, these authors propose mapping the flow into streamline potential coordinates then integrating the inviscid equations in characteristic form in this coordinate system. The system of equations are particularly straightforward to solve. At this time the λ approach of Moretti also appears to be most useful, since the method can be used to solve both the blunt-body flow as well as the supersonic sonic afterbody flow with one method and program. The method of Walkden et al. could probably be adapted to have similar possibilities, but unfortunately more testing and work is necessary. Both the Walkden–Caine–Laws and Moretti-type methods will yield accurate results because of the ease of incorporating boundary conditions. A study by Abbett (1973) clearly pointed out that by using characteristic-type schemes at boundaries one tended to obtain optimum accuracy. Abbett also found that the predictor-corrector finite-difference methods could be adapted to accurately incorporate solid-boundary conditions. The procedure is to use a one-sided predictor to obtain a boundary-point value even though the boundary condition may not be exactly satisfied. The boundary value is then corrected to satisfy the boundary condition. The procedure is as follows:

1. Predict the flow variables at a boundary point using the predictor of choice (for example, the first step of MacCormack's scheme).

2. Correct the predicted variables using the known surface entropy, enthalpy, and a simple wave expansion which adjusts p to turn the flow parallel to the wall.

For shock-wave boundaries Abbett found that a variety of methods worked equally well. These were the approaches of Kentzer (1971), Barnwell (1971), and Thomas et al. (1972) or the method of characteristics. The interesting fact is that the procedures of Kentzer and Barnwell for the shock waves encompass

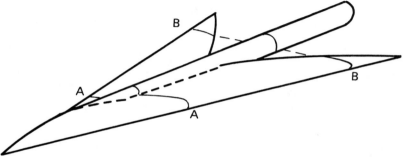

Fig. 10.2.9 Supersonic aircraft configuration.

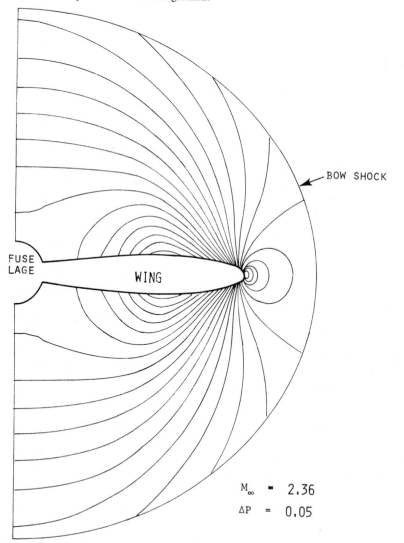

Fig. 10.2.10 Pressure distributions at station A as computed by Moretti using λ scheme, $M_\infty = 2.36$. (Courtesy of G. Moretti.)

ideas similar to the complete schemes employed by Walkden et al. or Moretti. As a consequence of Abbett's work it is clear that the best approach for computing supersonic flows is a scheme that encompass the ideas of characteristics but does not integrate along the characteristic directions. The schemes that accomplish this are those of Walkden et al. and Moretti. Moretti has demonstrated the use of the λ scheme in the calculation of flow past the complex supersonic aircraft shown in Fig. 10.2.9 for $M_\infty = 2.36$ and 4.63. Figures 10.2.10 and 10.2.11 show pressure isobars computed for stations A

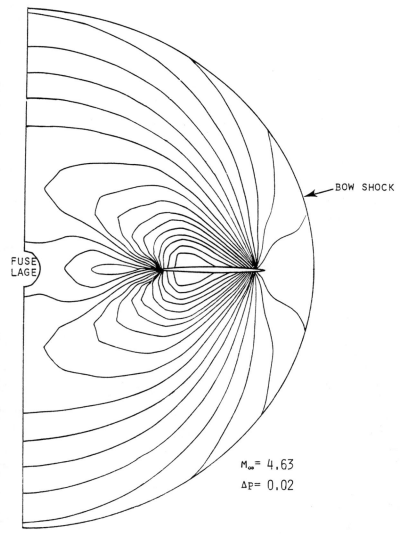

Fig. 10.2.11 Pressure distributions at station B as computed by Moretti using λ scheme, $M_\infty = 4.63$. (Courtesy of G. Moretti.)

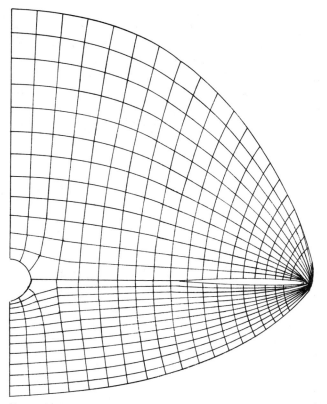

Fig. 10.2.12 Grid system employed at location B. (Courtesy of G. Moretti.)

and B. The isobars are spaced by $\Delta p = 0.05$ for location A and there is no angle of incidence. For $M_\infty = 4.63$ the isobars for location B are spaced by 0.2 and the angle of attack was taken to be 10°. The grid system employed for location B is shown in Fig. 10.2.12. As shown, 48 grids were employed circumferentially and 12 were utilized between the body and shock. For location A the grid was similar except only 24 were used circumferentially. Figure 10.2.13 shows comparison of the computed C_p at station A with experiments for both Mach numbers. Needless to say the results look quite good.

Results similar in complexity to Moretti have also been obtained by using a MacCormack-type method to capture shocks in the work of Kutler et al., Rakich and Kutler (1972), and Marconi and Koch (1979). For the details the reader is referred to the references.

From the results available, it is apparent that computation of three-dimensional steady supersonic flow is in rather good condition and appears limited only by one's ability to incorporate the complex geometries.

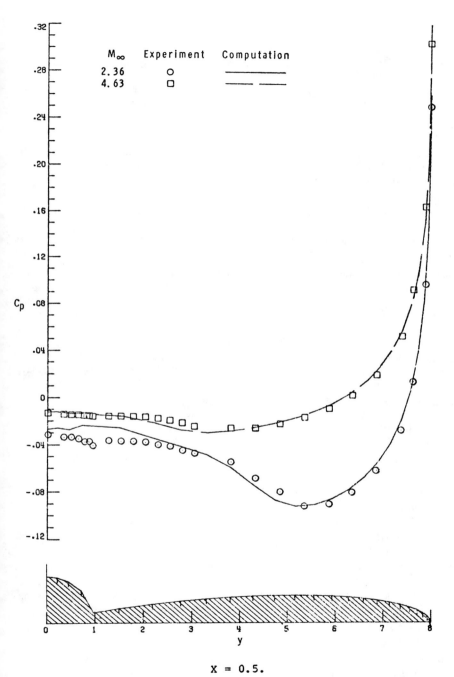

X = 0.5.

Fig. 10.2.13 Comparison of computed C_p at station A with experiment. (Courtesy of G. Moretti.)

References

Abbett, M. J. Proceedings of the AIAA Computational Fluid Dynamics Conference, pp. 153–172, Palm Springs, July (1973).

Babenko, K. I., Voskresenskii, G. P., Lyubimov, A. N., and Rusanov, V. V. Three-Dimensional Flow of Ideal Gases Around Smooth Bodies. NASA TTF-380 (1968). [Original published by Nauka Moscow (1964)].

Ballhaus, W. F., and Bailey, F. R. AIAA Paper 7-677, July (1972).

Ballhaus, W. F. Recent Progress in Transonic Flow Calculations. In: *Numerical Methods in Fluid Dynamics*, H. J. Wirz and J. J. Smolderen, Eds., pp. 155–236, Hemisphere Publishing, Washington, D. C. (1978).

Barnwell, R. AIAA Paper 71-56, January (1971).

Bauer, F., Garabedian, P., Korn, D., and Jameson, A. *Supercritical Wing Sections II*, Springer-Verlag, New York (1975).

Bohachevsky, I. O., and Rubin, E. L. *AIAA J.* **4,** 600–607 (1966).

Boppe, C. W., and Stern, M. A. AIAA Paper 80-0130, January (1980).

Boppe, C. W., and Aidala, P. V. Complex Configuration Analysis at Transonic Speeds. AGARD Conference on Subsonic/Transonic Configuration Aerodynamics, Preprint No. 285, May (1980).

Bristeau, M. O. Application of Optimal Control Theory to Transonic Flow Computations. In *Proceedings of the 3rd IRIA Symposium on Computational Methods in Applied Science and Engineering*, Versailles, France, December (1977).

Bristeau, M. O., et al. Application of Optimal Control and Finite Element Methods to the Calculation of Transonic and Incompressible Flow. IRIA Rept. No. 294, LeChesnay, France, April (1978).

Caughey, D. A., and Jameson, A. *Transonic Flow Problems in Turbomachinery*, T. C. Adamson and M. Platier, Eds., pp. 274–291, Hemisphere Publishing, Washington, D. C. (1977a).

Caughey, D. A. and Jameson, A. *AIAA J.* **15,** 1474–1480 (1977b).

Caughey, D. A. Lecture Notes, University of Tennessee Space Institute, Tullahoma, TN, December (1978).

Caughey, D. A., and Jameson, A. *AIAA J.* **17,** 175–181 (1979a).

Caughey, D. A., and Jameson, A. AIAA Paper No. 79-1513 (1979b).

Caughey, D. A.: "A (Limited) Perspective on Computational Aerodynamics." Presented at the 13th AIAA Fluid and Plasma Dynamics Conference, July (1980); also Cornell University Rept. FDA-80-07, July (1980).

Caughey, D. A. "A Review of Transonic Flow Computation," *Ann. Rev. Fluid Mech.* **14,** 261–283 (1982).

Chan, S. T. K., Brashears, M. R., and Young, V. Y. C. AIAA Paper 75-79, Pasadena, (1975).

Chang, I. S. AIAA Paper 80-0272, Pasadena, (1980).

Chen, L. T., and Caughey, D. A. *J. Aircraft* **17,** 167–174 (1980).

Chattot, J. J., and Coulombeix, C. Calculs D'Ecoulements Transsoniques Autour D'Ailes. ONERA Rept. T. P. No. 1978-125 Chatillon, France (1978).

Chorin, A. J. *J. Comput Phys.* **22,** 517–533 (1976).

Chorin, A. J. *J. Comput Phys.* **25**, 253–272 (1977).

Colella, P. The Effects of Sampling and Operator Splitting on the Accuracy of Glimm's Scheme. PhD Thesis, University of California, Berkeley (1978).

Euvrard, D., and Tournemine, G. *J. Méc.* **12**, 419–461 (1973).

Fletcher, C. A. J. Subsonic Inviscid Flow by the Finite-Element Method. Dept. of Defense, Weapons Research Establishment Rept. WRE-TR-1858 (W), Salisbury, S. Australia, August (1977).

Fletcher, C. A. J. *J. Comput. Phys.* **33**, 301–312 (1979).

Glaz, H. M. Development of Random Choice Numerical Methods for Blast Wave Problems. NSWC Report/WOL TR 78-211, March (1979).

Glaz, H. M., and Colella, P. Private communication results can be obtained from P. Colella at LLL, Livermore, CA (1980).

Godunov, S. K. *Mat. Sb.* **47**, 271–306 (1959).

Grossman, R., and Moretti, G. AIAA Paper 70–1322 (1970).

Holden, M. Studies of Transitional Flows, Unsteady Separation Phenomena, and Partial Induced Augmentation Heating on Ablated Noise Tips. AFOIL Rept TR-76-1066, October (1975).

Holt, M., and Masson, B. S. *Lecture Notes in Physics, Vol. 8*, pp. 207–214. Springer-Verlag, New York (1971).

Holt, M. *Numerical Methods in Fluid Dynamics*, Springer-Verlag, New York (1977).

Jameson, A. *Commun. Pure Appl. Math.* **27**, 283–309 (1974).

Jameson, A. In: *Proceedings of the 2nd AIAA Computational Fluid Dynamics Conference*, pp. 148–161, Hartford, CT, June (1975).

Jameson, A., and Caughey, D. A.: In: *Proceedings of the 3rd AIAA Computational Fluid Dynamics Conference*, p. 35–54, Albuquerque, N.M., June (1977c).

Jameson, A. In: *Proceedings of the 4th AIAA Computational Fluid Dynamics Conference*, pp. 122–146, Williamsburg, VA, July (1979).

Kentzer, C. P. *Lecture Notes in Physics, Vol. 8*, pp. 108–113, Springer-Verlag, New York (1971).

Kershaw, D. C. *J. Comput. Phys.* **26**, 43–65 (1978).

Khosla, P. K., and Rubin, S. G. *Comput. Fluids*, **1**, 109–121 (1981).

Kutler, P., and Lomax, H. *J. Spacecraft Rockets* **8**, 1175–1182 (1971).

Kutler, P., Lomax, H., and Warming, R. F. *AIAA J.* **11**, 196–204 (1973).

Kutler, P., Reinhardt, W., and Warming, R. F. *AIAA J.* **11**, 657–664 (1973).

Kutler, P. *AIAA J.* **12**, 557–578 (1974).

Kutler, P., and Sakell, L. AIAA Paper 75–49, January (1975).

Kutler, P., Chakravarthy, S. R., and Lombard, C. K. AIAA Paper 78–213 (1978).

Lax, P. D. *Commun. Pure Appl. Math.* **7**, 159–193 (1954).

Lax, P. D., and Wendroff, B. *Commun. Pure Appl. Math.* **13**, 217–237 (1960).

Lax, P. D., and Wendroff, B. *Commun. Pure Appl. Math.* **17**, 381–398 (1964).

Lerat, A., and Sides, J. In: *Proceedings of the Conference on Numerical Methods in Aeronautical Fluid Dynamics*, University of Reading, March 29–April 1 (1981).

Lomax, H. and Steger, J. L. *Ann. Rev. Fluid Mech.* **7**, 63–80 (1975).

Magnus, R., and Yoshihara, H. *AIAA J.* **8,** 2157–2162 (1970).

Magnus, R., and Yoshihara, H. *AIAA J.* **13,** 1622–1628 (1975).

Magnus, R. J. Some Numerical Solutions of Inviscid, Unsteady, Transonic Flows over the NLR 7301 Airfoil. Convair General Dynamics Rept. CASD/LVP 78-013, January (1978).

Marconi, F., and Koch, F. An Improved Supersonic 3-D External Inviscid Flow Field Code. NASA Contractor Rept. 3108, Grumman Aerospace, Bethpage, NY (1979).

Masson, B. S., Taylor, T. D., and Foster, R. M. *AIAA J.* **7,** 694–698 (1969).

Masson, B. S., and Taylor, T. D. *Polish Fluid Dyn. Trans.* **5,** 185–194 (1971).

Masson, B. S., and Friedman, G. Axisymmetric Transonic Flow Calculations. Picatinny Arsenal Tech. Rept. No. 4271, Dover, NJ (1972).

MacCormack, R. W. AIAA Paper 69-354, May (1969).

Moretti, G. In: *Proceedings of the 1974 Heat Transfer and Fluid Mechanics Institute,* pp. 184–201, Stanford University Press, Stanford (1974).

Moretti, G. Computation of Shock Layers about Ablated Blunt Nosed Bodies. Polytechnic Institute of New York Rept. 77-14, August (1977).

Moretti, G. An Old Integration Scheme Refurbished and put to Work. Polytechnic Institute of New York Rept. M/AE 78-22, September (1978).

Moretti, G. *Comput. Fluids* **7,** 191–205 (1979).

Moretti, G. A Numerical Analysis of Muzzle Blast-Precursor Flow. Polytechnic Institute of New York Rept. M/AE 80-10, May (1980).

Murman, E. M., and Cole, J. D. *AIAA J.* **9,** 114–121 (1971).

Periaux, J. *Int. J. Numer. Meth. Eng.* **9,** 775–831 (1975).

Rakich, J. V., and Kutler, P. AIAA Paper 72-191, January (1972).

Reddall, W. Private communication of result (1980).

Richtmyer, R. D., and Morton, K. W. *Difference Methods of Initial-Value Problems,* 2nd ed., Interscience, New York (1967).

Rusanov, V. V. *USSR. Comput. Math. Math. Phys.* **8,** 156–179 (1968).

Rusanov, V. V. *J. Comput. Phys.* **5,** 507–516 (1970).

Schmidt, W. Progress in Transonic Flow Computations; Analysis and Design Methods for 3-D Flows. In: *Numerical Methods in Fluid Dynamics,* H. J. Wirz and J. J. Smolderen, Eds., pp. 299–338, Hemisphere Publishing, Washington, D. C. (1978).

Schmidt, E. M., and Shear, D. D., The Flow Field About the Muzzle of an M-16 Rifle, BRL Report No. 1692, U.S. Army Ballistic Research Laboratory, Aberdeen Proving Ground, Md., January (1974).

Sod, G. A. *J. Comput. Phys.* **27,** 1–31 (1978).

Sod, G. A. Computational Fluid Dynamics with Stochastic Techniques. Princeton University Technical Rept. MAE 1479, March (1980a).

Sod, G. A. Automotive Engine Modeling with a Random Choice Hybrid Method, II. SAE Technical Paper 800288, February (1980b).

South, Jr., J. C., and Jameson, A. In: *Proceedings of the AIAA Computational Fluid Dynamics Conference,* pp. 8–12, July, AIAA, New York (1973).

Taylor, T. D., and Masson, B. S. *J. Comput. Phys.* **5,** 443–454 (1970).

Taylor, T. D. *AGARDograph,* No. 187, (1974).

Taylor, T. D., and Lin, T. C. *AIAA J*. **19,** 346–349 (1981).

Thomas, P. D., Vinokur, M. Bastianon, R. A., and Conti, R. J. *AIAA J*. **10,** 887–894 (1972).

Thommen, H. U. Z. *Angew. Math. Phys.* **17,** 369–384 (1966).

Thommen, H. U., and D'Attorre, L. AIAA Paper 65-25, January (1965).

Thompson, J. F., Thames, F. C., and Martin, C. W. *J. Comput. Phys.* **15,** 299–319 (1974).

VanLeer, B. *J. Comput. Phys.* **32,** 101–136 (1979).

Viviand, H., and Veuillot, J. P. Pseudo-Unstationary Methods for Calculation of Transonic Flows (in French). ONERA Publication No. 1978-4, Chatillon, France (1978).

Walkden, F., Caine, P. and Laws, G. T. *J. Comput. Phys.* **27,** 103–122 (1978).

CHAPTER 11

Viscous Compressible Flows

11.1 Introduction to Methods

The calculations of steady viscous flows based on the Navier–Stokes equations are generally conducted with the unsteady equations by considering the limit of large time. Until recently most of the solutions were developed with explicit finite-difference schemes such as those of Thommen (1966) and MacCormack (1969). Such explicit schemes are easy to implement and this is important because of the large number of dependent variables, the complexity of the equations, and, often, the complexity of the geometry. However, the limitation on the time step due to the explicit character of the scheme leads to expensive calculations. For this reason, schemes with less-restrictive stability conditions have always been a subject of interest. In this context, pioneer work in this direction was conducted by Crocco (1965) who considered a *nonconsistent* scheme based on the following approximation of the viscous derivatives:

$$\left(\frac{\partial^2 f}{\partial x^2}\right)_i^n = \frac{f_{i+1}^n - 2 f_i^{n+1} + f_{i-1}^n}{\Delta x^2} \tag{11.1.1}$$

The resulting finite-difference equations become consistent with the exact differential equations when the steady state is reached.

Such a nonconsistent approximation was later introduced by Allen and Cheng (1970) into the two-step scheme of Brailovskaya (1967). This latter scheme is analogous to the Thommen scheme (see Sections 2.7 and 2.8), but the provisional values are defined at the same points as the final values. The Brailovskaya scheme does not enter into the general class of schemes (2.7.11) or (2.8.9).

The Allen–Cheng scheme has good stability properties and is unconditionally stable if the mesh Reynolds number is smaller that 2 (Peyret and Viviand, 1975); for larger values of the mesh Reynolds number, the Courant–Friedrich–Lewy condition has been found to be sufficient. This means that with a nonconsistent approximation of type (11.1.1) the viscosity does not enter into the limitation on the time step. However, as shown by Peyret and Viviand (1975), the solution of the effective unsteady equation approximated by the Allen–Cheng scheme could have a slower decay toward the steady state than the solution of the exact equation; although, in fact, this relative slowness is counterbalanced by the large time steps that can be used.

Employing the same idea of using a nonconsistent scheme with good stability properties to convergence toward a steady state, Peyret and Viviand (1972, 1973) have considered the computation of low-Reynolds-number flows. A very simple one-step explicit scheme based upon the Gauss–Seidel technique characterized by

$$\left(\frac{\partial f}{\partial x}\right)_i^n = \frac{1}{2\,\Delta x}\left(f_{i+1}^n - f_{i-1}^{n+1}\right), \qquad \left(\frac{\partial^2 f}{\partial x^2}\right)_i^n = \frac{f_{i+1}^n - 2f_i^n + f_{i-1}^{n+1}}{\Delta x^2}$$

$$(11.1.2)$$

was employed.

Note also that the leapfrog DuFort–Frankel scheme, used in particular by Victoria and Widhopf (1973), could also be employed as a nonconsistent scheme with a time step larger than the one required for consistency (Section 2.6.3).

All of these unsteady techniques belong to the general class of *pseudo-unsteady methods* which employ either a nonconsistent scheme with respect to time or modify the original unsteady system of equations in order to obtain a faster or a less time-consuming convergence toward the steady state. A review of methods based upon the second concept and adapted to the computation of transonic inviscid flows has been carried out by Viviand (1980).

Another way to obtain less-restrictive stability conditions for multi-dimensional flows while preserving an explicit calculation is the method of splitting described in Section 2.8.4. The method has been introduced for the solution of the compressible Navier–Stokes equations by MacCormack (1971) and since has been employed in a number of applications, in particular, with a finite-volume method (Deiwert 1975, 1976; Baldwin et al., 1975; Hung and MacCormack, 1977).

In the last few years, a large and productive effort has occurred in the area of implicit schemes. Such implicit schemes are efficient if (i) they are based upon splitting or alternating direction techniques for replacing the multi-dimensional problem by successive one-dimensional problems, and (ii) they do not necessitate an iterative procedure for the calculation of the solution at each time step. The first use of an ADI technique without an iterative process seems to have been proposed by Polezhaev (1967). The generalized ADI procedure has been successfully applied to the compressible Navier–Stokes equations by Berezin et al. (1972, 1975a, 1975b), Briley and McDonald (1973, 1975, 1977), and more recently by Beam and Warming (1978) who extended the implicit scheme previously introduced for hyperbolic systems (Beam and Warming, 1976).

Finally, we note two methods of implicit nature devised by MacCormack.

The first method (MacCormack, 1976a, 1976b) considers splitting the equations in each space direction followed by a second splitting between inviscid and viscous parts. Each part is solved by a special technique: the method of characteristics combined with the explicit McCormack scheme for the inviscid part and the Crank–Nicolson scheme for the viscous part. The method has been used in several applications. However, the second method of MacCormack (1981a, 1981b) seems to be more interesting because of its simplicity of implementation (solutions of linear systems with bidiagonal matrices) and its good stability properties. The implicit part of the scheme makes successive use of forward and backward difference operators associated with the corresponding explicit MacCormack scheme, and the discretization of viscous terms is derived from the Saul'ev scheme (see Richtmyer and Morton, 1967). A study of the properties of the scheme applied to the solution of a hyperbolic equation has been made by Hollanders and Peyret (1981).

All the methods mentioned above belong to either the class of finite-difference or finite-volume methods. Until now, very limited work has been devoted to other types of methods discussed in this book: finite-element methods (Oden and Wellford, 1972; Baker, 1978; Cooke and Blanchard, 1979) or spectral methods (Bokhari et al., 1981). We do not intend to describe these efforts since they are very recent and will likely receive substantial improvements in the near future.

In the present chapter, general features of finite-difference explicit and implicit schemes will be discussed, along with the associated problems of boundary conditions and definition of computation domain. A large part of the chapter will be devoted to problems related to the use of nonuniform meshes.

We refer the reader to the original papers for more complete details concerning the methods or to review works (Peyret and Viviand, 1975; Viviand, 1978; Hollanders and Viviand, 1980; Baldwin et al., 1975; MacCormack and Lomax, 1979; Mehta and Lomax, 1981) for a general description and for references concerning the various applications.

11.2 Boundary Conditions

The unsteady Navier–Stokes equations for a compressible fluid form a hybrid parabolic–hyperbolic system (also called an *incompletely parabolic system*). The parabolic nature is related to the presence of second-order derivatives in the momentum and energy equations which characterize the dissipative effects. On the other hand, the absence of such derivatives in the continuity equation makes the system different since by itself it would be a hyperbolic equation. In the time-independent case, the Navier–Stokes equations form a hybrid elliptic–hyperbolic system.

Problems to be solved in practice are boundary-value problems, either time dependent with given initial values of the basic unknowns (ρ, $\rho\mathbf{V}$, ρE) or time independent. Only a few special results are known about the mathematical statement of boundary conditions to be imposed in order to ensure existence or uniqueness of the solution. This is contrary to the incompressible case for which the mathematical theory is well established. As a result, the question must be settled in a heuristic way by taking into account the physical meaning of the problem and the mathematical nature of the equations. Thus, it can be conjectured that the momentum and energy equations each require one condition at each boundary (or a scalar condition for each scalar equation), the simplest being $\mathbf{V}$ and T given on the boundary. The continuity equation, considered as an equation for ρ assuming a known velocity field, explicitly gives the change of density of a fluid particle along its trajectory. Therefore, if the fluid particle enters the domain of computation through the boundary, its density should be given as a boundary condition. On the contrary, if the particle leaves the computation domain, its density must be calculated as a part of the solution and it cannot be imposed arbitrarily.

In the case where the boundary is an impermeable wall and when rarefaction effects are not present, the usual no-slip condition is

$$\mathbf{V} = \mathbf{V}_w \tag{11.2.1}$$

where $\mathbf{V}_w$ is the velocity of the wall. For the temperature, two types of conditions are usually considered, either

$$T = T_w \tag{11.2.2}$$

where T_w is the wall temperature, or

$$\frac{\partial T}{\partial N} = 0 \tag{11.2.3}$$

where $\mathbf{N}$ is the normal unit vector to the wall. In the case of a perfect gas with constant specific heats, the above condition, (11.2.2) or (11.2.3), gives a condition for the internal energy $e = c_v T$. There is no boundary condition for the density ρ at an impermeable wall, and it therefore must be calculated.

A general discussion of the boundary conditions associated with a non-material boundary is presented by Hollanders and Viviand (1980). The question is considered mathematically by Gustaffson and Sundström (1978).

We describe next the procedure for defining the computation domain and associated boundary conditions for two typical examples: (i) the shock–boundary layer interaction problem and (ii) the problem of flow around a finite body.

In the shock–boundary layer interaction problem (Fig. 11.2.1) considered, for example, by Baldwin and MacCormack (1974), the domain in which the Navier–Stokes equations are solved overlaps inviscid supersonic flow regions (1) and boundary-layer-type regions (2) and (3). The flow quantities are fixed along AB and BC: a boundary-layer flow is given on AA′ and uniform flow conditions are imposed along A′S, SB, and BC compatible with the given

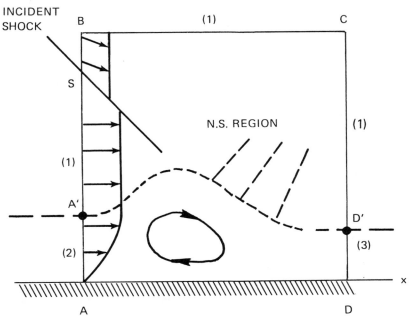

Fig. 11.2.1 Computation domain for shock boundary-layer interaction problem.

incident shock. Along CD, the flow quantities can be taken equal to the values computed at the previous column of discretization points (i.e., first-order approximation of $\partial f/\partial x = 0$). Such a simple treatment expresses the fact that the flow downstream of CD has no influence on the flow inside the Navier–Stokes region, because on the boundary CD the flow is either supersonic or of boundary-layer type. For the density, it would be more rigorous from a mathematical point of view to compute the density at the downstream limit CD from an upstream differencing of the unsteady continuity equation. The success of the simple extrapolation is directly related to the special nature of the flow at this boundary.

In some calculations (the example given in Section 11.4.2), the flow quantities along AB are obtained from a previous calculation including the leading edge of the plate.

The second boundary condition example concerns flow around a finite body with uniform-flow conditions imposed at infinity. In such a case, it is necessary to devise a technique to carry out the solution of the Navier–Stokes equations in a bounded domain. A technique that can be used is to perform a coordinate transformation which maps the entire physical plane into a finite domain in the transformed plane. A classical example is conformal mapping of the exterior of a body into the interior of a circle. Usually, however, simple stretched coordinate systems are used. A second technique, similar to the first, consists of introducing a coordinate transformation such that the external boundary of the transformed plane corresponds to a region in the physical plane on which uniform-flow conditions or, better, asymptotic behavior can be applied. In the latter case, the asymptotic solution at large distances is related to flow properties on the body and the overall solution usually requires an iterative procedure. This technique has been successfully applied in the case of incompressible viscous flow (Takami and Keller, 1969) or compressible inviscid transonic flows (Murman and Cole, 1971; Euvrard and Tournemine, 1973). However, due to its complexity, such a technique has not been applied to the compressible viscous case.

A third technique, and the most simple, is illustrated in Fig. 11.2.2. It consists of choosing an outer boundary at a finite distance and dividing this boundary into two part: (1) the upstream part BAD, through which the fluid with uniform conditions enters the domain of computation, and (2) the downstream part BCD, through which the fluid leaves the domain, subject to some type of empirical extrapolation condition. Here again, the success of such a simple treatment depends on the nature of the flow at the boundary. In the case of subsonic flows, it may be necessary to use a more elaborate technique such as the method of "nonreflecting" boundary conditions, based upon the elimination of disturbances propagating along the incoming characteristic lines. Such a procedure has been developed by Engquist and Majda (1977) and Hedstrom (1979) for the solution of hyperbolic equations and by Rudy and Strikwerda (1980) for the Navier–Stokes equations. An analogous idea was used by Peyret (1971) for the computation of a simple incompressible flow.

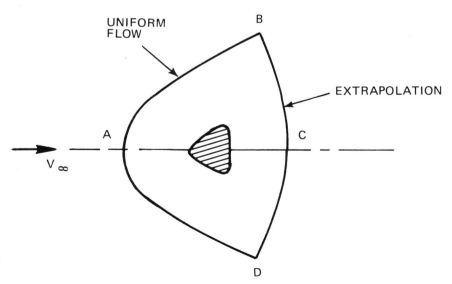

Fig. 11.2.2 Computation domain for flow around a finite body.

In the case of a supersonic flow, the flow field remains unperturbed upstream of a bow shock, and the computation domain can be limited upstream by a boundary located at a short distance ahead of the bow shock or by the bow shock itself, depending on whether the shock is treated as a sharp but continuous transition zone (shock-capturing method) or as a true discontinuity (shock-fitting method). In the first case, the uniform-flow conditions are imposed on the chosen boundary. In the second case, the location of the shock and the flow quantities behind it must be determined as a part of the solution (Moretti and Salas, 1969; Tannehill et al., 1975; Viviand and Ghazzi, 1976).

Concerning the boundary conditions on the material boundary (flat plate in Fig. 11.2.1, finite body in Fig. 11.2.2), the only problem is related to the determination of the density, or more precisely that of the pressure. As a matter of fact, when the conservative Navier–Stokes equations are approximated with finite differences in a mesh in which all the flow quantities are defined at the same points (as is usually done) and such that the wall is a mesh line, it can be seen that the knowledge of the pressure (and not the density) at the wall is needed only in the centered approximation of the pressure derivative occurring in the transversal momentum equation at the first line of discretization points adjacent to the wall.

Various techniques to handle this problem have been employed. The use of the continuity equation at the wall to obtain the density and then the pressure through the state law is delicate and may lead to strong oscillations or even to divergence if an artificial viscosity term is not added. In particular, in the case of separated flows (Roache, 1972; Peyret and Viviand, 1973) negative values of the density can be obtained. However, certain types of discretization of the continuity equation at the wall are reported to give good results (explicit: Victoria and Widhopf, 1973; implicit: Polezhaev, 1967; Scala and Gordon, 1970; Briley and McDonald, 1973). Another approach is simply the determination of p at the wall by a quadratic extrapolation (Carter, 1973; Tannehill et al., 1974), or a linear extrapolation in a mesh for which the wall is located at middistance between two mesh lines (Allen and Cheng, 1970).

A completely different approach is currently used with the best success. It consists of computing the wall pressure from a one-sided differencing of the normal momentum equation written at the wall. Assuming the wall to be located at $y = 0$, the normal momentum equation in Cartesian coordinates yields

$$\left(\frac{\partial p}{\partial y}\right)_{y=0} = \frac{1}{\mathrm{Re}}\left[\frac{\mu}{3}\left(\frac{\partial^2 u}{\partial x\,\partial y} + 4\frac{\partial^2 v}{\partial y^2}\right) + \frac{d\mu}{dT}\left(\frac{4}{3}\frac{\partial v}{\partial y}\frac{\partial T}{\partial y} + \frac{\partial u}{\partial y}\frac{\partial T}{\partial x}\right)\right]_{y=0}$$

(11.2.4)

In a high-Reynolds-number boundary-layer-type flow, the condition

$$\left(\frac{\partial p}{\partial y}\right)_{y=0} = 0$$

(11.2.5)

can be used in place of Eq. (11.2.4).

11.3 Finite-Difference Schemes in Uniform Cartesian Mesh

In this section, the application of difference schemes to the compressible Navier–Stokes equations is discussed. The discussion is restricted to the case of uniform Cartesian meshes. Such a restriction is made only for the sake of simplicity of presentation because, as mentioned in Chapter 1, the Navier–Stokes equations can be written in any arbitrary curvilinear system in a fully conservative form analogous to the conservative Cartesian form to be considered here. The equations for plane two-dimensional flow are written

$$\frac{\partial f}{\partial t} + \frac{\partial F}{\partial x} + \frac{\partial G}{\partial y} = 0 \tag{11.3.1}$$

where

$$f = (\rho, \rho u, \rho v, \rho E), \qquad F = F_{\mathrm{I}} - \frac{F_{\mathrm{V}}}{\mathrm{Re}}, \qquad G = G_{\mathrm{I}} - \frac{G_{\mathrm{V}}}{\mathrm{Re}}$$

The inviscid terms F_{I}, G_{I} and the viscous ones F_{V}, G_{V} were defined in Chapter 1.

11.3.1 Explicit schemes

The application of the two-step explicit schemes described in Section 2.8 to the system (11.3.1) is straightforward except for the treatment of mixed derivatives which were not present in the model equation. The fact that the equations are vectorial and no longer scalar does not change the formal construction of the scheme. However, it is necessary to consider the change in the stability introduced by these new terms.

We consider here the treatment of the mixed derivative

$$D = \frac{\partial}{\partial x} \left[\nu(f) \, \frac{\partial f}{\partial y} \right] \tag{11.3.2}$$

when added to the model equation (2.8.7). In the case of the "centered" version of scheme (2.8.9) obtained with the β, γ, and λ values equal to $\frac{1}{2}$, the mixed derivative D must be evaluated at point Q and R (see Fig. 11.3.1) for the computation of the predicted values and then at point P for the computation of the final value.

At point Q defined by $(i + \frac{1}{2}, j)$ the approximation is simply

$$D^n_{i+1/2,j} = \Delta^1_x (\nu^n_{i+1/2,j} \, \Delta^0_y f^n_{i+1/2,j}) \tag{11.3.3}$$

with the usual definition of the difference operators

$$\Delta^1_x \phi_{l,m} = \frac{\phi_{l+1/2,m} - \phi_{l-1/2,m}}{\Delta x}$$

$$\Delta^0_y \phi_{l,m} = \frac{\phi_{l,m+1} - \phi_{l,m-1}}{2 \, \Delta y}$$

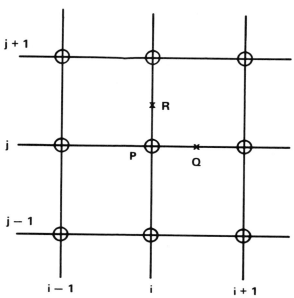

Fig. 11.3.1 Mesh for centered scheme.

and where $\nu_{i+1/2,j}$ is an approximation of $\nu(f)$ at point Q; for example, $\nu_{i+1/2,j} = \frac{1}{2}[\nu(f_{i+1,j}) + \nu(f_{i,j})]$. The approximation of D at point R $(i,j+1/2)$ is defined in the same way by

$$D^n_{i,j+1/2} = \Delta^0_x(\nu^n_{i,j+1/2} \Delta^1_y f^n_{i,j+1/2}) \tag{11.3.4}$$

and at point P (i,j) by

$$D^n_{i,j} = \Delta^0_x(\nu^n_{i,j} \Delta^0_y f^n_{i,j}) \tag{11.3.5}$$

In the case of schemes of the MacCormack type (2.8.8) [or (2.8.9) with values of the parameters other than α^x and α^y equal to 0 or 1], the mixed derivative D is usually approximated by a combination of noncentered and centered differences. Therefore, the approximation of D included in the first step (2.8.8a) is

$$D^n_{i,j} = \Delta^+_x(\nu^n_{i,j} \Delta^0_y f^n_{i,j}) \tag{11.3.6a}$$

and in the second step (2.8.8b):

$$\tilde{D}_{i,j} = \Delta^-_x(\tilde{\nu}_{i,j} \Delta^0_y \tilde{f}_{i,j}) \tag{11.3.6b}$$

where $\Delta^{\pm}_x \phi_{l,m} = \pm(\phi_{l\pm1,m} - \phi_{l,m})/\Delta x$

Another possibility for treating the mixed derivative D is considered by LeBalleur et al. (1980). It consists of using the following formulas

$$D^n_{i,j} = \Delta^+_x(\nu^n_{i,j} \Delta^-_y f^n_{i,j})$$
$$\tilde{D}_{i,j} = \Delta^-_x(\tilde{\nu}_{i,j} \Delta^+_y \tilde{f}_{i,j}) \tag{11.3.7}$$

which lead, as formulas (11.3.6), to second-order accuracy (proven in the linear case).

Exact criterion of stability of the explicit schemes (2.8.8) or (2.8.9) applied to the compressible Navier–Stokes equations are not known. In the case of a nondissipative fluid (Euler's equation), a necessary criterion of stability can be deduced from the condition (Courant et al., 1928; Lax and Wendroff, 1964) that the numerical domain of dependence must contain the exact domain of dependence. From this restriction, we obtain for schemes (2.8.8) or (2.8.9) the CFL condition

$$\Delta t \le \left(\frac{|u|}{\Delta x} + \frac{|v|}{\Delta y} + c \sqrt{\frac{1}{\Delta x^2} + \frac{1}{\Delta y^2}} \right)^{-1} \tag{11.3.8}$$

where c is the sound speed. Details of the derivation of such criterion can be found in the work of MacCormack (1971) who demonstrated that the condition was not sufficient for his own scheme.

For the Navier–Stokes equations, the maximal time step should be evaluated from numerical analysis of the eigenvalues of the amplification matrix. However, useful bounds on the time step are given by the approximate analysis carried out by MacCormack and Baldwin (1975) for a splitting technique (Section 2.8.4) based upon the one-dimensional MacCormack scheme. These results show that for the operator in the x direction the criterion is estimated to be

$$\Delta t_x \le \Delta x \left[|u| + c + \frac{1}{\operatorname{Re} \rho} \left(\frac{2 \gamma k}{\operatorname{Pr} \Delta x} + \sqrt{\frac{2}{3}} \frac{\mu}{\Delta y} \right) \right]^{-1} \tag{11.3.9a}$$

and for the operator in the y direction,

$$\Delta t_y \le \Delta y \left[|v| + c + \frac{1}{\operatorname{Re} \rho} \left(\sqrt{\frac{2}{3}} \frac{\mu}{\Delta x} + \frac{2 \gamma k}{\operatorname{Pr} \Delta y} \right) \right]^{-1} \tag{11.3.9b}$$

It should be noted that these criteria were estimated by assuming the Stokes law ($\lambda = -\frac{2}{3} \mu$) and that the limitation on the time step due to thermal effects is more restrictive than that due to purely viscous effects. Note also that the terms Δy in (11.3.9a) and Δx in (11.3.9b) are directly related to the presence of mixed derivatives.

Finally, by a heuristic argument based on the relationship between the stability criteria in one and two dimensions, a useful estimation for the time step can be obtained from (11.3.8) and (11.3.9) assuming $\Delta x = \Delta y$. The result is

$$\Delta t \le \Delta x^2 \left[\Delta x (|u| + |v| + c \sqrt{2}) + \frac{2}{\operatorname{Re} \rho} \left(\frac{2 \gamma k}{\operatorname{Pr}} + \sqrt{\frac{2}{3}} \mu \right) \right]^{-1} \tag{11.3.10}$$

Such a criterion can be used for the MacCormack scheme (2.8.8) or for the "centered" version of Thommen's type of scheme (2.8.9) (α^x, α^y arbitrary and all the other parameters equal to $\frac{1}{2}$).

11.3.2 Implicit schemes

Efficient implicit schemes for the solution of the compressible
Navier–Stokes equations are constructed by using the discretization scheme
(2.7.22) and the linearization technique (2.7.24) combined with the generalized ADI method described in Section 2.8.5. Such implicit schemes have
been developed by Briley and McDonald (1973, 1975, 1977) and Beam and
Warming (1978). Basically, both methods are similar and lead to the solution
of block-tridiagonal linear systems. The main difference between both methods
lies in the fact that in the Beam and Warming method the vector ϕ which
appears as unknown in the block-tridiagonal system is the vector f itself; i.e.,
$\phi = f = (\rho, \rho\mathbf{V}, \rho E)$ while Briley and McDonald introduce a supplementary
linearization so that the unknown vector is $\phi = (\rho, \mathbf{V}, T)$.

We briefly outline the method in the Beam and Warming formulation, by
explaining the treatment of viscous terms which is more complicated in the
Navier–Stokes equations than in the simple model equation (2.7.1) with constant viscosity and no mixed derivatives. In order to simplify the presentation,
we restrict the discussion to the three-level scheme deduced from (2.7.22) with
$\theta = 1$, $\epsilon = 2$, used in calculations reported by Beam and Warming. Thus, the
discretization of the vector equation (11.3.1) with respect to time is

$$\frac{1}{2\Delta t}(3f^{n+1} - 4f^n + f^{n-1}) + \left(\frac{\partial F}{\partial x}\right)^{n+1} + \left(\frac{\partial G}{\partial y}\right)^{n+1} = 0 \qquad (11.3.11)$$

Next, we describe the treatment of the term $(\partial F/\partial x)^{n+1}$, noting that the same
technique applies to the term $(\partial G/\partial y)^{n+1}$. The inviscid part F_1 of the term
$F = F_1 - F_V/\mathrm{Re}$ is linearized in the manner explained in Section 2.7.4. Let

$$F_1^{n+1} \cong F_1^n + A^n(f^{n+1} - f^n)$$

where A is the Jacobian matrix defined by $A = dF_1/df$. The viscous part F_V
contains derivatives of f with respect to x and y so that a mixed derivative
occurs which needs special treatment. The term F_V is decomposed into

$$F_V = F_{V1}\left(f, \frac{\partial f}{\partial x}\right) + F_{V2}\left(f, \frac{\partial f}{\partial y}\right).$$

The part F_{V1}^{n+1} is linearized according to the relationship

$$F_{V1}^{n+1} \cong F_{V1}^n + M^n(f^{n+1} - f^n) + P^n\left[\left(\frac{\partial f}{\partial x}\right)^{n+1} - \left(\frac{\partial f}{\partial x}\right)^n\right]$$

then

$$F_{V1}^{n+1} \cong F_{V1}^n + \left[M^n - \left(\frac{\partial P}{\partial x}\right)^n\right](f^{n+1} - f^n) + \frac{\partial}{\partial x}[P^n(f^{n+1} - f^n)]$$

$$(11.3.12)$$

where $M = \partial F_{V1}/\partial f$ and $P = \partial F_{V1}/\partial(\partial f/\partial x)$.

The term F_{V2}, which gives a mixed derivative leading to difficulty if it is treated in an implicit manner, is evaluated explicitly in the form

$$F_{V2}^{n+1} \cong 2 F_{V2}^n - F_{V2}^{n-1} .$$

This does not destroy the second-order accuracy in time of the scheme nor its unconditional stability (proven for a model scalar equation by Beam and Warming, 1978, 1980).

Finally, after carrying out the same treatment for the term G, approximating the spatial derivatives with finite-difference formulas, and substituting the resulting expressions into Eq. (11.3.11), we obtain the scheme written in the symbolic form (2.8.18):

$$(I + \Lambda_x + \Lambda_y)(f_{i,j}^{n+1} - f_{i,j}^n) = Rf_{i,j}^n \qquad (11.3.13)$$

where I is the identity operator and Λ_x, Λ_y, and R are the operators defined by the expressions:

$$\Lambda_x f_{i,j} = \frac{2\Delta t}{3}\left(\frac{\partial}{\partial x}\left\{\left[A^n - \frac{1}{Re}\left(M^n - \frac{\partial P^n}{\partial x}\right)\right]f\right\}\right)_{i,j} - \frac{2\Delta t}{3\,Re}\left[\frac{\partial^2}{\partial x^2}(P^n f)\right]_{i,j}$$

$$\Lambda_y f_{i,j} = \frac{2\Delta t}{3}\left(\frac{\partial}{\partial y}\left\{\left[B^n - \frac{1}{Re}\left(N^n - \frac{\partial Q^n}{\partial y}\right)\right]f\right\}\right)_{i,j} - \frac{2\Delta t}{3\,Re}\left[\frac{\partial^2}{\partial y^2}(Q^n f)\right]_{i,j}$$

$$Rf_{i,j}^n = -\frac{2\Delta t}{3}\left[\left(\frac{\partial F}{\partial x}\right)_{i,j}^n + \left(\frac{\partial G}{\partial y}\right)_{i,j}^n\right] + \frac{2\Delta t}{3\,Re}\left\{\left[\frac{\partial}{\partial x}\left(F_{V2}^n - F_{V2}^{n-1}\right)\right]_{i,j}\right.$$

$$\left. + \left[\frac{\partial}{\partial y}\left(G_{V1}^n - G_{V1}^{n-1}\right)\right]_{i,j}\right\} + \frac{1}{3}\left(f^n - f^{n-1}\right)_{i,j}$$

where

$$B = \frac{dG_I}{df}, \qquad N = \frac{\partial G_{V2}}{\partial f}, \qquad Q = \frac{\partial G_{V2}}{\partial(\partial f/\partial y)}$$

with

$$G = G_I - \frac{G_V}{Re}, \qquad G_V = G_{V1}\left(f, \frac{\partial f}{\partial x}\right) + G_{V2}\left(f, \frac{\partial f}{\partial y}\right)$$

In the above expression, the finite-difference operators for approximating the spatial derivatives are not stated. Various formulas can be used; however, it is necessary that the resulting equations lead at most to block-tridiagonal matrices. For example, Beam and Warming (1978) used the usual centered three-point second-order accurate finite-difference approximations.

Because the coefficients of viscosity, λ and μ, and thermal conductivity k depend generally on the temperature, the matrices M and N are rather complicated. However, a simplification occurs if the dependence on time of these coefficients can be neglected. Moreover, in this case, the sum $M - \partial P/\partial x$

depends homogeneously on the first derivatives of the dissipative coefficients, such that $M - \partial P/\partial x = 0$ if these derivatives can be neglected. Obviously this is the case if the coefficients λ, μ, and k are constant. The similar equality $N - \partial Q/\partial y = 0$ holds for the term G_{V2}. Consequently, the operators Λ_x and Λ_y are now defined by

$$\Lambda_x f_{i,j} = \frac{2}{3}\Delta t \left\{ \left[\frac{\partial}{\partial x}(A^n f) \right]_{i,j} - \frac{1}{Re}\left[\frac{\partial^2}{\partial x^2}(P^n f) \right]_{i,j} \right\}$$

$$\Lambda_y f_{i,j} = \frac{2}{3}\Delta t \left\{ \left[\frac{\partial}{\partial y}(B^n f) \right]_{i,j} - \frac{1}{Re}\left[\frac{\partial^2}{\partial y^2}(Q^n f) \right]_{i,j} \right\}$$

Note that, for a steady-state solution, these operators can be used without loss of accuracy whatever the variation of the dissipation coefficients since, at steady state $f_{i,j}^{n+1} = f_{i,j}^n = f_{i,j}^{n-1} = f_{i,j}$ and Eq. (11.3.13) reduces to $Rf_{i,j} = 0$; i.e.,

$$\left(\frac{\partial}{\partial x}F \right)_{i,j} + \left(\frac{\partial}{\partial y}G \right)_{i,j} = 0.$$

Finally, scheme (11.3.13) is solved by means of the generalized ADI method described in Section 2.8.5. Note that in applications in which only a steady solution is of interest the second-order accuracy in time is not necessary, and the construction outlined above can be applied to the first-order scheme resulting from (2.7.22) with $\theta = 1$ and $\epsilon = 1$, with the mixed derivatives evaluated at the level n.

11.3.3 Artificial viscosity

Although the schemes considered in the previous sections are stable in the von Neumann sense, this stability concerns only the solution of the linearized finite-difference systems without boundary conditions. In real-flow computations at high Reynolds number, instabilities can occur due to several sources including (i) nonlinear effects, (ii) rapid change of flow direction in a separated flow region, (iii) large pressure gradients, and (iv) the presence of walls and outer boundaries of the computational domain. The effect of such instabilities, which can be qualified as "weak," is not necessarily a fast divergence of the computation; but often they create oscillations that remain of finite amplitude.

The normal way to remove these oscillations as well as those occurring for other reasons (see Chapter 2) is to introduce an artificial viscosity or damping.

The fourth-order damping proposed by Richtmyer and Morton (1964) consists of adding the term $D/\Delta t$ with

$$D = -\alpha_1 \Delta x^4 \frac{\partial^4 f}{\partial x^4} - \alpha_2 \Delta y^4 \frac{\partial^4 f}{\partial y^4} \qquad (\alpha_1 > 0, \ \alpha_2 > 0) \qquad (11.3.14)$$

to Eq. (11.3.1). Such damping has been used by Barnwell (1971) and Chang

(1980) for inviscid calculations, and for viscous flows by Gnoffo (1974), Beam and Warming (1978), Steger (1978), and LeBalleur et al. (1980). This type of artificial viscosity is easily included in the general schemes (2.8.8) or (2.8.9) by adding the discretized version of $D_{i,j}$ [Eq. (10.1.2)] to the corrector term or by adding a third step which yields a smoothed value $\hat{f}_{i,j}^{n+1}$ of $f_{i,j}^{n+1}$ by the formula

$$\hat{f}_{i,j}^{n+1} = f_{i,j}^{n+1} + D_{i,j}^{n+1} \tag{11.3.15}$$

An analysis of the stability of the above scheme in a one-dimensional case leads to the condition $0 \le \alpha_1, \alpha_2 \le \frac{1}{8}$.

In the case of the implicit scheme of Section 11.3.2, the damping term (11.3.14) is split as $D = D_1 + D_2$ with

$$D_1 = -\alpha_1 \, \Delta x^4 \, \frac{\partial^4 f}{\partial x^4}$$

and

$$D_2 = -\alpha_2 \, \Delta y^4 \, \frac{\partial^4 f}{\partial y^4}$$

Then the discretized approximations of D_1 and D_2 are evaluated at level n and are included, respectively, into the right-hand sides of Eq. (2.8.17a and b), representing the two first steps of the generalized ADI method used for the solution of (11.3.13) as described in Section 2.8.5.

Another type of damping is used by MacCormack and Baldwin (1975) with the explicit splitting scheme (Section 2.8.4) and is based upon the backward scheme (2.7.10) where $\alpha = 1$, $\beta = \gamma = 1$.

The damping term in the x direction, for example, is

$$D = \alpha \Delta t \, \Delta x^3 \, \frac{\partial}{\partial x} \left[\frac{|u| + c}{p} \left| \frac{\partial^2 p}{\partial x^2} \right| \frac{\partial f}{\partial x} \right]$$

and its discretized approximation is added to each of the two steps of the scheme. $D_{i,j}^n = \Delta t (S_{i,j}^n - S_{i-1,j}^n)$ with

$$S_{i,j}^n = 4\alpha (|u_{i,j}^n| + c_{i,j}^n) \frac{|p_{i+1,j}^n - 2 p_{i,j}^n + p_{i-1,j}^n|}{p_{i+1,j}^n + 2 p_{i,j}^n + p_{i-1,j}^n} (f_{i+1,j}^n - f_{i,j}^n)$$

for the first step and $\tilde{D}_{i,j} = \Delta t (\tilde{S}_{i+1,j} - \tilde{S}_{i,j})$ with

$$\tilde{S}_{i,j} = 4\alpha (|\tilde{u}_{i,j}| + \tilde{c}_{i,j}) \frac{|\tilde{p}_{i+1,j} - 2 \tilde{p}_{i,j} + \tilde{p}_{i-1,j}|}{\tilde{p}_{i+1,j} + 2 \tilde{p}_{i,j} + \tilde{p}_{i-1,j}} (\tilde{f}_{i,j} - \tilde{f}_{i-1,j})$$

for the second step. Note that $0 \le \alpha \le \frac{1}{8}$ for stability.

Moreover, as pointed out by MacCormack and Baldwin (1975), an instability can be created by the convective flux term when there is an expansion in which the velocity changes sign. Such phenomenon occurs in a reversed-flow region. The instability can be removed by replacing any convective term of the type $u_{i,j}^n \, \phi_{i,j}^n$ in the first step by $\frac{1}{2}(u_{i+1,j}^n + u_{i,j}^n) \, \phi_{i,j}^n$ and $\tilde{u}_{i+1,j} \, \tilde{\phi}_{i+1,j}$ in the second step by $\frac{1}{2}(\tilde{u}_{i,j} + \tilde{u}_{i+1,j}) \, \tilde{\phi}_{i+1,j}$.

11.4 Finite-Difference Schemes in Non-Cartesian Configurations

The problems related to the computation of flows in complex geometries have been largely discussed in Chapter 10 which was devoted to inviscid flows. The problems for viscous flows are identical. Moreover, the presence of thin regions where the solution presents large gradients due to viscous effects (boundary layer and "viscous shock") necessitates adapting the mesh not only to the geometry but also to the solution. Note that the need of an adaptation to the solution is not restricted to viscous flows but appears also in inviscid calculations, when large gradients exist. The problem of adapting the mesh to the solution is a complex one when considered in all its generality. Interesting approaches for constructing self-adapted meshes as functionals of the solution have been proposed by Gough et al. (1975), Yanenko et al. (1979), Schönauer et al. (1980), and by Dwyer et al. (1980) for special cases and should lead to fruitful developments.

Up to now, the methods used for practical applications with complex configurations are less general in their principle and usually appeal to some *a priori* knowledge of the solution. The generation of meshes can be performed in various ways according to the complexity of the problem. These can include (i) a curvilinear coordinate system known analytically in a closed form, (ii) a system constructed by conformal mapping or by more general numerical techniques (Moretti, 1976, 1977; Chu, 1971; Thompson et al., 1974; Eiseman, 1978, 1980) with possible stretching (Sorenson and Steger, 1977), and (iii) supplementary mesh refinement (Viviand and Ghazzi, 1976; LeBalleur et al., 1980).

In general, when flows in nonrectangular configurations have to be computed, two types of approaches are possible according to the space (transformed or physical) in which the equations are approximated. In the present section both approaches will be discussed and illustrated with examples.

11.4.1 Discretization in transformed space

In transformed space the solution approach consists of writing the Navier–Stokes equations in curvilinear coordinates and discretizing the resulting equations by means of a finite-difference scheme. In this approach, one has a choice of two bases for expressing the vectors and tensors.

In the first, the local basis can be chosen to be simple orthogonal systems, i.e., polar coordinates (Moretti and Salas, 1969; Tannehill et al., 1975), parabolic coordinates (plane: Peyret and Viviand, 1972, 1973; axisymmetric: Weilmuenster and Graves, 1981), and spherical coordinates (Victoria and

Widhopf, 1973). Results obtained by Peyret and Viviand (1973) by using this approach are shown in Fig. 11.4.1. The figure exhibits the computation domain, the computed edges of the viscous shock, and the sonic line in the case of a supersonic flow ($M_\infty = 2$), around a plane body at low Reynolds number (Re = 25 and 50, based upon the length of the body). The Prandtl number Pr = 0.72 and the adiabatic index $\gamma = 1.4$. The Navier–Stokes equations (with constant viscosity) were solved by means of the pseudo-unsteady method based upon the differencing (11.1.2) as mentioned in the introduction to this chapter. Uniform-flow conditions were imposed on the upstream boundary and linear extrapolations were used on the downstream boundary. The wall temperature with respect to the free-stream temperature is equal to 1.20. The need for the wall pressure evaluation is avoided by using a noncentered discretization of the normal pressure gradient at lines adjacent to the body.

Instead of using a local basis, it is generally simpler to express the vectors

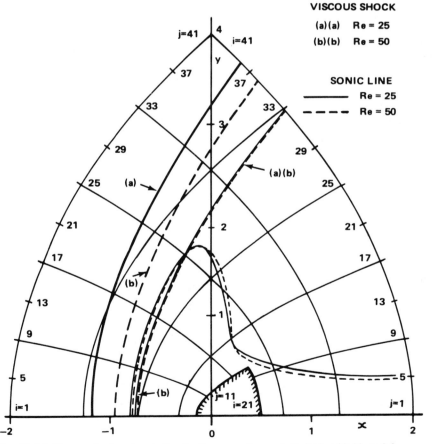

Fig. 11.4.1 Viscous shock and sonic line for flow around a parabolic body with $M_\infty = 2.0$.

and tensors in a fixed Cartesian coordinate system while conserving the equations written in the curvilinear system. Such a formulation has been used in particular by Viviand and Ghazzi (1976) for the calculation of the supersonic flow around a hemisphere-cylinder by using the explicit MacCormack scheme. The same formulation, but with the thin-layer approximation of the Navier–Stokes equations, has been employed to compute (i) flows around airfoils by Steger (1978) and Steger and Bailey (1980), and (ii) flow around an ogive cylinder and hemisphere-cylinder by Schiff and Steger (1980), Pulliam and Lomax (1979), and Pulliam and Steger (1980). In these works, the equations were solved by means of the implicit scheme of Beam and Warming (1978) described in Section 11.3.2. Note also that in some of these calculations, the turbulent nature of the flow is taken into account by using eddy–viscosity models.

Results obtained by Pulliam and Steger (1980) and Pulliam and Lomax (1979) for flow past a hemisphere-cylinder at an angle of attack are shown in Fig. 11.4.2. These calculations were made using a general curvilinear coordinate formulation with warped spherical coordinates. The thin-layer approximation was also employed. This approximation was implemented by writing the Navier–Stokes equations in curvilinear coordinates (ξ, η, ζ) with (ξ, η) oriented in the direction of the body so that the surface of the body corresponded to $\zeta = $ const. All viscous terms in the (ξ, η) directions were then neglected. For the details the reader is referred to the original papers. The calculated results for pressure and separation along with comparison with experimental results are shown in the figure. Note that even with the thin-layer approximation the results are quite good.

Finally, for the full Navier–Stokes equations, it has been pointed out by Viviand (1978) as well as by Shamroth and Gibeling (1979) that the quasiconservative form [Eq (1.31a), Section 1.2.3] is preferred for the case of large mesh deformations in the physical plane to the fully conservative form [Eq. (1.31)] in order to reduce the discretization errors. These errors are present if the derivatives of the transformation are exactly evaluated, but they can be eliminated by using special finite differences to approximate these derivatives in the fully conservative equation (1.31) as proposed by Viviand and Veuillot (1978). Calculations reported by Shamroth and Gibeling (1979) concerning flows around a cylinder and a NACA 0012 airfoil were performed with the Navier–Stokes equation in quasiconservative form using the Briley and McDonald (1973) method, in a grid generated by the method of Eiseman (1978).

11.4.2 Discretization in the physical space

The second approach for solution the Navier–Stokes equations consists of writing the equations in Cartesian coordinates and discretizing directly in the physical space on a nonuniform mesh adapted to the geometry and possibly to the solution. Various approximations can be used: finite-difference methods (Viviand and Veuillot, 1978; Le Balleur et al., 1980), finite-volume methods

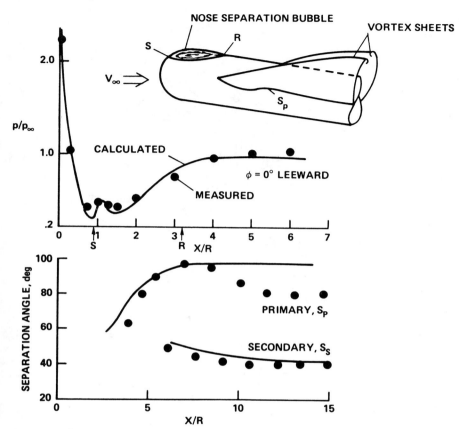

Fig. 11.4.2 Three-dimensional calculation of flow past a hemisphere-cylinder for $M_\infty = 1.2$, $\alpha = 19°$ $Re_D = 445,000$. (Courtesy of T. Pulliam.)

(Deiwert, 1975), or finite-element methods (Baker, 1978; Cook and Blan-chard, 1979). Note that all these discretization techniques are close to each other in their behavior.

11.4.2.1 *A finite-difference scheme*: In this section, we describe the finite-difference method first introduced by Viviand and Veuillot (1978) for the calculation of inviscid flows and extended to viscous flows by Hollanders and Viviand (1980) and LeBalleur et al. (1980). The scheme is a generalization of the explicit MacCormack scheme (2.8.8).

Let us consider the nonuniform mesh of Fig. 11.4.3, where the nodes P are identified by means of two indices (i,j). Although the mesh is usually con-structed through a coordinate transformation, it is not necessary to explicitly introduce such a transformation—only the location of the mesh points (i,j) with respect to the original Cartesian coordinate system (x,y) are needed. Approximation of the derivatives $\partial\phi/\partial x$, $\partial\phi/\partial y$ at point P of any function

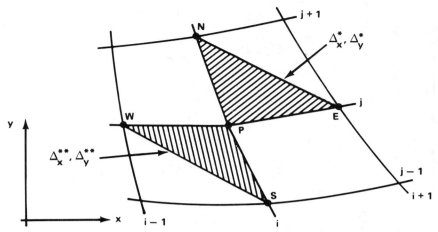

Fig. 11.4.3 Definition of mesh for difference operators.

$\phi(x, y)$ are defined in terms of the values of ϕ at P and at two neighboring points Q and R, such that P, Q, and R are not aligned (Fig. 11.4.4). These approximations are written

$$\left(\frac{\partial \phi}{\partial x}\right)_P \cong (\Delta_x \phi)_P = \frac{\Delta \phi_{PR}\, \Delta y_{PQ} - \Delta \phi_{PQ}\, \Delta y_{PR}}{\Delta y_{PQ}\, \Delta x_{PR} - \Delta y_{PR}\, \Delta x_{PQ}} \qquad (11.4.1a)$$

$$\left(\frac{\partial \phi}{\partial y}\right)_P \cong (\Delta_y \phi)_P = \frac{\Delta \phi_{PR}\, \Delta x_{PQ} - \Delta \phi_{PQ}\, \Delta x_{PR}}{\Delta x_{PQ}\, \Delta y_{PR} - \Delta x_{PR}\, \Delta y_{PQ}} \qquad (11.4.1b)$$

with

$$\Delta \phi_{PR} = \phi_R - \phi_P, \qquad \Delta x_{PR} = x_R - x_P, \qquad \Delta y_{PR} = y_R - y_P, \text{ etc.}$$

where ϕ_P is the value of ϕ at point P of coordinates (x_P, y_P), etc.

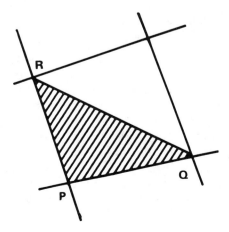

Fig. 11.4.4 Mesh definition.

The above approximations can be obtained (i) by Taylor expansions to first order of ϕ at Q and at R with respect to P; or (ii) by assuming $\phi(x,y)$ to be a linear function of x and y over the triangle PQR; or (iii) by using the formulas (3.7.11) in which the curvilinear coordinates (ξ,η) correspond to the mesh lines, then by approximating the various derivatives with respect to ξ and η with noncentered differences.

Now, we consider the point P and the four neighboring points N, S, E, W as shown in Fig. 11.4.3. We define the following finite-difference operators of the type (11.4.1): Δ_x^*, Δ_y^* using the points P, N, E and Δ_x^{**}, Δ_y^{**} using the points W, P, S. Then, the discretization of the Navier–Stokes equations (11.3.1) in Cartesian coordinates is carried out by the two-step scheme:

$$\tilde{f}_{i,j} = f_{i,j}^n - \Delta t (\Delta_x^* F_{i,j}^n + \Delta_y^* G_{i,j}^n) \tag{11.4.2a}$$

$$f_{i,j}^{n+1} = \frac{1}{2}(f_{i,j}^n + \tilde{f}_{i,j}) - \frac{\Delta t}{2}(\Delta_x^{**} \tilde{F}_{i,j} + \Delta_y^{**} \tilde{G}_{i,j}) \tag{11.4.2b}$$

where $f_{i,j}$ refers to the value of f at (i,j) and $\tilde{F}_{i,j} = F(\tilde{f}_{i,j})$, $\tilde{G}_{i,j} = G(\tilde{f}_{i,j})$. Moreover, the dissipative part F_v and G_v of F and G includes first-order derivatives of the velocity components u, v and of the internal energy e. These derivatives are approximated by means of the operators Δ_x^{**}, Δ_y^{**} at the first step and by means of the operators Δ_x^*, Δ_y^* at the second step. In the case of a uniform rectangular mesh, scheme (11.4.2) reduces to the MacCormack scheme (2.8.8) except the differencing of mixed derivatives which is then given by (11.3.7).

The truncation error or scheme (11.4.2) has been found, in the case of a scalar equation, to be $O(\Delta E_1, \Delta t \, \Delta E_2, \Delta t^2, \Delta E_2^2)$ where

$$\Delta E_1 = \text{Max}(|\mathbf{PE} + \mathbf{PW}|, |\mathbf{PN} + \mathbf{PS}|)$$

$$\Delta E_2 = \text{Max}(|\mathbf{PE}|, |\mathbf{PW}|, |\mathbf{PN}|, |\mathbf{PS}|)$$

so that the scheme is of second-order accuracy if $\Delta E_1 = O(\Delta E_2^2)$. For stability the time step Δt is chosen to satisfy condition (11.3.10) in which

$$\Delta x = \text{Min}(|\mathbf{PE}|, |\mathbf{PW}|, |\mathbf{PN}|, |\mathbf{PS}|)$$

11.4.2.2 *A mesh refinement technique*: In discretization of the Navier–Stokes equations in the physical plane, a mesh refinement technique often is needed to obtain an adequate description of a viscous region such as a boundary layer or "viscous shock." In addition to the technique of stretching or in conjunction with it, the following technique (Viviand and Ghazzi, 1975; LeBalleur et al., 1980) is based upon a division of the computation domain into several zones in which the mesh size differs by a factor of 2. Such a technique is associated with a systematic procedure for matching the solution between adjacent zones while the solution is advanced in time.

We describe the technique for the case where the refinement is needed in the y direction and $y=0$ corresponds to a wall on which a boundary layer develops. The computation domain is divided into a number of zones defined by $y_1^{(r)} \leq y \leq y_2^{(r)}$ for the zone r where $r = 1, 2, \ldots, R$.

In each zone, the mesh size $\Delta y^{(r)}$ is a constant. From one zone to the next, the mesh size varies as $\Delta y^{(r+1)} = 2 \, \Delta y^{(r)}$ so that $\Delta y^{(r+1)} = 2^r \, \Delta y^{(1)}$. Zone 1 corresponds to the viscous layer near the wall $y = 0$. Zone R corresponds to the outer region discretized with the basic mesh Δy. Hence $\Delta y = \Delta y^{(R)} = 2^{R-1} \, \Delta y^{(1)}$. The ratio $\Delta y^{(1)}/\Delta y = 2^{1-R}$ characterizes the degree of refinement in the viscous layer.

A matching procedure must be set out to link the solution in two adjacent zones. For this, it is convenient to make adjacent zones overlap as shown in Fig. 11.4.5 where horizontal lines in the same column represent the mesh lines $y = y_l^{(r)} = y_1^{(r)} + (l-1) \, \Delta y^{(r)}$ with $l = 1, \dots, L_r$ for zone r. Considering scheme (11.4.2), it is apparent that the solution at time $t + \Delta t$ and at a mesh point P involves the solution at time t and at mesh points located on two lines on each side of P. Therefore, considering each zone, the two lower and the two upper mesh lines in the zone r must be considered as boundary lines, and the solution is advanced in time from t to $t + \Delta t^{(r)}$ only on the inner lines. The solution $f_l^{(r)}$ at time $t + \Delta t^{(r)}$ on the boundary lines $l = 1, 2, L_{r-1}, L_r$ in zone r is obtained from the adjacent zones by the following matching conditions:

$$f_1^{(r)} = f_{L_{r-1}-4}^{(r-1)}, \qquad f_2^{(r)} = f_{L_{r-1}-2}^{(r-1)} \tag{11.4.3a}$$

and

$$f_{L_r}^{(r)} = f_3^{(r+1)}$$

$$f_{L_r-1}^{(r)} = \frac{-f_1^{(r+1)} + 9\,(f_2^{(r+1)} + f_3^{(r+1)}) - f_4^{(r+1)}}{16} \tag{11.4.3b}$$

The time step $\Delta t^{(r)}$ and the number $N^{(r)}$ of steps to be performed in zone r to

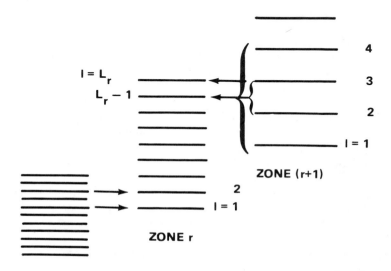

Fig. 11.4.5 Mesh refinement and matching of adjacent zones.

advance the solution from t to $t + \Delta t$ with $\Delta t = \Delta t^{(R)}$ are defined by

$$\Delta t^{(r)} = 2^r \, \Delta t^{(1)}, \qquad r > 1$$
$$N^{(r)} = 2^{R-r} \tag{11.4.4}$$

where the time step $\Delta t^{(1)}$ in zone 1 must satisfy the stability criterion (11.3.10). Note that the variation of time step defined by (11.4.4) corresponds approximately to the maximum value in each zone $r > 1$ if stability is governed by the inviscid CFL criterion (11.3.8).

The solution is advanced in time from t to $t + \Delta t^{(R)}$ by performing two steps in zone 1, and one in zone 2. Then by matching the solution between these two zones by means of formulas (11.4.3), one obtains new boundary conditions at time $t + 2 \, \Delta t^{(1)} = t + \Delta t^{(2)}$. The process is continued until the solution in the outer region R is computed. It is important to note that the resulting method is not consistent in time with the unsteady equations. For steady calculations this inconsistency is not important provided it does not introduce a divergence of the procedure. If a time-accurate solution is needed, it is necessary to devise a special treatment of the boundary lines.

11.4.2.3 *Example problem solutions*: We now present two example applications of scheme (11.4.2) used with the mesh refinement technique to compute steady flows. These calculations were performed by Viviand's Group at ONERA and they concern shock–boundary-layer interaction and flow around an airfoil. In both applications, the fluid is a perfect gas with $\gamma = 1.4$ and the viscosity is calculated according to Sutherland's law. The wall is assumed to be adiabatic and a no-slip condition is imposed on the velocity. The internal energy e_w and the pressure p_w at the wall are obtained from one-sided differencing, respectively, of the adiabatic condition $\partial e / \partial N = 0$ and of the normal momentum equation at the wall.

The shock–boundary-layer interaction on a flat plate is solved for the flow conditions $M_\infty = 2$ and $\mathrm{Re}_{\infty, x_s} = 2.96 \times 10^5$ where x_s is the distance from the leading edge of the flat plate to the point of impact of the incident shock on the plate assuming inviscid flow. The strength of the incident shock is such that $p_f / p_\infty = 1.4$ where p_f is the theoretical pressure downstream of the reflected shock in an inviscid flow. The Prandl number is $\mathrm{Pr} = 0.72$. The incident shock is imposed as a discontinuity in the free-stream conditions. In the calculation, the leading edge shock as well as the incident shock are captured by using scheme (11.4.2) with the addition of a third step for artificial damping (11.3.15), where $\alpha_1 = \alpha_2 = 10^{-3}$. The interaction of the incident shock with the boundary layer is computed in the domain shown on Fig. 11.4.6. In a preliminary calculation, not presented here, the flow past the flat plate is determined in a domain extending downstream to $x = 0.7$ and outwardly to the free stream. All the flow conditions on the upstream boundary are known from the preliminary calculation; the upper boundary of the computation domain is located on the free-stream and linear extrapolation is used on the downstream boundary.

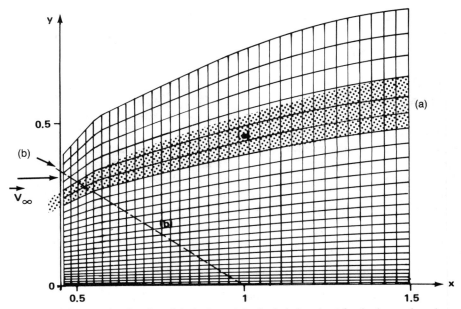

Fig. 11.4.6 Computation domain and the basic mesh in physical plane (x, y) for shock-wave boundary layer interaction. (a) = Leading Edge Shock Region, (b) = Incident (Inviscid) Shock

The manner in which the basic mesh, shown in Fig. 11.4.6, is established is now described. First, the mesh is approximately adapted *a priori* to the leading edge shock and to the boundary layer using the coordinate transformation $(x, y) \rightarrow (x, \bar{y})$ defined by

$$\bar{y}(x, y) = \frac{y}{y_e(x)} \left[\beta_1(x) + \beta_2(x) \frac{y}{y_e(x)} \right]$$

For each value of x, the function $y_e(x)$, defining the upper boundary of the computation domain, and the functions $\beta_1(x)$, $\beta_2(x)$ are determined by requiring the following conditions:

1. The outer limit $y = y_\delta(x)$ of the boundary layer, estimated by $y_\delta(x) = 7x/\sqrt{\mathrm{Re}_x}$, corresponds to the line $\bar{y} = 0.025$.
2. The leading edge shock $y = y_s(x)$, estimated by $y_s(x) = x/8$, corresponds to the line $\bar{y} = 0.8$.
3. The outer boundary $y = y_e(x)$ corresponds to the line $\bar{y} = 1.07$.

Next a second transformation $(x, \bar{y}) \rightarrow (X, Y)$ is carried out in order to determine the uniform basic mesh $\Delta X, \Delta Y$. The purpose of such a transformation, defined by

$$\frac{x - x_0}{x_1 - x_0} = \frac{X(X+1)}{2}, \qquad \bar{y} = \frac{Y(3Y+1)}{4}$$

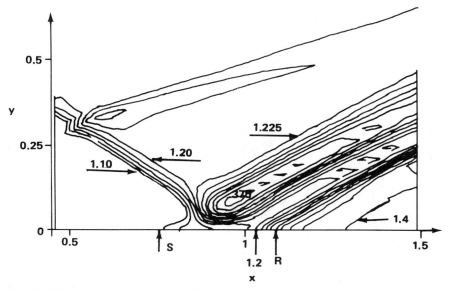

Fig. 11.4.7 Isobar contour map for shock boundary-layer interaction on a flat plate.

(where x_0 and x_1 correspond to the lateral boundaries of the domain), is to introduce a dilatation in the x direction and a refinement in the y direction. Consequently, it is sufficient to use three zones to refine the mesh further near the wall by the method which has already been described. The number of mesh lines in each zone is $L_1 = 17$, $L_2 = 13$, and $L_3 = 20$—zone 3 being a part of the basic mesh. From the above transformations and the mesh refinement, one determines the Cartesian coordinates $(x_{i,j}, y_{i,j})$ of the mesh point P used in the discretization (11.4.2). Figure 11.4.7 shows an isobar contour map which exhibits the separation shock and the reattachment shock. The streamlines in the separated region are shown in Fig. 11.4.8.

The second example deals with supersonic flow $(M_\infty = 2.0)$ around an NACA 0012 airfoil at Reynolds number Re $= 106$ based on the cord of the

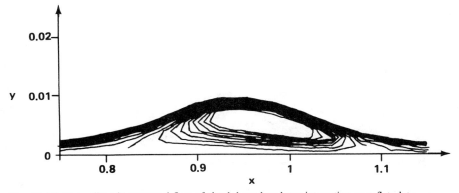

Fig. 11.4.8 Streamlines in separated flow of shock boundary-layer interaction on a flat plate.

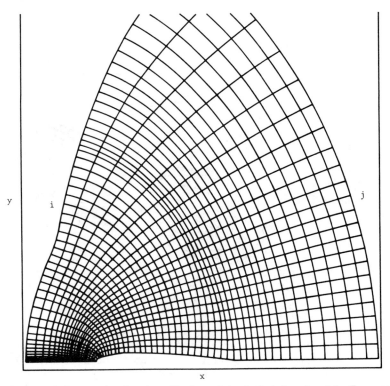

Fig. 11.4.9 Computation domain and basic mesh in physical plane (x, y) for flow around an NACA 0012 airfoil.

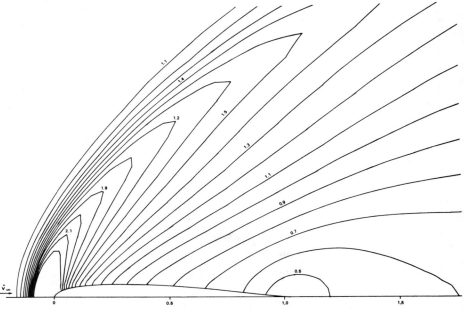

Fig. 11.4.10 (a) Computed lines of constant density (ρ/ρ_∞) for flow around an NACA 0012 airfoil with $\alpha = 0°$, Re $= 106$ and $M_\infty = 2.0$. (Courtesy of ONERA.)

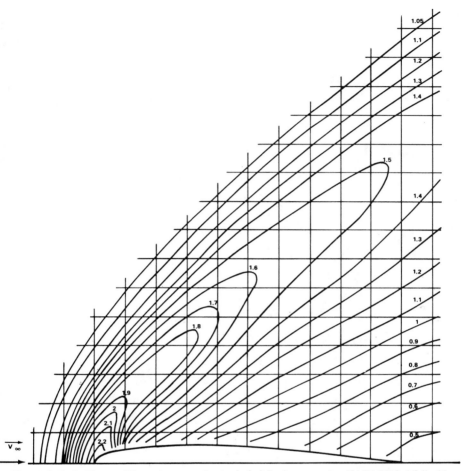

Fig. 11.4.10 (b) Measured lines of constant density (ρ/ρ_∞) for flow around an NACA 0012 airfoil with $\alpha = 0°$, Re $= 106$, and $M_\infty = 2.0$. (Courtesy of Dr. Allegre.)

airfoil. The Prandl number is Pr $= 0.723$. The basic mesh (half-domain) is shown in Fig. 11.4.9. It is constructed numerically by solution of ordinary differential equations determining the orthogonal trajectories ($i = $ const) of a family of curves ($j = $ const) defined by interpolation–extrapolation based on the airfoil contour ($j = 1$) and the estimated location of the leading edge shock. The total basic mesh consists of 101×23 points, and the method of refinement makes use of three zones with $L_1 = 9$, $L_2 = 7$, and $L_3 = 3$—zone 3 being a part of the basic mesh.

Uniform-flow conditions are imposed on the upstream boundary of the computation domain and linear extrapolation is used at the downstream boundary. Figure 11.4.10a and b show the constant density lines ($\rho/\rho_\infty = $ const)

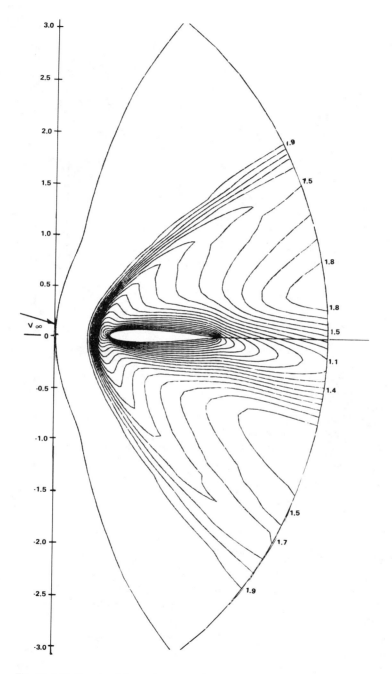

Fig. 11.4.11 Computed lines of constant Mach number for flow around an NACA 0012 airfoil with $\alpha = 10°$, Re = 106, and $M_\infty = 2.0$.

which were computed and also measured by experiments performed by Dr. Allegre at the Laboratoire d'Aérothermique, C.N.R.S. Meudon. The computations were carried out in a half-domain with a condition of symmetry, since the angle of attack α was zero. Figure 11.4.11 shows the iso-Mach lines ($M = $ const) for the flow around the same airfoil with an angle of attack equal to $10°$. For this case, computations were completed in the whole domain since symmetry no longer existed.

References

Allen, J. S., and Cheng, S. I., *Phys. Fluids* **18**, 37–52 (1970).

Baker, A. J. Finite-Element Analysis of Turbulent Flows. In: *Proceedings of the 1st International Conference on Numerical Methods in Laminar and Turbulent Flow*, pp. 203–229, Pentech Press, Swansea (1978).

Baldwin, B. S., and MacCormack, R. W. AIAA Paper No. 74-558 (1974).

Baldwin, B. S., MacCormack, R. W., and Deiwert, G. S. Numerical Techniques for the Solution of the Compressible Navier–Stokes Equations and Implementation of Turbulent Models. In: AGARD-LS-73, *Computational Methods of Inviscid and Viscous Two- and Three-Dimensional Flow Fields*, pp. 2.1–2.24, Von Karman Institute, Rhode-St.-Genèse, Belgium (1975).

Barnwell, R. W. A Time-Dependent Method for Calculating Supersonic Angle of Attack Flow about Axisymmetric Blunt Bodies with Sharp Shoulders and Smooth Nonaxisymmetric Blunt Bodies. NASA-TND-6283, Wash., D. C. August (1971).

Beam, R. M., and Warming, R. F. *J. Comput. Phys.* **22**, 87–110 (1976).

Beam, R. M., and Warming, R. F. *AIAA J.* **16**, 393–402 (1978).

Beam, R. M., and Warming, R. F. *SIAM J. Sci. Statis. Comput.* **1**, 131–159 (1980).

Berezin, Yu. A., Kovenja, V. M., and Yanenko, N. N. Numerical Methods in Continuum Mechanics, **3**, pp. 3–18, (in Russian). A. N. SSSR Siberian Computer Center, Novosibirsk (1972).

Berezin, Yu. A., Kovenja, V. M., and Yanenko, N. N. *Lecture Notes in Physics*, Vol. 35, pp. 85–90, Springer-Verlag, New York (1975a).

Berezin, Yu. A., Kovenja, V. M., and Yanenko, N. N. *Comput. Fluids* **3**, 271–281 (1975b).

Bokhari, S., Hussaini, M. Y., Lambiotte, J. J., and Orszag, S. A. A Navier–Stokes Solution on the CYBER-203 by a Pseudospectral Technique. ICASE Rept. No. 81-12 (1981).

Brailovskaya, I. Yu. *Fluid Dyn.* **2**, 49–55 (1967).

Briley, W. R., and McDonald, H. An Implicit Numerical Method for the Multidimensional, Nonstationary Navier–Stokes Equations. United Aircraft Research Laboratory Rept. M911363-6, November (1973).

Briley, W. R., and McDonald, H. *Lecture Notes in Physics*, Vol. 35, pp. 105–110, Springer-Verlag, New York (1975).

Briley, W. R., and McDonald, H. *J. Comput. Phys.* **24**, 372–397 (1977).

Carter, J. E. *Lecture Notes in Physics*, Vol. 19, pp. 69–78, Springer-Verlag, New York (1973).

Chu, W. H. *J. Comput. Phys.* **8**, 392–408 (1971).

Chang, I. S. AIAA Paper No. 80-0272 (1980).

Cooke, C. H., and Blanchard, D. K. *Int. J. Numer. Meth. Eng.* **14**, 271–286 (1979).

Courant, R., Friedrichs, K. O., and Lewy, H. *Math. Ann.* **100**, 32–74 (1928). [English translation: *IBM J. Res. Dev.* **11**, 215–234 (1967)].

Crocco, L. *AIAA J.* **10**, 1824–1832 (1965).

Deiwert, G. S. *AIAA J.* **13**, 1354–1359 (1975).

Deiwert, G. S. *Lecture Notes in Physics,* Vol. 59, pp. 159–163, Springer-Verlag, New York (1976).

Dwyer, H. A., Kee, R. J., and Sanders, B. R. *AIAA J.* **10**, 1205–1212 (1980).

Eiseman, P. R. *J.Comput. Phys.* **26**, 307–338 (1978).

Eiseman, P. R. Coordinate Generation with Precise Controls over Mesh Properties. ICASE Rept. No. 80-30 (1980).

Engquist, E., and Majda, A. *Math. Comput,* **31**, 629–651 (1977).

Euvrard, D., and Tournemine, G. *J. Méc.* **12**, 419–461 (1973).

Gnoffo, P. A. Solutions of the Navier–Stokes Equations for Supersonic Flow Over Blunt Bodies in a Generalized Orthogonal System. M.S. Thesis, Polytechnic Institute of Brooklyn, New York (1974).

Gough, D. O., Spiegel, E. A., and Toomre, J. *Lecture Notes in Physics,* Vol. 35, pp. 191–196, Springer-Verlag, New York (1975).

Gustaffson, B., and Sundström, A. *SIAM J. Appl. Math.* **35**, 343–357 (1978).

Hedstrom, G. W. *J. Comput. Phys.* **30**, 222–237 (1979).

Hollanders, H., and Peyret, R. *Rech. Aérosp.* No. 1981-4, 287–294 (1981).

Hollanders, H., and Viviand, H. The Numerical Treatment of Compressible High Reynolds Number Flows. In: *Computation Fluid Dynamics*, Von Karman Institute Book, Vol. 2, pp. 1–65, Hemisphere Publishing, Washington, D.C. (1980).

Hung, C. M., and MacCormack, R. W. *AIAA J.* **15**, 410–416 (1977).

Lax, P. D., and Wendroff, B. *Commun. Pure Appl. Math.* **17**, 381–398 (1964).

LeBalleur, J. C., Peyret, R., and Viviand, H. *Comput. Fluids* **8**, 1–30 (1980).

MacCormack, R. W. AIAA Paper No. 69-354 (1969).

MacCormack, R. W. *Lecture Notes in Physics,* Vol. 8, pp. 151–161, Springer–Verlag, New York (1971).

MacCormack, R. W. An Efficient Numerical Method for Solving the Time-Dependent Compressible Navier–Stokes Equations at High Reynolds Number. NASA TM.73129; also In: *Computing in Applied Mechanics,* Applied Mechanics Division, Vol. 18, ASME, New York (1976a).

MacCormack, R. W. *Lecture Notes in Physics,* Vol. 59, pp. 307–317, Springer–Verlag, New York (1976b).

MacCormack, R. W. *Lecture Notes in Physics,* Vol. 148, pp. 254–267, Springer-Verlag, New York (1981a).

MacCormack, R. W. AIAA Paper No. 81-0110 (1981b).

MacCormack, R. W., and Baldwin, B. S. AIAA Paper No. 75-1 (1975).

MacCormack, R. W., and Lomax, H., Numerical Solution of Compressible Viscous Flow. In: *Annual Reviews of Fluid Mechanics,* Vol. 11, pp. 289–316, Annual Reviews Inc., Palo Alto, CA (1979).

Mehta, U., and Lomax, H. Reynolds Averaged Navier–Stokes Computations of Transonic Flows: The State of the Art. In: *Symposium on Transonic Perspective*, Ames Research Center, NASA, February (1981).

Moretti, G. Conformal mappings for computations of steady, three-dimensional, supersonic flows. In: *Numerical/Laboratory Computer Methods in Fluid Mechanics*, A. A. Pouring and V. I. Shah, Eds., pp. 13–28, ASME, New York (1976).

Moretti, G. Computation of Shock Layers About Ablated Blunt-Nosed Bodies, Polytechnic Institute of New York, Rept. 77-14, August (1977).

Moretti, G., and Salas, M. D. AIAA Paper No. 69-139 (1969).

Murman, E. M., and Cole, J., *AIAA J.* **9**, 121–141 (1971).

Oden, J. T., and Wellford, L. C. *AIAA J.* **10**, 1590–1599 (1972).

Peyret, R. *Rech. Aérosp.* No. 1971-3, 133–138 (1971).

Peyret, R., and Viviand, H. *Rech. Aérosp.* No. 1972-3, 123–131 (1972).

Peyret, R., and Viviand, H. *Lecture Notes in Physics,* Vol. 19, pp. 222–229, Springer–Verlag, New York (1973).

Peyret, R., and Viviand, H. *AGARDograph,* No. 212, September (1975).

Polezhaev, V. I. *Fluid Dyn.* **2**, 70–74 (1967).

Pulliam, T. H., and Lomax, H. AIAA Paper 79-0206 (1979).

Pulliam, T. H., and Steger, J. L. *AIAA J.* **18**, 159–167 (1980).

Richtmyer, R. D., and Morton, K. W. Stability Studies with Difference Equations. (1) Nonlinear Stability, (2) Coupled Sound and Heat Flow. Courant Institute of Mathematical Sciences, New York University Rept. NYO-1480-5 (1964).

Richtmyer, R. D., and Morton, K. W. *Difference Methods for Initial Value Problems,* Interscience, New York (1967).

Roache, P. J. *Computational Fluid Dynamics,* Hermosa Publishing, Albuquerque, NM (1972).

Rudy, O., and Strikwerda, J. *J. Comput. Phys.* **36**, 55–70 (1980).

Scala, S. M., and Gordon, P. Solution of the Navier–Stokes Equations for Viscous Supersonic Flows Adjacent to Isothermal and Adiabatic Surfaces. In: *Proceedings of the 1969 Symposium on Viscous Interaction Phenomena in Supersonic and Hypersonic Flow,* pp. 319–391, University of Dayton Press, Dayton, OH (1970).

Schiff, L. B., and Steger, J. L. *AIAA J.* **18**, 1421–1430 (1980).

Schönauer, W., Glotz, G., and Raith, K. The solution of the laminar boundary layer equations by variable self-adaptive difference methods. In: *Recent Advances in Numerical Methods in Fluids,* C. Taylor and K. Morgan, Eds. Vol. 1, pp. 81–118, Pineridge, Swansea, U. K. (1980).

Shamroth, S. J., and Gibeling, H. J. A Compressible Solution of the Navier–Stokes Equations for Turbulent Flow about an Airfoil. NASA-CR-3183 (1979).

Sorenson, R., and Steger, J. L. Simplified Clustering of Nonorthogonal Grids Generated by Elliptic Partial Differential Equations. NASA-TM-73252 (1977).

Steger, J. L. *AIAA J.* **16**, 679–686 (1978).

Steger, J. L., and Bailey, H. E. *AIAA J.* **18**, 249–255 (1980).

Takami, H., and Keller, H. B. *Phys. Fluids* **12**, Supp. II, II51–II56 (1969).

Tannehill, J. C., Mohling, R. A., and Rakich, J. V. *AIAA J.* **12**, 129–130 (1974). [See also AIAA Paper No. 73-200 (1973)]

Tannehill, J. C., Holst, T. L., and Rakich, J. V. AIAA Paper No. 75–154 (1975).

Thommen, H. U., Z. Angew. Math. Phys. **17,** 369–384 (1966).

Thompson, J. F., Thames, F. C., and Mastin, C. M. J. Comput. Phys. **15,** 299–319 (1974).

Victoria, K. J., and Widhoff, G. F. Lecture Notes in Physics, Vol. 19, pp. 254–267, Springer–Verlag, New York (1973).

Viviand, H. Traitement des Problèmes d'Interaction Fluide Parfait-Fluide Visqueux en Ecoulement Bidimensionel Compressible à partir des Equations de Navier–Stokes. AGARD-LS-94 on Three-Dimensional and Unsteady Separation of High Reynolds Numbers, pp. 3.1–3.21, Von Karman Institute, Rhode-St-Genèse, Belgium (1978).

Viviand, H., Lecture Notes in Physics, Vol. 141, pp. 44–54, Springer–Verlag, New York (1981).

Viviand, H., and Ghazzi, W. Lecture Notes in Physics, Vol. 59, pp. 434–439, Springer–Verlag, New York (1976).

Viviand, H., and Veuillot, J.-P. Méthodes Pseudo-Instationnaires pour le Calcul d'Ecoulements Transsoniques. Publication ONERA No. 1978-4 (1978).

Weilmuenster, K. J., and Graves, Jr., R. A. AIAA J. **19,** 1089–1090 (1981).

Yanenko, N. N., Lisseikin, V. D., and Kovenja, V. M. Lecture Notes in Physics, Vol. 91, pp. 48–61, Springer–Verlag, New York (1979).

Concluding Remarks

In the material presented, the authors have attempted to provide a reasonable summary of numerical methods in fluid mechanics which are in use in both the United States and Europe. We did not attempt to review technology in the area which was generated in the U.S.S.R. due to the difficulty in obtaining the results. We did include those methods and results which were available and did not overlap the book of M. Holt in this series.

Due to the vast and rapid progress in the field of computational fluid mechanics we are fully aware that there may be a development of which we are unaware. We tried to avoid this by extensive communication. Certainly in the areas of finite-element and spectral approaches there will be new advances coming forth rapidly and we look forward to this progress.

R. Peyret
T. Taylor

Appendix A: Stability

Assume a scheme to be of the form

$$C_1 u_h^{n+1} + C_0 u_h^n + C_{-1} u_h^{n-1} = 0 \tag{A.1}$$

where C_1, C_0, and C_{-1} are finite-difference operators, $u_h^n \cong u(\mathbf{x}_h, n\, \Delta t)$ with $u = (u_1, \ldots, u_m)$, and $\mathbf{x} = (x_1, \ldots, x_d)$ so that $\mathbf{x}_h$ refers to the discretization in space: $\mathbf{x}_h = (\gamma_1\, \Delta x_1, \ldots, \gamma_d\, \Delta x_d)$, $\gamma_1, \ldots, \gamma_d$ being integers; the index n refers to the time-discretization $t_n = n\, \Delta t$, n = integer.

The stability of scheme (A.1) is examined by considering a Fourier analysis of the solution. One assumes a solution of the form

$$u_h^n = U^n(\mathbf{k}) \exp(i\, \mathbf{k} \cdot \mathbf{x}_h) \tag{A.2}$$

where $i = \sqrt{-1}$, and $\mathbf{k}$ is the vector wave number $\mathbf{k} = (k_1, \ldots, k_d)$. Substituting (A.2) into (A.1) and dividing by the common factor $\exp(i\, \mathbf{k} \cdot \mathbf{x}_h)$ yields the vector equation

$$G_1 U^{n+1} + G_0 U^n + G_{-1} U^{n-1} = 0 \tag{A.3}$$

This three-level equation is transformed into a two-level equation by introducing the auxiliary unknown $V^n = U^{n-1}$. Equation (A.3) then becomes

$$G_1 U^{n+1} + G_0 U^n + G_{-1} V^n = 0 \tag{A.4}$$

$$V^{n+1} - U^n = 0$$

which can be written

$$\Phi^{n+1} = G \Phi^n \tag{A.5}$$

where $\Phi = (U, V)$ is a $2m$-component vector, G a $2m \times 2m$ matrix deduced from (A.4).

Stability of the scheme (*strict* Von Neumann condition) requires that the spectral radius of the amplification matrix G is not larger than unity; i.e.,

$$\rho(G) = \operatorname*{Max}_l |\lambda_l| \le 1, \qquad 0 \le k_j\, \Delta x_j \le 2\pi, \; j = 1, \ldots, d \tag{A.6}$$

where λ_l are the eigenvalues of G. These eigenvalues are the roots of the algebraic equation

$$\det(G - \lambda I) = 0$$

where I is the identity matrix, or equivalently they are the zeros of a polynomial of degree q (with $q = 2m$)

$$f(\lambda) = a_0 + a_1\lambda + \cdots + a_q\lambda^q = \sum_{j=0}^{q} a_j\lambda^j \qquad (A.7)$$

where the coefficients can be complex numbers.

Note that if the scheme is a two-level scheme ($C_{-1} \equiv 0$) and if u is a scalar, the matrix G reduces to a scalar g (called the factor of amplification) so that condition (A.6) reduces to

$$|g| \leq 1, \qquad 0 \leq k_j\,\Delta x_j \leq 2\pi,\ j = 1, \ldots, d \qquad (A.8)$$

A study of the zeros of $f(\lambda)$ is simplified by using results obtained by Miller (1971) concerning the zeros of polynomials. Here, we shall give only the theorem ensuring that the zeros of the polynomial $f(\lambda)$ satisfy condition (A.6); the reader will find the proof as well as other interesting theorems in Miller's paper.

First, we define the polynomial $\tilde{f}(\lambda)$ constructed from (A.7) where it is assumed that $a_0 \neq 0$ and $a_q \neq 0$; i.e.,

$$\tilde{f}(\lambda) = a_0^*\lambda^q + \cdots + a_q^* = \sum_{j=0}^{q} a_{q-j}^*\,\lambda^j \qquad (A.9)$$

where a_j^* is the complex conjugate of a_j. Then we consider the polynomial $f_1(\lambda)$ defined by

$$f_1(\lambda) = \frac{1}{\lambda}[\tilde{f}(0)f(\lambda) - f(0)\tilde{f}(\lambda)] \qquad (A.10)$$

which is of a degree lower than q (at most equal to $q-1$).

Now, the *theorem* states that:

The zeros, λ, of $f(\lambda)$ are such that $|\lambda| \leq 1$ if *either* (i) $|\tilde{f}(0)| > |f(0)|$ and the zeros λ of $f_1(\lambda)$ are such that $|\lambda| \leq 1$, *or* (ii) $f_1 \equiv 0$ and the zeros λ of $df/d\lambda$ are such that $|\lambda| \leq 1$.

By using this theorem, the study of the zeros of the polynomial f is reduced to the study of the zeros of a polynomial with a lower degree. This can be repeated until this polynomial is sufficiently simple. Obviously, if the degree of $f(\lambda)$ is large, the successive conditions obtained by a repeated application of the theorem could be too complicated to be useful and numerical calculations are then needed.

An alternate possibility is to compute directly the eigenvalues of G by numerical methods. The calculations are carried out by varying the parameters (physical or numerical) of the problem and constructing graphs of the results. For such an approach it is recommended that a preliminary analytical study of the stability be made in simplified cases; for instance, by (i) considering problems in less space dimensions, (ii) zeroing out some physical parameters, or (iii) considering some particular relationships between the components of the wave-number vector or even some particular values of it. Such a preliminary analysis often allows determination of a rough form of the analytical stability criterion which can be refined by numerical calculations of the eigenvalues.

Reference

Miller, J. J. H. *J. Inst. Math. Appl.* **8,** 397–406 (1971).

Appendix B: Multiple-Grid Method

In the application of finite-difference and finite-element methods one may be faced with the problem of developing an accurate solution on a small-scale mesh with limited computer availability. A method has been proposed by Federenko (1964) which, in principle, allows development of an accurate solution on a small grid by employing solutions on grids of increasing size. The result is that a solution can be constructed with decreased computation time and cost.

This approach has been examined by Bakhvalov (1966) and applied by Brandt (1977), Jameson (1979), and South and Brandt (1977). The method has been used principally for transonic flows, but it has application to other problems as well.

B.1 The Linear Problem

The multiple-grid procedure is applied to linear problems in the following manner. First, assume that one is solving the equation

$$LU = F \tag{B.1.1}$$

by a finite-difference approach on a grid of spacing h_k. For this case L and F become matrices. Next, assume u is an approximation to the solution and that u can then be defined by

$$U = u + v \tag{B.1.2}$$

where v is the correction to u. One then solves the problem

$$Lv = f \tag{B.1.3}$$

where

$$f = F - Lu \tag{B.1.4}$$

on a grid that ranges from the smallest spacing, which we denote by M, to the largest spacing, which we denote by N. The value of U is then computed from $U = u + v$ once v has converged to the final solution. The procedure for solving for v can be described as a general strategy of escalating a fine-grid,

M, solution to a coarse-grid, N, solution and then back to the fine grid. The strategy for constructing the solution in this manner becomes somewhat of an art. Brandt (1977) and Jameson (1979) have presented two separate approaches. First we describe the strategy suggested by Brandt.

The steps of the Brandt approach are as follows:

1. Assume or calculate by relaxation an approximation to U on the smallest grid M. Call this u^M.

2. Compute the values of f at the grid points from the values of u using Eq. (B.1.4). Call these values f^M.

3. Next, solve for v^M from the equation

$$Lv^M = f^M \qquad\qquad\qquad (B.1.5)$$

   by a relaxation procedure of your choice. Note the solution for v^M should proceed as long as convergence is rapid. If the convergence slows, then one should begin the grid escalation process in step 4.

4. When convergence of v^M slows, escalate to a larger grid $(M+1)$ and solve

$$Lv^{M+1} = f^{M+1} \qquad\qquad\qquad (B.1.6)$$

   as in step 3. Note that f^{M+1} must be obtained by interpolation of f^M onto the $M+1$ grid system. Follow this procedure until convergence of v to your specified criteria or until convergence slows. If convergence slows, escalate the grid again. This procedure can be employed until one escalates to the final grid size N. Note that N is the option of the user. At level N, v^N should be made to converge.

5. When convergence of v is obtained at any level in the escalation of grids one then must try to deescalate the solution. This is accomplished by computing

$$v_{\text{new}}^k = v^k + I_{k+1}^k v^{k+1} \qquad\qquad\qquad (B.1.7)$$

   where k is the level the user is going to and $k+1$ is the level the user is leaving. I_{k+1}^k denotes that one must interpolate from the coarser grid to the finer grid. The new value of v^k is then used as an estimate of v at the level k to start a new iteration to a solution on level k. The solution is relaxed toward convergence, but at this point a decision has to be made. If the solution begins to converge slowly, does one escalate up or down? At this point the procedure is not absolute. Brandt indicates that one can proceed in either way. The choice seems to be the user's, since Jameson (1979) has pointed out that a simplified approach may work equally well. The simplified approach is to iterate the solution only once per level, k, going both up and down the scale. Jameson interpolates the solution at the final stage going from the $M+1$ grid to the final grid M to obtain the converged solution. Due to the significant difference in strategies between the studies of Brandt (1977) and Jameson (1979), it is not clear what the optimum strategy should be. As a result research in this area seems needed.

B.2 The Nonlinear Equation

The multiple-grid approach just discussed needs to be modified for nonlinear problems. For the nonlinear problem, Brandt (1977) suggests that the equations for interaction and escalation take the form for grid level k from level $k-1$

$$L^k u^k = L^k (I^k_{k-1} u^{k-1}) + I^k_{k-1}(F^{k-1} - L^{k-1} u^{k-1}) \tag{B.2.1}$$

Employing this equation one marches up and down the grid scale as for the linear problem. The value of u at a level $k-1$ then is obtained in downward march from k by

$$u_{\text{new}}^{k-1} = u^{k-1} + I^{k-1}_k (u^k - I^k_{k-1} u^{k-1}) \tag{B.2.2}$$

It is not at all clear that this procedure is unique, but both Brandt (1977) and Jameson (1979) have demonstrated that the procedure yields practical results. In summary, the procedure seems useful, but more standardization is required.

References

Bakhvalov, N. S. *USSR Math. Math. Phys.* **6,** 101–135 (1966).

Brandt, A. *Math. Comput.* **31,** 333–390 (1977).

Federenko, R. P. *USSR Math. Math. Phys.* **4,** 227–235 (1964).

Jameson, A. AIAA Paper No. 79-1458 (1979).

South, J. C., and Brandt, A. In: *Transonic Flow Problems in Turbomachinery,* T. C. Adamson and M. F. Platzer, Eds., pp. 180–207, Hemisphere Publishing, Washington, D.C. (1977).

Appendix C: Conjugate-Gradient Method

In the text there are numerous reference to the encouraging improvements in speed and efficiency with which the conjugate-gradient method can be applied to solve algebraic equations generated by finite-difference or spectral approximations. The most recent work of Koshla and Rubin (1981) has brought this point out most clearly. Koshla and Rubin show that, for finite-differenced Laplace equations, the conjugate-gradient method converged to a solution an order of magnitude faster than the more commonly used point successive relaxation, successive line relaxation, alternating direction implicit, and strongly implicit procedure (Stone, 1968).

A similar improvement has been noted by Kershaw (1978) who indicates solutions are obtain 6000 times faster than by the point Gauss–Seidel method, 200 times faster than by ADI methods, and 30 times faster than block successive relaxation.

Glowinski et al. (1980) also demonstrate similar improvements in applying the approach to solution of the matrices from finite-element formulations. Progress in the spectral-method area has begun and improvements should occur there also. For the prospective user we outline the method in this section. A derivation of the method is given in Beckman (1960).

The conjugate-gradient method is employed to solve the matrix equation

$$Ax = k$$

for the x vector. The method is iterative and proceeds as follows:

1. Assume a starting value of x_0.
2. Compute $p_0 = r_0 = k - Ax_0$.
3. Compute $\alpha_i = |r_i|^2/(p_i, Ap_i)$.
4. Compute $x_{i+1} = x_i + \alpha_i p_i$.
5. Compute $r_{i+1} = r_i - \alpha_i Ap_i$.
6. Compute $\beta_i = |r_{i+1}|^2/|r_i|^2$.
7. Compute $p_{i+1} = r_{i+1} + \beta_i p_i$.

In these expressions (p_i, Ap_i) represents the inner product of the vectors p_i and Ap_i. Note also that alternate expressions can be used for α_i and β_i which give greater accuracy at the expense of more computation. These are

$$\alpha_i = \frac{(p_i, r_i)}{(p_i, Ap_i)}$$

$$\beta_i = -\frac{(r_{i+1}, Ap_i)}{(p_i, Ap_i)}$$

Basically this solution technique is the use of the Gram–Schmidt orthogonalization process to form sequences of vectors. The method traditionally is derived for the symmetric and positive definite matrix, but it can be made to work on most finite-difference-type equations by using preconditioning. This accomplished premultiplying by a matrix A^T to obtain*

$$(A^TA)x = A^Tk \quad \text{or} \quad A^T(Ax - k) = A^Tr_i$$

The matrix (A^TA) should satisfy the conditions of symmetric and positive definiteness. For this case the solution algorithm is

$$r_0 = k - Ax_0, \quad p_0 = A^Tr_0$$

$$\alpha_i = \frac{|A^Tr_i|^2}{|Ap_i|^2}, \quad \beta_i = \frac{|A^Tr_{i+1}|^2}{|A^Tr_i|^2}$$

$$x_{i+1} = x_i + \alpha_ip_i$$

$$r_{i+1} = r_i - \alpha_iAp_i$$

$$p_{i+1} = A^Tr_{i+1} + \beta_ip_i$$

The approach outlined provides a good approach for solving the algebraic equations generated by the application of finite-difference, finite-element or spectral methods to the solutions of various formulations of the flow equations.

*Note A^T may be the transpose of A or an alternate form.

References

Beckman, F. S. In: *Mathematical Methods for Digital Computers,* A. Ralston and H. S. Wilf, Eds., pp. 62–72, John Wiley & Sons, New York (1960).

Glowinski, R., Periaux, J., and Pironneau, O. *Appl. Math. Modelling* **4,** 187–192 (1980).

Koshla, P. K., and Rubin, S. G. *Comput. Fluids* **9,** 109–122 (1981).

Kershaw, D. S. *J. Comput. Phys.* **26,** 43–65 (1978).

Stone, H. L. *SIAM J. Numer. Anal.* **5,** 530–558 (1968).

Index

RETURN TO: PHYSICS LIBRARY
351 LeConte Hall   510-642-3122

LOAN PERIOD 1 **1-MONTH**	2	3
4	5	6

ALL BOOKS MAY BE RECALLED AFTER 7 DAYS.
Renewable by telephone.

DUE AS STAMPED BELOW.

MAY 1 7 2005		
AUG 2 2 2005		
FEB 1 3 2009		
		•

FORM NO. DD 22
500   4-03

UNIVERSITY OF CALIFORNIA, BERKELEY
Berkeley, California 94720–6000